Tunneling Field Effect Transistors

This book will give insight into emerging semiconductor devices from their applications in electronic circuits, which form the backbone of electronic equipment. It provides desired exposure to the ever-growing field of low-power electronic devices and their applications in nanoscale devices, memory design, and biosensing applications.

Tunneling Field Effect Transistors: Design, Modeling and Applications brings researchers and engineers from various disciplines of the VLSI domain together to tackle the emerging challenges in the field of nanoelectronics and applications of advanced low-power devices. The book begins by discussing the challenges of conventional CMOS technology from the perspective of low-power applications, and it also reviews the basic science and developments of subthreshold swing technology and recent advancements in the field. The authors discuss the impact of semiconductor materials and architecture designs on TFET devices and the performance and usage of FET devices in various domains such as nanoelectronics, Memory Devices, and biosensing applications. They also cover a variety of FET devices, such as MOSFETs and TFETs, with various structures based on the tunneling transport phenomenon.

The contents of the book have been designed and arranged in such a way that Electrical Engineering students, researchers in the field of nanodevices and device-circuit codesign, as well as industry professionals working in the domain of semiconductor devices, will find the material useful and easy to follow.

Materials, Devices, and Circuits: Design and Reliability

Series Editor: Shubham Tayal, K. K. Paliwal and Amit Kumar Jainy

Tunneling Field Effect Transistors

Design, Modeling and Applications

Edited by T.S. Arun Samuel, Young Suh Song, Shubham Tayal, P. Vimala and Shiromani Balmukund Rahi

For more information about this series, please visit:
https://www.routledge.com/Materials-Devices-and-Circuits/book-series/MDCDR

Tunneling Field Effect Transistors
Design, Modeling and Applications

Edited by
T.S. Arun Samuel, Young Suh Song,
Shubham Tayal, P. Vimala and
Shiromani Balmukund Rahi

CRC Press
Taylor & Francis Group
Boca Raton London New York

CRC Press is an imprint of the
Taylor & Francis Group, an **informa** business

First edition published 2023
by CRC Press
6000 Broken Sound Parkway NW, Suite 300, Boca Raton, FL 33487-2742

and by CRC Press
4 Park Square, Milton Park, Abingdon, Oxon, OX14 4RN

CRC Press is an imprint of Taylor & Francis Group, LLC

ISBN: 978-1-032-34876-6 (hbk)
ISBN: 978-1-032-35469-9 (pbk)
ISBN: 978-1-003-32703-5 (ebk)

DOI: 10.1201/9781003327035

Typeset in Times
by MPS Limited, Dehradun

Contents

Editor Biographies

T.S. Arun Samuel received B.E degree in Electronics and Communication Engineering from Syed Ammal Engineering College (2004) and M.E degree in Computer and communication engineering from National Engineering College (2006). He was awarded a Ph.D. in Nanoelectronic Devices (2014) from Thiagarajar College of Engineering, Tamilnadu, India, under Anna University Chennai. He is currently working as Professor at National Engineering College, Kovilpatti, India. He has authored more than 60 research articles in SCI and Scopus-indexed International Journals, six book chapters, and 15 International Conferences. He is a life member of the Institute of Engineering (IE), India, and a member of IEEE. His research interest includes Modeling and Simulation of Multigate transistors and Tunnel Field-effect Transistors.

Prof. Young Suh Song is assistant professor in the department of Computer Science (CS) at Korea Military Academy. His current research interests have included semiconductor reliability (self-heating effect and retention characteristics), AI semiconductors, NAND Flash and NOR Flash, low-temperature logic device for CPU, low-power logic device (Tunnel FET, TFET) for cell phones and laptops, and germanium (Ge)-based logic device (2030 ~ 2050), which is one of the promising candidates for replacing current silicon (Si) based CPU technology. He has authored and co-authored over 40 research papers in journals and conferences, including IEEE Transactions on Electron Devices (IEEE TED) and IEEE Journal of the Electron Devices Society (IEEE JEDS). He served on the technical committee at IEEE International Conference on Circuits, Systems and Simulations (IEEE ICCSS), in 2022. He has also served as a guest speaker in universities including VIT-AP University and VIT-Chennai University (India), in 2022. He received the "Best Paper" Award from "the Institute of Electronics and Information Engineers (IEIE)" in 2021 and "IEEE ICCSS" in 2022.

Dr. Shubham Tayal is assistant professor in the Department of Electronics and Communication Engineering at SR University, Warangal, India. He has more than 6 Years of academic/research experience of teaching at UG and PG level. He has received his Ph.D. in Microelectronics & VLSI Design from National Institute of Technology, Kurukshetra, M.Tech (VLSI Design) from YMCA University of Science and Technology, Faridabad and B.Tech (Electronics and Communication Engineering) from MDU, Rohtak. He has qualified GATE (2011,2013,2014) and UGC-NET (2017).

He has published more than 40 research papers in various International journals and conferences of repute and many papers are under review. He is on the editorial and reviewer panel of many SCI/SCOPUS-indexed international journals and conferences. He is editor/co-editor of 8 books in total from CRC Press (Taylor & Francis Group, USA) and Springer Nature. He acted as a keynote speaker and delivered professional talks on various forums. He is a member of various professional bodies like IEEE, IRED, etc. He is on the advisory panel of many international conferences. He is a recipient of Green ThinkerZ International Distinguished Young Researcher Award 2020. His research interests include the simulation and modeling of multi-gate semiconductor devices, Device-Circuit co-design in the digital/analog domain, machine learning, and IoT.

P. Vimala received her B.E and M.E degrees, both in electronics and communication engineering, from Anna University, Chennai. She completed her Ph.D. in Semiconductor device modeling and simulation at Anna University, Chennai, in 2014. She is awarded the Women Scientist Fellowship for three years for her research work from the Department of Science and Technology (DST), New Delhi, Government of India. Currently, she is working as a Professor in the Department of ECE, Dayananda Sagar College of Engineering, Bangalore, Karnataka. She had three funded projects from various government funding agencies like DRDO and DST. She published more than 100+ research papers, including SCI and Scopus-indexed journals. Also, she had one granted patent and published seven book chapters (CRC Press:5) based on her research work. Her research interests focused on Modeling of Nanoscale semiconductor device modeling and Simulation using the Sentaurus Atlas TCAD Simulation tool. She received the IEI- Young engineer award in 2017. She has been selected as a senior IEEE member in 2018. Also, she is a life member of ISTE and a member of IEI.

Shiromani Balmukund Rahi received B.Sc. (Physics, Chemistry & Mathematics) in 2002, M.Sc. (Electronics) from DeenDyal Upadhyaya Gorakhpur University, Gorakhpur, in 2005, GATE (2009), M. Tech. (Microelectronics) from Panjab University Chandigarh in 2011 and a Doctorate of Philosophy in 2018 from the Indian Institute of Technology, Kanpur, India. He has completed his master's project (M.Sc.) at Central Electronics Engineering Research Institute (CEERI, 2005), Pilani Rajasthan, under the supervision of Dr. P C Panchariya (Director and Chief Scientist, CEERI, Pilani) and master's thesis (M. Tech.) under Prof. Renu Vig (director and Professor, UIET Panjab University Chandigarh). He has 25 international publications and eight book chapters. He has edited one book for CRC publication. He is associated with research with the Indian Institute of Technology Kanpur, India, and the Electronics Department University MostefaBenboulaid of Algeria to develop ultra-low power devices such as Tunnel FETs, NC TFET, Negative Capacitance, and Nanosheet FETS.

Contributors

Bibhudendra Acharya
National Institute of Technology Raipur
Chhattisgarh, India

J. Ajayan Associate
SR University Warangal
Telangana, India

Andrew Robert
Francis Xavier Engineering College
Tamil Nadu, India

Anju
National Institute of Technology
    Raipur
Chhattisgarh, India

Naushad Alam
Z. H. College of Engineering
    and Technology
Aligarh Muslim University
Aligarh, UP, India

A.V. Arun
Model Engineering College
APJ Abdul Kalam Technological
    University
Thrikakkara, India

T.S. Arun Samuel
National Engineering College
Nallatinputhur, India

Shiromani Balmukund Rahi
IIT Kanpur
India

Manisha Bharti
National Institute of Technology
Delhi, India

Saurabh Chaudhury
NIT Silchar
India

Diganta Das
VIT-AP University
Amaravati, Andhra Pradesh

Debashish Dash
VIT, Vellore
India

Ritam Dutta
Siksha 'O' Anusandhan University
Bhubaneswar, India

Dr. Mohammad Ehteshamuddin
Indian Institute of Technology - Roorkee
Roorkee, India

G. Gifta
Karunya Institute of Technology
    and Science
Coimbatore, India

Jobymol Jacob
College of Engineering Poonjar
APJ Abdul Kalam Technological
    University
Thiruvananthapuram, India

L. Jerart Julus
National Engineering College
Nallatinputhur, India

Mr. B. Karthikeyan
Velammal College of Engineering
    and Technology
Madurai, Tamilnadu, India

Abhishek Kumar Upadhyay
R&D Engineer
X-FAB Semiconductor Foundries
Germany

Ashish Kumar Sharma
Z. H. College of Engineering &
   Technology
Aligarh Muslim University
Aligarh, UP, India

Chandan Kumar Pandey
VIT-AP University
Amaravati, India

Pramod Kumar
National Institute of Technology
Delhi, India

G. Lakshmi Priya
School of Electronics Engineering
VIT University
Vellore, India

Dr. S. Manikandan
Indian Institute of Technology-Roorkee
Roorkee, India

Ilho Myeong
Seoul National University
Seoul, South Korea

Umakant Nanda
VIT-AP University
Amaravati, Andhra Pradesh, India

Neha Paras
National Institute of Technology
Delhi, India

Eswaran Parthasarathy
SRM Institute of Science
   and Technology
Chennai, India

Dr. Adhithan Pon
Indian Institute of Technology - Roorkee
Roorkee, India

Guru Prasad Mishra
National Institute of Technology Raipur
Chhattisgarh, India

K. Ramkumar
Vellore Institute of Technology
Vellore, India

M. Saravanan
Sri Eshwar College of Engineering
Coimbatore, India

Prabhat Singh
National Institute of Technology
Hamirpur, Himachal Pradesh, India

Dharmendra Singh Yadav
National Institute of Technology
Hamirpur, Himachal Pradesh, India

Youngjae Song
Konkuk University
Seoul, South Korea

Dr. D. Sriram Kumar
National Institute of Technology
Trichy, Tamilnadu, India

Young Suh Song
Korea Military Academy
Seoul, South Korea

Dr. P. Suveetha Dhanaselvam
Velammal College of Engineering
   and Technology
Madurai, Tamilnadu, India

Shubham Tayal
SR University
Telangana, India

C. Usha
Cambridge Institute of Technology
Bengaluru, India

Dr. P. Vanitha
SRMIST Ramapuram Campus
Chennai, India

M. Venkatesh
CMR Institute of Technology
Bengaluru, India

P. Vimala Professor
Dayananda Sagar College
   of Engineering
Bengaluru, India

Mohd Yasir
Z. H. College of Engineering
   and Technology
Aligarh Muslim University
Aligarh, UP, India

1 Challenges of Conventional CMOS Technology in Perspective of Low-Power Applications

Andrew Robert, M. Venkatesh, G. Lakshmi Priya,
G. Gifta, and L. Jerart Julus

CONTENTS

1.1 INTRODUCTION

Junction Field Effect Transistors (JFETs) with a Schottky-Junction gate, which is made up of Silicon (Si) substrate, are called MOSFET (Metal Oxide Semiconductor Field Effect Transistor). Two types of MOSFETs are used in the Complementary Metal-Oxide-Semiconductor (CMOS) [1]. They are named NMOS (N-Channel MOSFET) and PMOS (P-Channel MOSFET). In CMOS technology, the performance of the NMOS transistor is better than the PMOS transistor. Thus, the NMOS transistors are preferred in the RFIC. Especially the NMOS transistor has high transconductance and current drive due to the mobility difference between the electrons and holes. In the same width and length, the NMOS transistor has high output resistance, leading to an increase in the gain of the amplifier.

PMOS and NMOS devices are pictured in Figures 1.1 and 1.2, respectively. The Gate, Source, Drain, and Bulk terminals are represented as G, S, D, and B in Figures 1.1

DOI: 10.1201/9781003327035-1

(a)   (b)

(c)

(d)

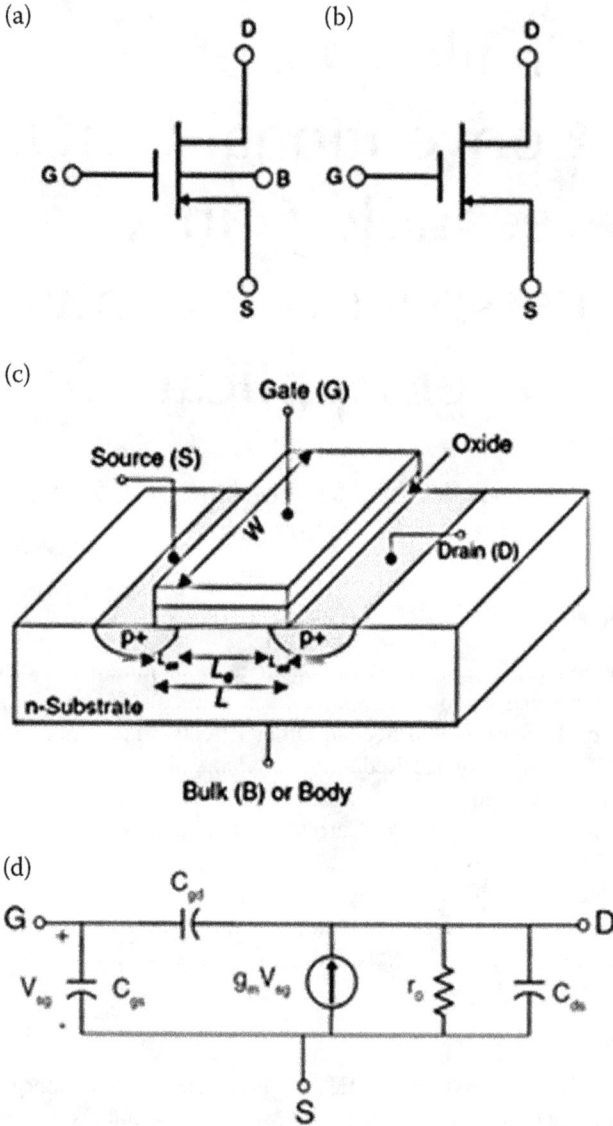

FIGURE 1.1   P-type metal oxide semiconductor (PMOS): (a) the symbol with external bulk terminal, (b) the symbol with a bulk terminal connected to the supply, (c) internal structure, (d) small signal model.

and 1.2. The W, L, and L_e are the channel's width, length, and effective length, respectively. The effective length $L_e = L - 2L_{diff}$ where, L_{diff} is the diffused length due to the lateral diffusion. C_{gs}, C_{gd}, and C_{ds} are the parasitic capacitance between the gate and source, gate and drain, and drain and source, respectively. V_{sg}, and V_{gs} are the voltages between the source and gate, and gate and source, respectively. Moreover, r_0 is the output resistance of the MOS device [2].

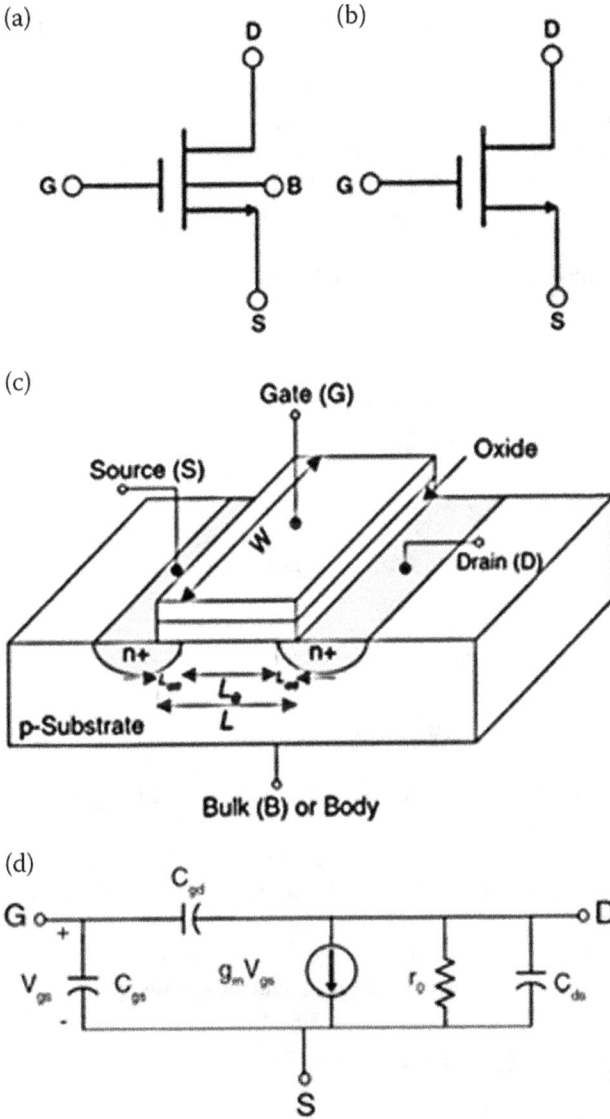

FIGURE 1.2 N-type metal oxide semiconductor (NMOS): (a) the symbol with external bulk terminal, (b) the symbol with a bulk terminal connected to the ground, (c) internal structure, (d) small signal model.

CMOS technology is used in most integrated circuits. There are several reasons for the dominance of CMOS technology. The area required for the CMOS IC is small. CMOS consumes less power, and the operating speed is high. Apart from this, the manufacturability of the CMOS is more important. Defects that occur during the fabrication process are also fewer in CMOS circuits. The mass fabrication cost of the CMOS IC gets reduced with the scale-down process. Significantly, the layout designed for a particular nanometre technology in digital circuits can be scaled down in simple steps [3].

In the initial stages, the CMOS technology was used for digital design; the lower cost and improved functionality of the CMOS ICs have motivated the designers to use CMOS technology for analog and mixed-signal design. When the CMOS is used in analog or mixed signal design, the matching gets essential [4]. The electrical characteristics of the transistors need to be matched for perfect signal power transfer in these circuits [5]. The quality of the design depends on the matching. In CMOS technology, multiple methods are used for matching the transistors.

In digital circuits, several advantages are provided by CMOS technology. They are,

    i. Requires minimum devices/gates
    ii. Minimum static power dissipation
    iii. MOS devices are easily scaled down for higher density
    iv. Lower fabrication cost with high volume

While using the CMOS in analog/RF circuits, the following advantages are observed,

    i. Possibility of SoC (System on Chip) to get a single chip with digital, mixed-signal, and RF modules
    ii. Reduced cost
    iii. Minimum power consumption

However, the following issues are present in the advanced CMOS technologies [6],

    i. The cost increases with the number of masks used in the design and fabrication process. Moreover, the lithography gets expensive due to the scale down.
    ii. In the nanoscale CMOS devices, the subthreshold leakage, gate leakage, band-to-band tunneling leakage, and static power dissipation are higher [7].

1.2 CMOS INVERTER DESIGN

The CMOS inverter complements the digital input signal. NOT gate of the digital circuits also doing the same Boolean operation. In the ideal case, the input and output resistance of the inverter are infinite and zero. The inverter can be used in analog amplification if the operating point is set in the transition region [8]. Recently, the digital IC design has been significant due to the reduced design overhead.

The response of the ideal inverter and the practical inverter are analyzed in this section. The schematic of the CMOS inverter is shown in Figure 1.3. It has the PMOS on the pull-up side and the NMOS on the pull-down side.

The input and output response of the inverter is shown in the upper section of Figure 1.4, where the input and output are complementary. In the transfer characteristics of the ideal inverter, we can refer that for the input value of 0 to $V_{DD}/2$,

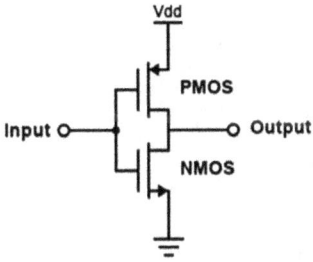

FIGURE 1.3   Schematic of the CMOS inverter.

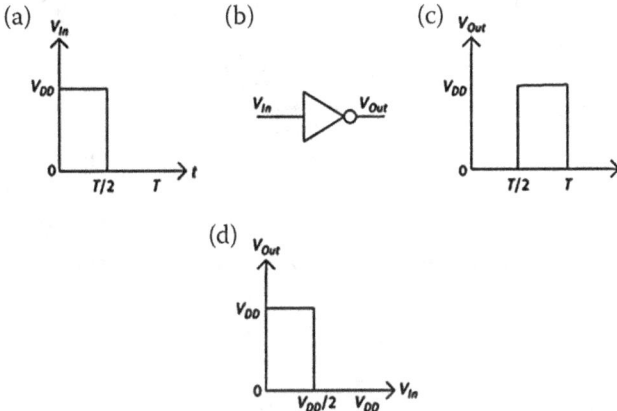

FIGURE 1.4   Response of the ideal inverter: (a) input signal, (b) symbol of the inverter, (c) output signal, (d) transfer characteristics.

the output of the inverter is high (V_{DD}). When the input value increases higher than $V_{DD}/2$, the output becomes zero.

In practical cases, the transfer characteristics of the inverter will be shown in Figure 1.5. The CMOS inverter has different regions of operation based on the condition of the PMOS and NMOS devices. The five regions of operations are mentioned in Table 1.1 and marked in the transfer curve [9].

The ratio between the width and length of the PMOS and NMOS devices is named aspect ratio $\left(\frac{W}{L}\right)_p$ and $\left(\frac{W}{L}\right)_n$, respectively. In general, electron mobility is greater than holes' mobility. The speed of the NMOS device is higher than PMOS device. In the CMOS inverter, the pull-up side is made up of a PMOS device, and the pull-down side is made up of an NMOS device. Thus, the pull-down operation will be quicker than the pull-up operation. In the input-output response of the CMOS inverter circuit shown in Figure 1.6, the high-to-low transition is expected when the input voltage is at $V_{DD}/2$, as in Figure 1.5. In practical cases, the aspect ratio of the PMOS device is increased than the NMOS device to make the transition at $V_{DD}/2$.

The average propagation delay of the CMOS inverter concerning the supply voltage variation is shown in Figure 1.7. It shows that the supply voltage and the

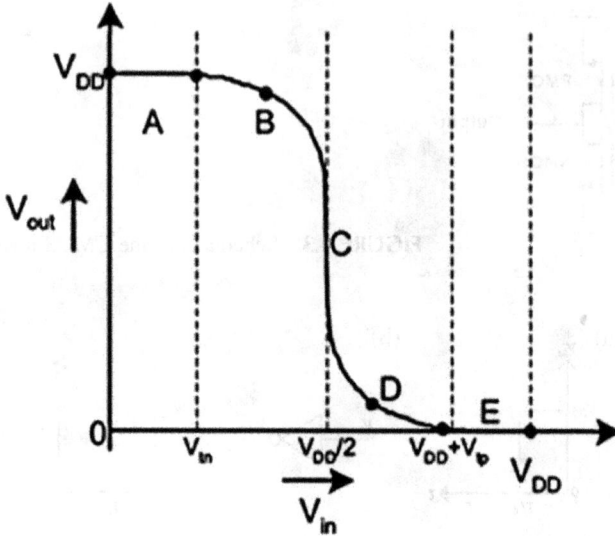

FIGURE 1.5 Transfer characteristics of practical inverter with different regions of PMOS and NMOS.

TABLE 1.1

Different Operating Regions of PMOS and NMOS in CMOS Inverter

Region	PMOS	NMOS
A.	Linear region	Cut-off region
B.	Linear region	Saturation region
C.	Saturation region	Saturation region
D.	Saturation region	Linear region
E.	Cut-off region	Linear region

propagation delay have an inverse relation. When the supply voltage is reduced below 0.5 V, the propagation delay is highly increased.

VLSI circuits have different logic configurations like CMOS, pseudo, pass, transmission, and domino logic. CMOS logic has less leakage current among the mentioned logic configurations due to the complementary structure. When the channel length of the transistor is reduced in the scale-down process, the leakage current gets increased. Figure 1.8 shows the sources of leakage current in an NMOS device of the CMOS logic [10]. Subthreshold leakage, gate leakage, and reverse-biased junction band-to-band tunneling leakage are mentioned in Figure 1.8. In the above-mentioned leakage, the subthreshold and gate leakage are the major sources in the 50 nm or higher nanometre technology nodes. When the transistors of the lower nanometre technology nodes (lower than 25 nm) are used in the circuit design, the band-to-band tunneling leakage gets increased, as in Figure 1.9.

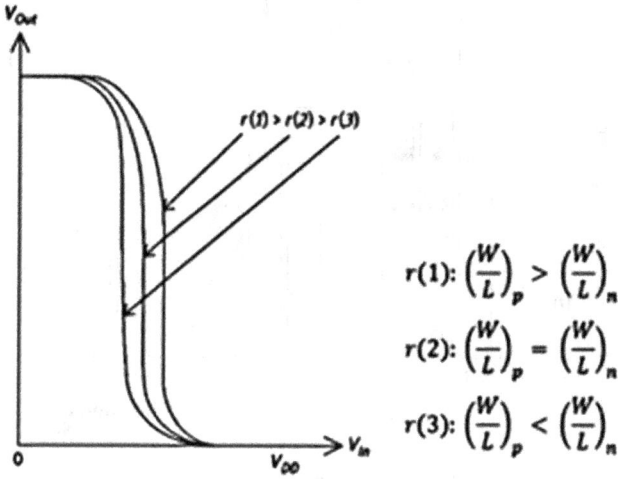

$$r(1): \left(\frac{W}{L}\right)_p > \left(\frac{W}{L}\right)_n$$

$$r(2): \left(\frac{W}{L}\right)_p = \left(\frac{W}{L}\right)_n$$

$$r(3): \left(\frac{W}{L}\right)_p < \left(\frac{W}{L}\right)_n$$

FIGURE 1.6 Impact of the aspect ratio of PMOS and NMOS on transfer characteristics.

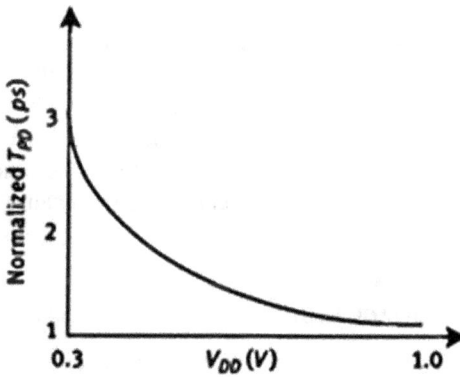

FIGURE 1.7 The normalized propagation delay of the inverter with a supply voltage variation.

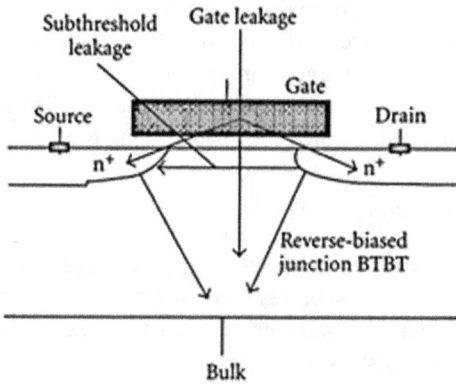

FIGURE 1.8 Leakage sources of NMOS device in static CMOS.

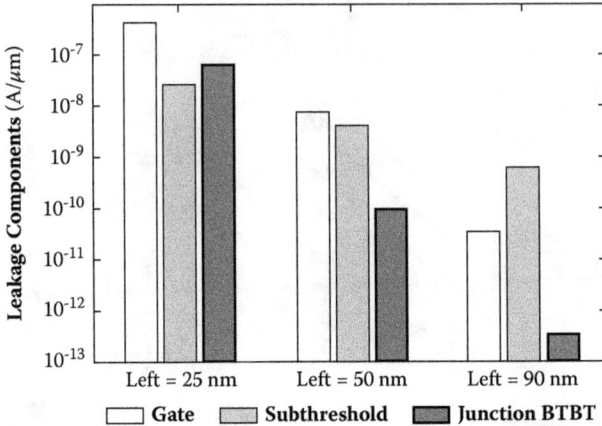

FIGURE 1.9 Gate leakage, subthreshold leakage, and junction leakage of NMOS devices at various technology nodes.

1.3 SCALING OF TRANSISTORS AND SECOND-ORDER EFFECTS

The performance of the CMOS can be improved by the scale-down operation of the NMOS and PMOS devices. The stray capacitances in the transistors are also reduced in the scale-down operation. Thus, the speed of the transistors gets improved. When the channel length is reduced to the order of the depletion layer, the non-ideal response is observed in the MOS transistors. These non-ideal effects are called second order effects, and the short-channel effects cause these second order effects.

1.3.1 THRESHOLD VOLTAGE REDUCTION

The minimum voltage required between the gate and source region for creating the inversion layer at the channel region below the gate terminal is called the threshold voltage [11]. During the scaling operation, the channel length of the NMOS and PMOS devices is decreased. When the gate voltage is applied, the minority careers accumulation under the gate terminal increases with respect to the reduced channel length. As a result, the reduced effective threshold voltage is observed with the scaling of the transistor. The charge-sharing mechanism is shown in Figure 1.10. It shows that the gate voltage can easily deplete the channel. Thus, the charge-sharing mechanism is more important when the channel length is reduced [12].

1.3.2 SUBTHRESHOLD SWING

The threshold voltage reduction increases the transistor leakage. When the transistor is in the OEE state, there is a subthreshold current to cause the transistor leakage. In integrated circuits (ICs), the number of transistors integrated into a unit area or chip increases as the transistors are scaled down. Thus, the subthreshold leakage current of all transistors creates an impact on the IC level. For example, 10 nA of

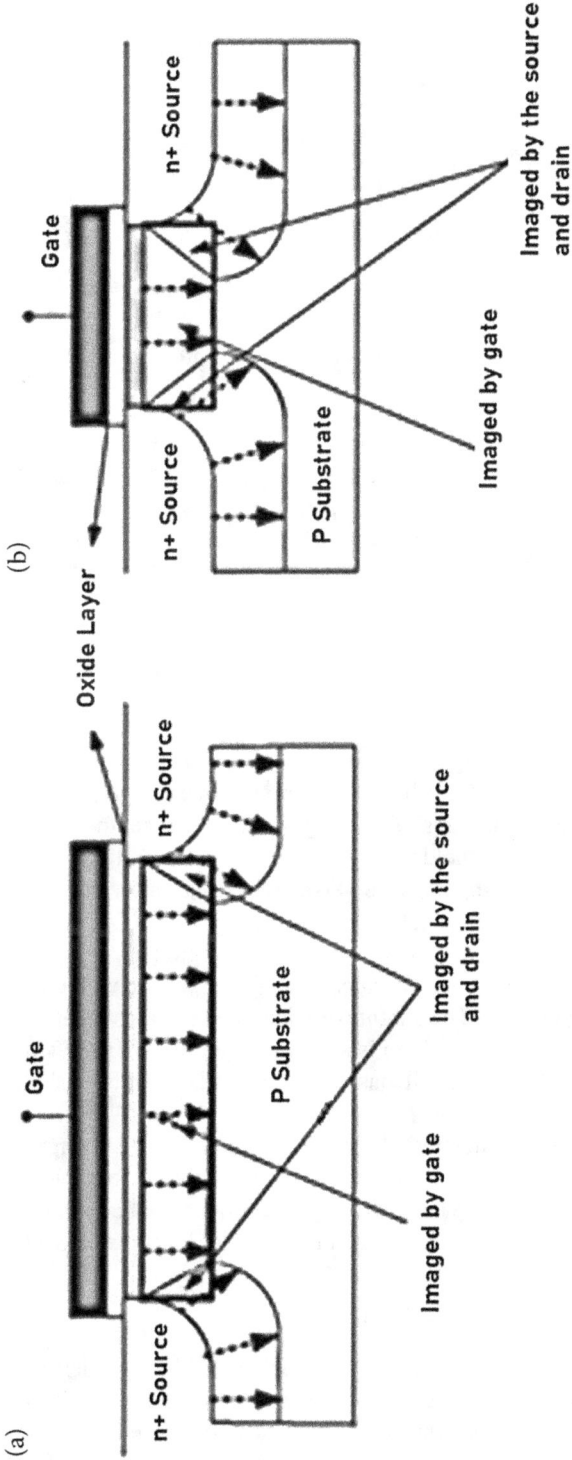

FIGURE 1.10 Illustration of the threshold voltage related to short channel effects and charge sharing between the source/drain depletion regions and the channel depletion region.

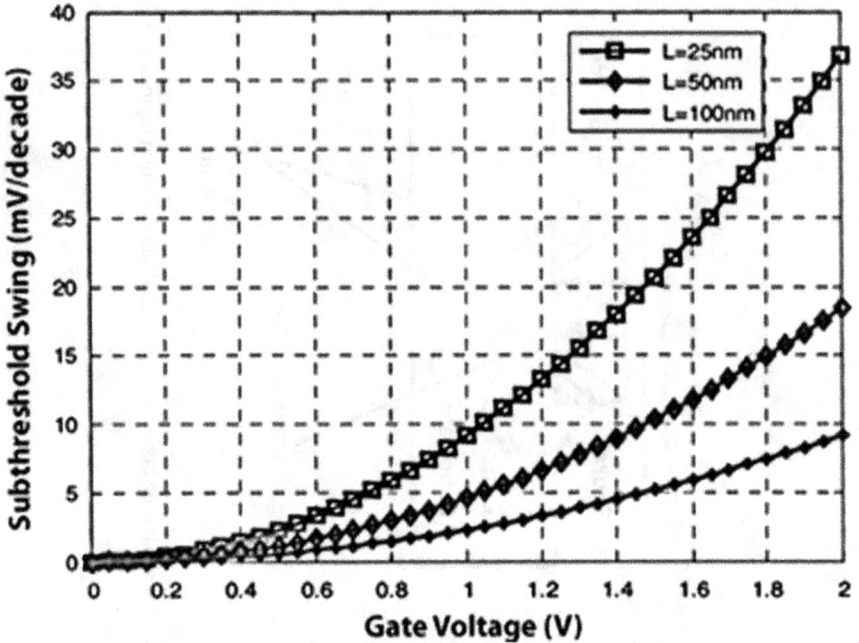

FIGURE 1.11   Subthreshold swing vs gate voltage for different channel length.

subthreshold current of a single transistor at VGS = 0 is smaller, but the power consumption in ICs with 100 million transistors is much larger. In some ICs, the MOSFET with two different threshold voltages is used to meet the speed and power requirements. When the threshold voltage is low, the speed and power consumption of the transistor will be high, and vice versa. So, based on the requirement of the specific circuit, the transistor selection can be made.

The transistor's behavior in the subthreshold region is determined by the subthreshold swing, which can be calculated using the rate of change in gate voltage to the drain current. The subthreshold swing increases when the transistor is scaled down below 130 nm. There is a limitation on the MOSFET performance by thermal voltage (kT/q), and it has 60 mV/decade of subthreshold swing at room temperature.

The dependency of the subthreshold swing on different gate voltage and channel lengths is depicted in Figure 1.11. It shows that the subthreshold swing is directly proportional to the gate voltage and inversely proportional to the channel length. The subthreshold swing rapidly increases when the channel length is below 60 nm [13].

1.3.3   SUBTHRESHOLD LEAKAGE CURRENT

The CMOS circuit's overall leakage power is calculated by considering the leakage current of each transistor. There are two essential sources of leakage current in CMOS. The first one is subthreshold leakage current, and the second one is gate tunneling leakage current.

With the advancement in technology, in the scale-down process, the supply voltage has to be scaled down to minimize the dynamic power consumption. The threshold voltage is also scaled down to keep the high drive current. The scaling of threshold voltage results in subthreshold leakage current. When the transistor is in a weak inversion region ($V_{GS} < V_{TH}$), the subthreshold leakage current occurs between the drain and the source.

The drift and diffusion currents are present in the drain to source current. During the vital inversion region ($V_{GS} > V_{TH}$), the drift current has a higher value. The concentration of the minority careers is approximately zero in the weak inversion region; the drain to source voltage makes a small electric field longitudinally. Now, the carriers move between the source and drain due to diffusion. The diffusion current highly contributes to the subthreshold current. Gate-to-source voltage and threshold voltage have exponential relation with diffusion current.

The subthreshold leakage current of the MOSFET device can be expressed as in equation (1.1),

$$I_{Subthreshold} = I_0 e^{\frac{V_{GS}-V_{TH}}{nV_T}} \left[1 - e^{-\frac{V_{DS}}{V_T}} \right] \tag{1.1}$$

Here, $I_0 = \frac{W \mu_0 C_{ox} V_T^2 e^{1.8}}{L}$, the thermal voltage $V_T = \frac{KT}{q}$, the threshold voltage V_{TH}, the drain to source voltage V_{DS}, the gate to source voltage V_{GS}, transistor width W, Length L, gate oxide capacitance C_{ox}, carrier mobility μ_0, and subthreshold swing coefficient n.

The subthreshold leakage happens when the transistor is in the OFF state. The above equation will suit the leakage current of an individual transistor. In a complete circuit, the leakage current is not equal to the sum of the leakage current of all individual devices. The transistors connected in a series or stacked manner have less total leakage current than the sum of individual transistor's leakage current. As depicted in Figure 1.12, the leakage current can be effectively reduced by stacking more stages of transistors [14].

As a negative effect, the stacking of transistors results in increased delay [15], shown in Figure 1.13.

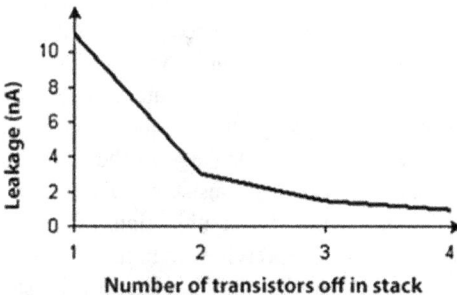

FIGURE 1.12 Subthreshold leakage current vs. off-transistors stacking.

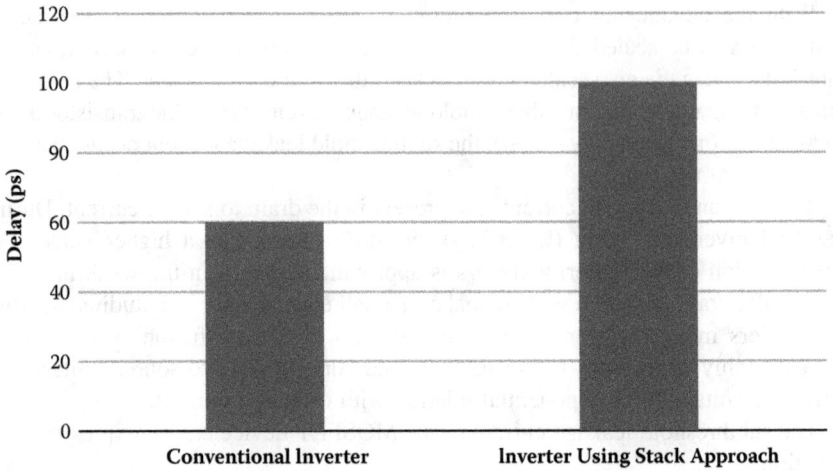

FIGURE 1.13 The comparison between traditional inverter and inverter using stack approach (a) The comparison of propagation delay.

1.3.4 GATE LEAKAGE POWER

In the scale-down process, the vertical dimension reduction is more complex than the horizontal one. The gate oxide thickness has to be reduced to get enough current drive at reduced supply voltage to reduce the short channel effects like drain-induced barrier lowering. The field through the oxide layer gets increased by the scaling of the oxide layer. From the gate to the channel and source or drain over-lapping region, the gate leakage current passes, which can also occur from the source or drain overlapping region to the gate. The scenarios mentioned above are shown in the following figure

When the thickness of the oxide layer is decreased, the gate leakage current increases exponentially. When a 3 nm or lower value of oxide thickness is reached, the gate leakage current is comparable with the subthreshold leakage current. The potential across the oxide layer and the gate leakage current density (A/m^2) of an NMOS transistor for various oxide thicknesses are compared in Figure 1.14.

When the transistor is turned OFF, the subthreshold leakage is calculated. Gate leakage occurs when the transistor is turned on or turned OFF, as shown in Figure 1.15. The gate leakage occurs through the source and drain overlap regions when the transistor is OFF (Figure 1.15(b). Gate leakage occurs along with the source and drain overlap regions through the channel (Figure 1.15(a). Thus, the gate leakage is higher than the subthreshold leakage in the transistor's ON state.

Since the gate leakage depends on the transistor's ON or OFF state, the transistors' bias conditions are essential. The possible combinations of bias methods are present in Figure 1.16. The transient states are represented in Figures 1.16(f) and (g); these biased conditions can be ignored because they are unrelated to the steady state. The same potential is present in all the terminals of Figure 1.16(a) and (h), and the gate leakage is not present in this case. Gate leakage is present in all other conditions in Figure 1.16.

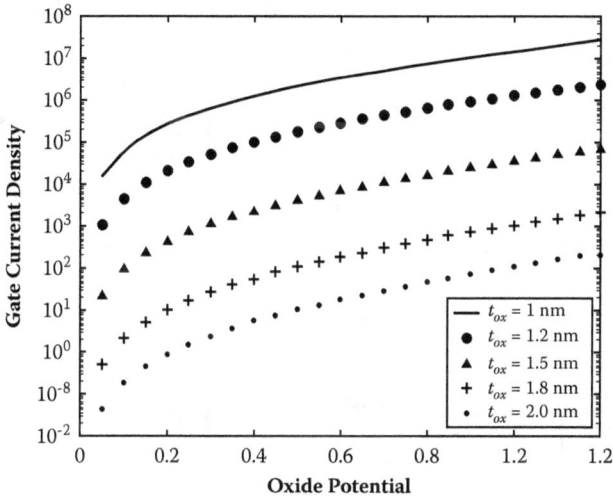

FIGURE 1.14 Variation of tunneling current density with potential drop across the oxide.

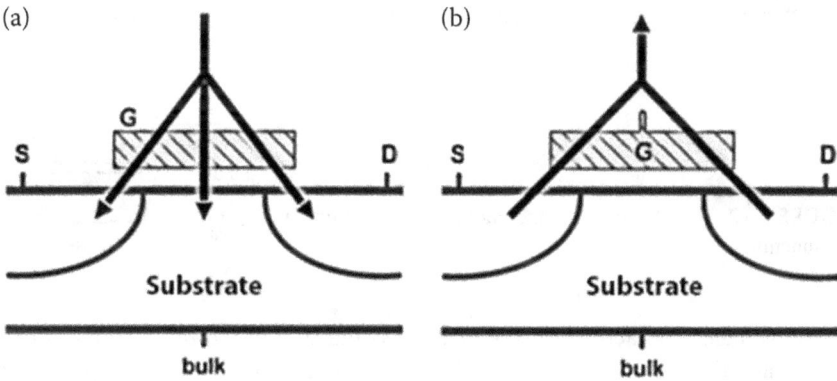

FIGURE 1.15 Gate oxide leakage (a) through source or drain overlap region and gate to channel, (b) through source or drain overlap region.

1.3.5 BAND-TO-BAND TUNNELING LEAKAGE

Two PN junctions are available in the MOS transistor. One is between the source and the well, and another is between the drain and the well. These reverse-biased PN junctions cause junction leakage. The doping concentration and junction area influence the junction leakage, and they get increased and decreased respectively in the scale-down operation. Band-to-band tunneling leakage is high when the n-type and p-type materials are highly doped [16].

When the high electric field is applied across the PN junction, tunneling of electrons occurs through the p-regions valance band to the n-regions conduction band in reverse biasing.

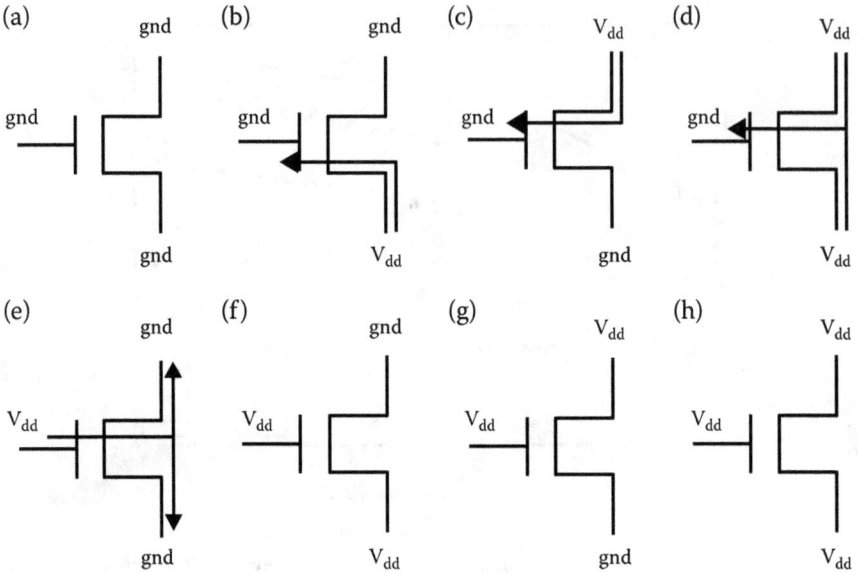

FIGURE 1.16 Possible bias condition of NMOS transistors in CMOS logic circuits.

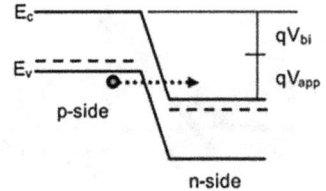

FIGURE 1.17 Band-to-band tunneling in reverse biased
PN junction.

The voltage drop across the PN junction is the sum of the reverse bias voltage applied and the built-in potential. When it is increased than the band gap, the tunneling current occurs, as shown in Figure 1.17. The built-in potential is a non-scaling factor [17]. Thus, the tunneling current increases with the scale-down operation. In lower nanometre technology node transistors with less than 25 nm of channel length, the band-to-band tunneling current causes more leakage in the transistor.

1.4 SUMMARY

The threshold voltage and subthreshold swing are essential in low-power, high-performance CMOS design. The threshold voltage is reduced in the scale-down process to improve the performance, leading to subthreshold leakage and static power dissipation. In addition to this, the gate leakage power increases as the scaling down of CMOS devices. Moreover, band-to-band tunneling leakage occurs at the reverse-biased PN junctions of the MOSFET.

The subthreshold swing indicates the transistor's behavior in the subthreshold regime. In MOSFET, 60 mV of gate voltage is required to create a decade of change in drain current. Thus, the subthreshold swing puts a restriction on low-power applications.

The scaling of transistors has ended, and the more Moore trends cannot improve efficiency and performance [18,19]. The CMOS technology has difficulties fulfilling the performance requirements of big data, the Internet of Things, edge sensors, deep learning, artificial intelligence, supercomputing, robotics, and autonomous systems. Thus, advanced transistors are required to meet the requirements of the systems mentioned above [20].

REFERENCES

[1] Ã, F. Ramundo, P. Nenzi, and M. Olivieri, "First Integration of MOSFET Band-to-Band-Tunneling Current in BSIM4," *Microelectronics Journal*, vol. 44, no. 1, Elsevier: pp. 26–32, 2013, 10.1016/j.mejo.2011.07.016

[2] A. Sarkar, S. De, M. Chanda, and C. Kumar Sarkar, *Low Power VLSI Design*. DE Gruyter Oldenbourg, 2016.

[3] P. F. Butzen, and R. P. Ribas, "Leakage Current in Sub-Micrometer CMOS Gates," *Universidade Federal Do Rio Grande Do Sul*, pp. 1–30, 2007, http://www.inf.ufrgs.br/logics/docman/book_emicro_butzen.pdf.

[4] D. Das, *VLSI DESIGN*. Second Edi. Oxford University Press, 2015.

[5] R. Das, and S. Baishya, "Analytical Model of Surface Potential and Threshold Voltage in Gate-Drain Overlap FinFET," *Microelectronics Journal*, vol. 75, pp. 153–159, April 2018, 10.1016/j.mejo.2018.04.005

[6] S. Dhar, M. Pattanaik, and P. Rajaram, "Advancement in Nanoscale CMOS Device Design En Route to Ultra-Low-Power Applications," *VLSI Design*, vol. 2011, pp. 1–19, 2011, 10.1155/2011/178516

[7] D. J. Frank, R. H. Dennard, E. Nowak, P. M. Solomon, Y. Taur, and H.-S. Philip Wong, "Device Scaling Limits of Si MOSFETs and Their Application Dependencies," *Proceedings of the IEEE*, vol. 89, pp. 259–288, 2001, IEEE, 10.1109/5.915374

[8] Z. Han, "The Power-Delay Product and Its Implication to CMOS Inverter The Power-Delay Product and Its Implication to CMOS Inverter," *Journal of Physics*, vol. 1754, 2021, 10.1088/1742-6596/1754/1/012131

[9] D. Harris, "Introduction to CMOS VLSI Design," *Harvey Mudd College*, 2004.

[10] IRDS. *INTERNATIONAL ROADMAP FOR DEVICES AND SYSTEMS: MORE MOORE*, 2021a.

[11] IRDS. *INTERNATIONAL ROADMAP FOR DEVICES AND SYSTEMS: BEYOND CMOS*, 2021b.

[12] M. Sanaullah, and M.H. Chowdhury, "Subthreshold Swing Characteristics of Multilayer MoS2 Tunnel FET Muhammad," *International Midwest Symposium on Circuits and Systems (MWSCAS)*, pp. 10–14, 2015, IEEE, 10.1109/MWSCAS.2015.7282101

[13] G. Naima, and S. Balmukund Rahi, "Low Power Circuit and System Design Hierarchy and Thermal Reliability of Tunnel Field Effect Transistor," *Silicon*, vol. 14, no. 7, Silicon: pp. 3233–3243, 2022, 10.1007/s12633-021-01088-2

[14] G. Rajakumar, and A. A. Roobert, "Design of Low Power VLSI Architecture of Line Coding Schemes," *Wireless Personal Communications*, 2018 January, 10.1007/s11277-018-5286-4

[15] A. A. Roobert, P. Sherly Arunodhayamary, D. Gracia Nirmala Rani, M. Venkatesh, and L. Jerart Julus, "Design and Analysis of 28 GHz CMOS Low Power LNA with 6,4 DB Gain Variability for 5G Applications," *Transactions on Emerging Telecommunications Technologies*, vol. 33, pp. 1–11, January 2022, 10.1002/ett.4486

[16] A. A. Roobert, and D. Gracia Nirmala Rani, "Survey on Parameter Optimization of Mobile Communication Band Low Noise Amplifier Design," *International Journal of RF and Microwave Computer-Aided Engineering*, pp. 1–16, 2019a, 10.1002/mmce.21720

[17] A. A. Roobert, and D. Gracia Nirmala Rani, "Design and Analysis of 0, 9 and 2, 3 GHz Concurrent Dual Band CMOS LNA for Mobile Communication," *International Journal of Circuit Theory and Applications*, pp. 1–14, July 2019b, 10.1002/cta.2688

[18] A. Roobert, "Design, Analysis and Optimization of CMOS Low Noise Amplifier for Wireless," Anna University, 2020, http://hdl.handle.net/10603/333501.

[19] A. Sarkar, *Low Power VLSI Design Fundamentals*, 2016.

[20] J. Shalf, "The Future of Computing beyond Moore' s Law," *The Royal Society Publishing*, 2020.

2 Basic Science and Development of Subthreshold Swing Technology

P. Suveetha Dhanaselvam, D. Sriram Kumar,
B. Karthikeyan, and P. Vanitha

CONTENTS

2.1 INTRODUCTION

In the existing scenario, advances in CMOS technology have made significant miniaturization of MOSFETs [1–5]. However, in addition to advances in production technology, the power consumed by integrated circuits is becoming increasingly important. Power dissipation is more pronounced in nanoelectronic circuits. In a field-effect transistor (FET), the minimum voltage needed to make the transistor switch ON is a dominant factor that ultimately determines the range of input supply voltage and power dissipation of device technology. Reducing the supply voltage minimizes the switching power. However, field-effect transistors (FETs) in recently developed integrated circuits require a minimum gate voltage of 60 mV in order to increase the current by one order at room temperature. "The subthreshold swing (SS) is defined as the difference in the gate voltage needed for increasing the drain current by a factor of 10 and is measured in mV/decade". Reducing subthreshold swing (SS) is needed to resume high ON-state current with acceptable OFF-state leakage, which has been a critical technical issue [6].

Subthreshold swings of MOSFETs are restricted by the Boltzmann distribution of carriers to the range of 60 mV/dec at room temperature, and subthreshold swing further increases with scaling. This increase in subthreshold swing with a decrease in gate length represents a fundamental limitation of traditional FETs. This constraint of SS is that all traditional FETs are based on the modulation of charge carriers' thermal radiation generated at the source contacts. Therefore, a minimum SS of 60 mV/decade at room temperature results from the carrier's thermally expanded Fermi distribution and thus occurs in any conventional FET regardless of size, materials used, etc. Thus, to obtain subthreshold swings below 60 mV/decade, the current from the source terminal must be modified to become independent of the thermally extended Fermi distribution function [7].

In the nanoscale regime, miniaturization is the key factor in today's electronic device applications [8]. As MOSFETs become smaller and smaller [9] and face fundamental performance limitations, the novel devices used on band-to-band tunneling are focused. In particular, the emphasis is on devices that act as field effect transistors (FETs), in which the variations in gate voltage switch the device but practice band-to-band tunneling in the ON state and transitions of both states. These devices have the prospective for very low OFF-current and offer the opportunity to minimize the subthreshold swing beyond traditional MOSFETs' 60 mV/dec limit. Hence, they appear to be a perfect replacement for minimized quasi-ideal switches that improve subthreshold swing and drain current. The aforesaid device is the Tunnel FET that contains a SiGe delta layer at the edge of the p+ region, reducing the barrier width and thus enhancing the subthreshold swing and ON-current. Tunnel FETs sidestep the limitations of traditional MOSFET by using band-to-band tunneling instead of heat injection to let the charge carriers into the device channel. A Tunnel Field Effect Transistor (TFET) is a gated p-i-n diode that operates in reverse bias. When OFF, the width of the potential barrier measured between the source and the channel is so high that tunneling does not occur. The leakage current is negligible. When ON, while the gate voltage increases beyond the threshold voltage, the width of the potential barrier between the channel and source is lesser and allows significant tunnel current. Tunnel FETs based on ultrathin semiconductor films or nanowires can reduce power 100 times on CMOS transistors; thus, integrating tunneling FETs with CMOS technology can improve low-power microchips. One of the significant characteristics of a TFET that distinguishes it from traditional MOSFETs is its excellent subthreshold swing, especially near the gate voltage where the drain current begins to increase. A 3D tunnel FET can achieve an inverse subthreshold slope of less than 60 mV/dec. However, there is a trade-off between high inrush current and small subthreshold swing values [7]. The Tunnel FET has become more efficient because of its ability to achieve sub-60 mV/dec subthreshold swing at room temperature, proving it to be a promising ultra-low power consumption device.

2.2   SUBTHRESHOLD BEHAVIOR

This section deals with the basics of subthreshold behavior. The transfer characteristics of a transistor are classified into two based on its drain current flow, i.e., the subthreshold and superthreshold regions. This is illustrated in Figure 2.1. Generally, a transistor conducts when its input voltage equals the threshold voltage. Until that voltage, it is in the OFF state, and that region below the threshold voltage is referred to

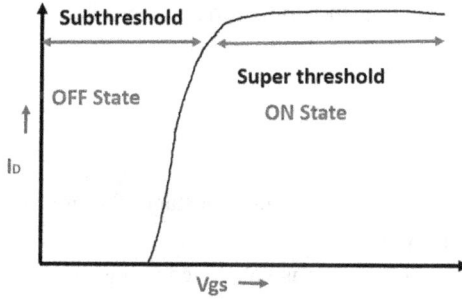

FIGURE 2.1 Showing the drain characteristics of MOSFETs – subthreshold regions.

as the subthreshold region. Similarly, when the input voltage extends beyond the threshold voltage, it is in the ON state and is called the superthreshold region. The current which flows from the source to drain in the subthreshold part of the cut-off region is called subthreshold current. Usually, there is no current in the subthreshold region since there is no channel formation. The drain junction is reverse-biased, assuming the source and body are tied together. The minority carriers are diffused into the source junction. The field applied to the transistor produces a drift current, which opposes the diffusion current. The minority carriers are injected into the drain junction before the recombination process. This increases the minority carriers and it even exceeds the majority carriers present. This results in subthreshold current due to the presence of minority carriers. It is also referred to as leakage current. In the super-threshold region, the situation reverts, i.e., the majority of carriers increase, and this causes an increase in drain current. The subthreshold current is given by:

$$I_{DS} = \mu_{eff} C_{ox} \frac{W}{L} (m - 1) \left(\frac{kT}{q}\right)^2 e^{q(V_G - V_T)/mkT} \left(1 - e^{-qV_{DS}/kT}\right) \tag{2.1}$$

The drain characteristics (drain current Vs gate voltage) of MOSFETs are exhibiting log linear behavior in their operating regions. The slope of this logarithmic curve is referred to as the subthreshold slope and it is shown in Figure 2.2.

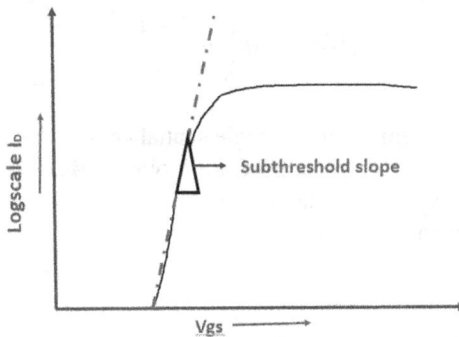

FIGURE 2.2 Subthreshold slope.

In the subthreshold region, the drain current depends exponentially on the gate voltage [10].

$$I_d \propto \exp\left(\frac{V_{gs}}{nV_T}\right) \tag{2.2}$$

Where $V_T = \frac{kT}{q}$, K represents Boltzmann Constant, T represents absolute temperature and q is the electron charge.

The subthreshold slope factor n is calculated using the formula

$$n = 1 + \frac{C_b}{C_g} \tag{2.3}$$

Where C_b represents bulk capacitance and C_g represents gate capacitance

2.3   PARAMETERS OF SUBTHRESHOLD SWING

The subthreshold swing (SS) of a device is the difference in gate voltage (V_{GS}) required to increase the output drain current (I_D) by a factor of 10 [11,12].

$$SS = \frac{d\,V_G}{d\,(log I_D)}\,[mV/dec] \tag{2.4}$$

With considerable biasing of the transistor, the drain current (I_D) in a MOSFET is proportionate to the number of electron states per unit volume per unit energy D(E), the probability that a particle will have energy is Fermi–Dirac $f_s(E)$, and the carrier momentum $v(E)$, E is the electron energy.

$$I_D \propto \int dE \cdot D(E) \cdot v(E) \cdot f_s(E) \tag{2.5}$$

The source Fermi–Dirac distribution function $f_s(E)$ in the exponential tail region, can be given by a Boltzmann approximation:

$$f_s(E) \approx exp\left(-\frac{E - E_F^S}{kT}\right) \tag{2.6}$$

For ease, supposing movement in the single spatial course, the product of D(E) and $v(E)$ is considered constant. Furthermore, for a typical MOSFET, it is inferred that the difference in the conduction band is in direct relation to the difference in the gate voltage. Thus, we may derive

$$\frac{\partial I_D}{\partial V_{GS}} = q\frac{\partial I_D}{\partial E} \tag{2.7}$$

The subthreshold swing of a MOSFET can be computed considering equations (2.6) and (2.7), such that the least probable swing in a typical device is given as,

$$SS = \left(\frac{\partial \log(I_D)}{\partial V_{GS}}\right)^{-1} = \ln 10 \left(\frac{1}{I_D}\frac{\partial I_D}{\partial V_{GS}}\right)^{-1} \approx \ln 10 \frac{kT}{q} \qquad (2.8)$$

A low SS would concede an improved ON-OFF (I_{ON}/I_{OFF}) ratio, and leads to the lowest power loss in the OFF state. It is noteworthy that equation (2.9) denotes the ideal subthreshold swing achievable in a traditional MOSFET device. At room temperature, the Subthreshold swing of the device is approximately 60 mV/decade.

$$SS_{MOSFET} = \ln(10)\frac{kT}{q}[mV/dec] \qquad (2.9)$$

Equation (2.9) describes the minimum bound on the SS imposed by the dynamics of the current movement in an ideal MOSFET device. Alternatively, equation (2.7) defines the subthreshold swing of MOSFET in terms of depletion region capacitance C_d and gate oxide capacitance $C_{ox} = \varepsilon_{ox}/t_{ox}$, where ε_{ox} is the permittivity of the oxide and t_{ox} is the gate oxide thickness.

$$SS_{MOSFET} = \left(\frac{\partial \log(I_D)}{\partial V_{GS}}\right)^{-1} = 2.3\frac{kT}{|e|}\left(1 + \frac{C_d}{C_{ox}}\right) \qquad (2.10)$$

Evidently, from equation (2.10), the lowermost value of SS_{MOSFET} is 2.3 kT/q, which is 60 $mV/decade$ at room temperature. Suppose that, to achieve an $I_{ON}/I_{OFF} = 10^5$, a gate voltage of 5 × 60 mV = 0.3V has to be applied. Thus, scaling down the supply voltage is arduous to achieve a higher ratio of I_{ON}/I_{OFF}. In addition to the restriction inflicted by the SS, various short channel effects (SCEs), such as DIBL-drain-induced barrier lowering, impact ionization, threshold voltage roll-off, hot electron effect, charge sharing amidst drain and gate also degrades the performance of MOSFET when scaling down the channel length below 50 nm. To overwhelm these basic constraints of higher I_{OFF}, poor SS_{MOSFET}, and SCEs of MOSFETs, TFETs can serve as an ideal alternate device.

Tunnel FET, in contrast, does not suffer from this physical constraint since the current mechanism depends on the tunneling barrier width instead of the formation of an inversion channel. In contrast to the traditional MOSFET, the subthreshold swing of a TFET is a strong function of the gate voltage V_{GS}. Below the subthreshold, the drain current (I_D) falls faster than exponential. In the Subthreshold regime of TFET, the drain current starts to increase only when the gate voltage reaches $V_{OFF} \approx 0.3V$, where V_{OFF} is the voltage at which the reverse-biased p-i-n diode characteristics make a transition to tunnel FET characteristics. When the gate voltage exceeds V_{OFF}, I_D begins to rise swiftly. In this region, the energy levels in the source and channel are in line enabling band-to-band tunneling (BTBT). As a result, the tunneling current begins to flow in this operating range. The drain current increases sharply near $V_{GS} = V_{OFF}$, and this increase in drain current with the gate voltage gradually reduces. The proportionate increase in the drain current with increasing gate voltage is computed by the subthreshold slope or its reciprocal subthreshold swing (SS). The transistor should have a petite value of SS. With small SS, I_D can be increased to the heights of ON-condition from the OFF-condition by a reduced gate voltage fashioning the device to operate at trifling bias voltages.

TFET conducts in the subthreshold region through BTBT, which takes place amongst the intrinsic and p+ regions. Equation (2.11) indicates that the magnitude of tunneling generation is an inverse exponential of the lateral electric field (E) in the channel, which in turn is a function of the gate voltage (V_{GS}).

$$I_D \propto e^{-1/E} \tag{2.11}$$

Therefore, the Subthreshold swing of TFET changes with V_{GS} and is not a constant contrast to the instance of a MOSFET [6]. For the Tunnel FET, the tunnel current can be expressed roughly as given below:

$$I = aV_{eff}\xi \exp\left(-\frac{b}{\xi}\right) \tag{2.12}$$

where ξ is the maximum electric potential at the junction, a and b are coefficients determined by the properties of the junction of the material and the cross-sectional area of device A. Precisely, $a = Aq^3\sqrt{2m_R^*/E_G}/8\pi^2\hbar^2$ and $b = 4\sqrt{2m_R^*}E_G^{\frac{3}{2}}/3q\hbar$, where E_G is the effective bandgap, m is the carrier effective mass and $\hbar$ is the reduced Plank constant [6].

In view of the fact that the output current of the Tunnel FET is the tunneling current between the source and channel determined by the gate voltage, its SS value depends on several parameters, including the thickness of the gate oxide (t_{ox}), the thickness of the SOI layer (t_{SOI}), and the steepness of the doping profile of the source. The SS value becomes smaller as the gate oxide layer and SOI layer thickness decrease. Indeed, the Subthreshold swing value is usually governed by the interdependency of the gate and channel voltages and the effect of the drain potential on the tunnel barrier. With the reduction in t_{ox} and t_{SOI}, the channel voltage turns out to be more directly linked to the gate potential than to the drain potential, with a subsequent decrease in the Subthreshold value. When the t_{ox} and t_{SOI} were insignificant, the SS value dropped below 60 mV/dec [13]. Thin gate oxide, increased doping at the source, and abrupt doping profile at the tunneling junction can enhance the TFET operation. Simulation results [14] show that increased doping concentration of the source region results in decreased SS. Also, SiGe material could enhance the TFET operation for the reason that the concentration of boron impurities is significant in SiGe when compared to silicon.

The derivative of the tunneling current equation of (2.12) relating to the gate-to-source voltage can then be utilized to govern an expression for the TFET subthreshold swing.

$$SS_{TFET} = \ln 10\left[\frac{1}{V_R}\frac{dV_R}{dV_{GS}} + \frac{\xi+b}{\xi^2}\frac{d\xi}{dV_{GS}}\right]^{-1} \tag{2.13}$$

The denominator of the above equation (2.13) which is not explicitly limited by kT/q, should be increased to reduce the subthreshold swing. Rendering to the first term, the transistor should be designed in such a way that the voltage at the gate terminal directly influences the bias of the tunnel junction. This term infers that the geometry of the transistor is influenced by a gate with a strong field so that the gate directly modifies the reverse bias of the junction. Suppose gate electrostatics uses a thin dielectric made of high-k material and an ultrathin body; the gate bias directly regulates the band overlap $dV_R/dV_{GS} \approx 1$ making the first term in the denominator of equation (2.13) inversely proportional to V_{GS}. Accordingly, the SS reduces as gate-to-source voltage decreases. Alternatively, that SS is lessened by increasing the denominator of equation (2.13). This happens when the gate is held to line up the applied field with the field of the tunnel junction. Thus, the gate potential complements the internal field to enhance the tunneling possibility [6].

Zhang [15] derived an analytical formula in which the gate regulates the band-to-band overlap as well as the tunnel junction potential, resulting in less than 60 $mV/decade$ swing at 300 K. The subthreshold swing (SS) is expressed as,

$$SS_{POINT} = (dlogI_D/dV_{GS})^{-1} \text{ mV/decade} \qquad (2.14)$$

The SS_{POINT} defined by (2.14) is known as the point subthreshold swing, which represents the inverse of the actual slope of the I-V characteristics at a specific gate voltage (V_{GS}). For imminent low-power CMOS applications, it is required that the TFETs display improved subthreshold swing related to traditional MOSFETs over a few decades of drain current. Unlike MOSFET, the Subthreshold slope of TFET is a strong function of V_{GS}. Thus, the average SS must be extricated for the TFETs instead of SS_{POINT}. The average subthreshold swing (SS_{AVG}) is stated in terms of threshold voltage V_{TH}, the gate voltage V_{OFF} at which I_D begins to rise swiftly, I_{VT} which is the value of I_D when $V_{GS} = V_{TH}$ and I_{OFF} which indicates the value of drain current at $V_{GS} = V_{OFF}$ is as follows;

$$SS_{AVG} = \frac{(V_{TH} - V_{OFF})}{\log(I_{VT}) - \log(I_{OFF})} \qquad (2.15)$$

The threshold voltage (V_{TH}) is referred to as the gate voltage at which the drain current reaches $10^{-7} A/\mu m$ at $V_{DS} = Bias\ voltage\ V_{DD}$ [14,16]. SS_{AVG} represents the amount of V_{GS} required (on average) to increase the device current ten folds in the subthreshold region. SS_{AVG} is a vital factor that affects the device performance as a switch [16].

2.4   A SURVEY ANALYSIS OF SUBTHRESHOLD SWING

In this section, let us discuss the subthreshold swing characteristics of diverse structures in detail with reference to the dependent factors. The analysis of subthreshold swing for various devices is summarized in Figure 2.3

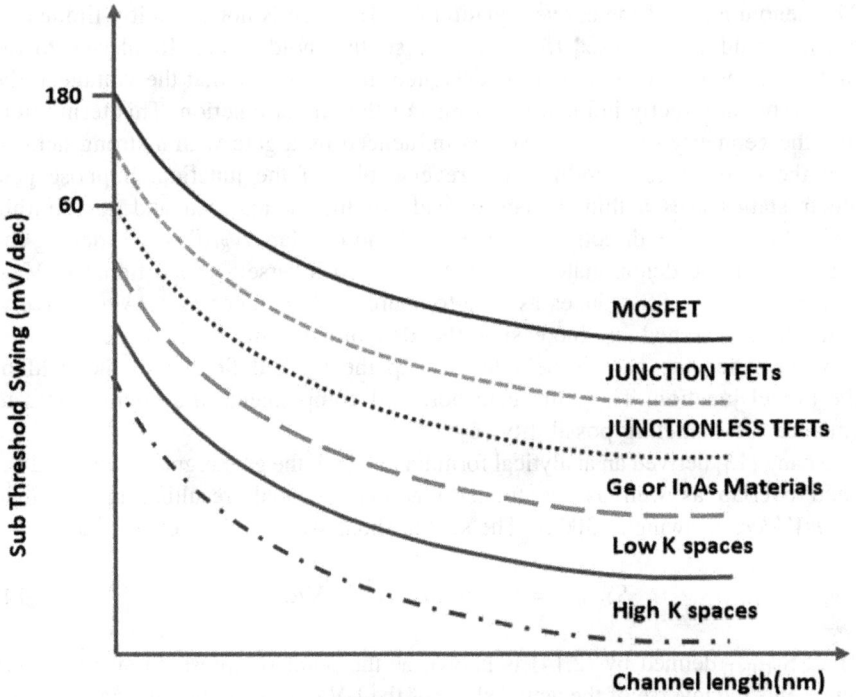

FIGURE 2.3 Variation of subthreshold swing.

2.4.1 GATE MATERIAL

Graphene nanoribbon-based TFET is proposed by Wei Cao et al. [17]. The dimensions of the Graphene nano ribbon-based channel are 2.34 nm × 0.34 nm. It is detected that the SS of this Graphene nanoribbon-based TFET points even to zero. The nominal value of SS is 180 mV and it is inferred to increase faster than silicon-based TFETs. The next device for discussion is silicon-based Interband TFET [6], proposed by Qin Zhang et.al. The subthreshold swing depends on the gate voltage and it is evident from this work that the SS can be achieved even less than 60 mV/dec for this device when the input value is nearly zero voltage. The channel materials Ge or InAs are responsible for achieving the lower value of SS. The least value of SS is 7 mV/dec for the gate to source voltage 0.07 V.

Junctionless TFETs are also proposed in the literature [18–20], which are more efficient than junction-based devices. When the doping concentration increases, it poses more OFF current at a faster rate than ON current. Hence it reduces the I_{ON}/I_{OFF} ratio. This also has the effect of an increase in the subthreshold slope. Using high K dielectric and spacers between the gates, the ON current can be increased, resulting in a low subthreshold slope. Low K spacers are also used in this structure which helps in decreasing the gate coupling and the threshold voltage. The simulation results of this work prove that the higher the dielectric constant value, the lower the OFF current. This enables an increase in I_{ON}/I_{OFF} ratio. This lowers the subthreshold slope even below 60 mV/dec. The major reason for this lower SS in junction-less TFETs is the current conduction due to tunneling.

2.4.2 Effect of Temperature

The effect of temperature in varying the subthreshold swing is studied for MOSFETs [18]. In the proposed experimental work, the temperature is varied from 10 to 300 K, and the subthreshold swing is observed. The SS is found to be initially low at 10 mV/dec and then gradually increases with temperature to 60 mV/dec at 300 K. In [21,22], a heterostructure junctionless double gate TFET is proposed. In this structure, AlGaAs and Si are used as hetero materials. The increase in temperature increases the ON current gradually. Obviously, the higher the temperature, the lower the OFF or leakage current. The effect of temperature is obtained by the analysis of the following equation in the junction-based double gate MOSFET [23]

$$I_{ds} = \frac{n_i \mu_n W k T \left[1 - \exp\left(\frac{-qV_{ds}}{kT}\right) \right]}{\int_0^{L_g} \frac{dy}{\int_{-\frac{t_{si}}{2}}^{\frac{t_{si}}{2}} \exp\left(\frac{-q\varphi(x,y)}{kT}\right) dx}} \qquad (2.16)$$

This equation emphasizes the dependence of the subthreshold swing on the temperature. The temperature is inversely proportional to the drain current; hence, the increase in temperature causes a decrease in the leakage or subthreshold current.

2.4.3 Effect of Scale Length

From the literature [23], it is evident that the scale length and the proportionate oxide thickness determine the value of subthreshold swing. The range of the SS varies based on the geometric mean of the silicon and oxide thickness, along with the scale length factor. The following equation [23] deserves to be the condition for calculating SS for devices with a scale length less than 10 nm:

$$\frac{L_g - 0.215}{6.38} > t_{0x} \qquad (2.17)$$

The channel length should be within $L_g > 1.5\lambda$ in order to reduce the deviation of the subthreshold swing. Interpretation of the scale theory outlines that the Silicon's geometric mean and oxide thickness is directly proportional to the channel length. Using this, it is also inferred from the simulation results [23] that when the channel length satisfies (2.17), the deviation is optimum and the range is also acceptable for the subthreshold swing. This is depicted in Figures 2.4a and 2.4b. It shows the percentage of deviation of subthreshold swing for varying channel lengths.

(a)

(b)

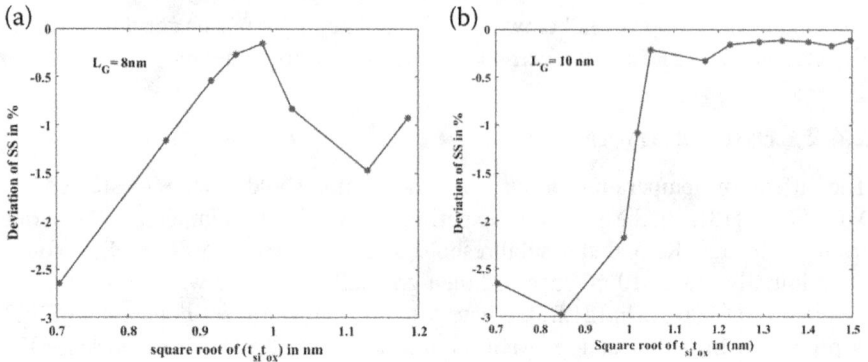

FIGURE 2.4   (a) Deviation of subthreshold swing for varying channel lengths. (b) Deviation of subthreshold swing for varying channel lengths.

2.5   CONCLUSION

The basics of subthreshold swing are explained in this chapter. It also emphasizes the origin of subthreshold swing from its subthreshold behavior and compares the SS of diverse structures, including MOSFETs and TFETs. The SS variation for various factors like gate material, temperature, and scale length factor is also studied and analyzed with reference to the existing literature.

REFERENCES

[1] P. Suveetha Dhanaselvam, and N. B. Balamurugan, "Analytical approach of a nanoscale triple-material surrounding gate (TMSG) MOSFETs for reduced short-channel effects," *Microelectronics Journal*, vol. 44, no. 5, pp. 400–404, May 2013.

[2] P. Suveetha Dhanaselvam, and N. B. Balamurugan, "A 2D transconductance and sub-threshold behavior model for triple material surrounding gate MOSFETs," *Microelectronics Journal*, vol. 44, no. 12, pp. 115–116, December 2013.

[3] P. Suveetha Dhanaselvam, and N. B. Balamurugan, "A 2D subthreshold current model single halo triple material surrounding gate (SHTMSG) MOSFET," *Microelectronics Journal*, vol. 45, no. 6, pp. 574–577, June 2014.

[4] P. Suveetha Dhanaselvam, N. B. Balamurugan, G. C. Vivek Chakaravarthi, R. P. Ramesh, and B. R. Sathish Kumar, "A 2D analytical modelling of single halo triple material surrounding gate (SHTMSG) MOSFET," *Journal of Electrical Engineering and Technology*, vol. 9, no. 4, pp. 1355–1359, July 2014, 10.5370/ JEET.2014.9.4.1355

[5] P. Suveetha Dhanaselvam, and N. B. Balamurugan, "Performance analysis of fully depleted triple material surrounding gate (TMSG) SOI MOSFET," *Journal of Computational Electronics*, vol. 13, pp. 449–455, 2014.

[6] Q. Zhang, W. Zhao, and A. Seabaugh, "Low-subthreshold-swing tunnel transistors," *IEEE Electron Device Letters*, vol. 27, no. 4, pp. 297–300, April 2006, 10.1109/ LED.2006.871855

[7] J. Knoch, S. Mantl, and J. Appenzeller, "Impact of the dimensionality on the performance of tunneling FETs: Bulk versus one-dimensional devices," *Solid-State Electron.*, vol. 51, no. 4, pp. 572–578, April 2007, 10.1016/j.sse.2007.02.001

[8] P. S. Dhanaselvam, P. Vimala, and T. S. A. Samuel, "A 2D analytical modeling and simulation of double halo triple material surrounding gate (DH-TMSG) MOSFET," *Silicon*, vol. 13, no. 8, pp. 2631–2637, August 2021, 10.1007/s12633-020-00617-9

[9] P. Suveetha Dhanaselvam, P. Vimala, and T. S. Arun Samuel, "A 2D analytical modelling and simulation of double halo triple material surrounding gate (DH-TMSG) MOSFET," *Silicon*, vol. 13, pp. 2631–2637, 2021.

[10] Y. Tsividis, *Operation and modeling of the MOS transistor*, 2nd edition. Boston: WCB/McGraw-Hill, 1999.

[11] K. Boucart, and A. Ionescu, "Double-gate tunnel FET with high- k gate dielectric," *Electron Devices IEEE Transactions On*, vol. 54, pp. 1725–1733, August 2007, 10.1109/TED.2007.899389

[12] M. J. Kumar, R. Vishnoi, and P. Pandey, *Tunnel field-effect transistors (TFET): Modelling and simulations*, 1st edition. Hoboken: Wiley, 2016.

[13] W. Y. Choi, B.-G. Park, J. D. Lee, and T.-J. K. Liu, "Tunneling field-effect transistors (TFETs) with subthreshold swing (SS) less than 60 mV/dec," *IEEE Electron Device Letters*, vol. 28, no. 8, pp. 743–745, August 2007, 10.1109/LED.2007.901273

[14] P.-F. Wang et al., "Complementary tunneling transistor for low power application," *Solid-State Electronics*, vol. 48, no. 12, pp. 2281–2286, December 2004, 10.1016/j.sse.2004.04.006

[15] A. C. Seabaugh, and Q. Zhang, "Low-voltage tunnel transistors for beyond CMOS logic," *Proceedings IEEE*, vol. 98, no. 12, pp. 2095–2110, December 2010, 10.1109/JPROC.2010.2070470

[16] S. Saurabh, and M. J. Kumar, *Fundamentals of tunnel field-effect transistors*, 1st edition. Boca Raton: CRC Press, 2016.

[17] W. Cao, D. Sarkar, Y. Khatami, J. Kang, and K. Banerjee, "Subthreshold-swing physics of tunnel field-effect transistors," *AIP Advances*, vol. 4, no. 6, p. 067141, June 2014, 10.1063/1.4881979

[18] B. Ghosh, P. Bal, and P. Mondal, "A junctionless tunnel field effect transistor with low subthreshold slope," *Journal of Computational Electronics*, vol. 12, no. 3, pp. 428–436, September 2013, 10.1007/s10825-013-0450-2

[19] S. Manikandan, P. Suveetha Dhanaselvam, and M. Karthigai Pandian, "A 2D unified subthreshold drain current investigation for junctionless cylindrical surrounding gate(JCSG) silicon nanowire transistor," *Silicon*, 2021.

[20] S. Manikandan, P. Suveetha Dhanaselvam, and M. Karthigai Pandian, "A quasi 2-D electrostatic potential and threshold voltage model for junctionless triple material cylindrical surrounding gate si nanowire transistor," *Journal of Nanoelectronics and Optoelectronics*, vol. 16, no. 2, pp. 318–323.

[21] G. Ghibaudo, M. Aouad, M. Casse, S. Martinie, T. Poiroux, and F. Balestra, "On the modelling of temperature dependence of subthreshold swing in MOSFETs down to cryogenic temperature," *Solid-State Electronics*, vol. 170, August 2020.

[22] M. Suguna, V. A. Nithya sree, R. Kaveri, M. Hemalatha, N. B. Balamurugan, D. Sriramkumar, and P. Suveetha Dhanaselvam, "Analytical modelling and simulation based investigation of triple material surrounding gate heterojunction tunnel FET," *Silicon*, vol. 14, pp. 10729–10740, March 2022.

[23] H. Jung, "Subthreshold swing model using scale length for sub-10 nm junction-based double-gate MOSFETs," *International Journal of Electrical and Computer Engineering*, vol. 10, no. 2, April 2020.

3 Historical Development of MOS Technology to Tunnel FETs

S. Manikandan and Adhithan Pon

CONTENTS

3.1 INTRODUCTION

The conventional Metal Oxide Semiconductor Field Effect Transistors (MOSFETs) can act as a gated switch that controls the current flow between the source and drain terminals. This invention later became the heart of Integrated circuits (ICs).

DOI: 10.1201/9781003327035-3

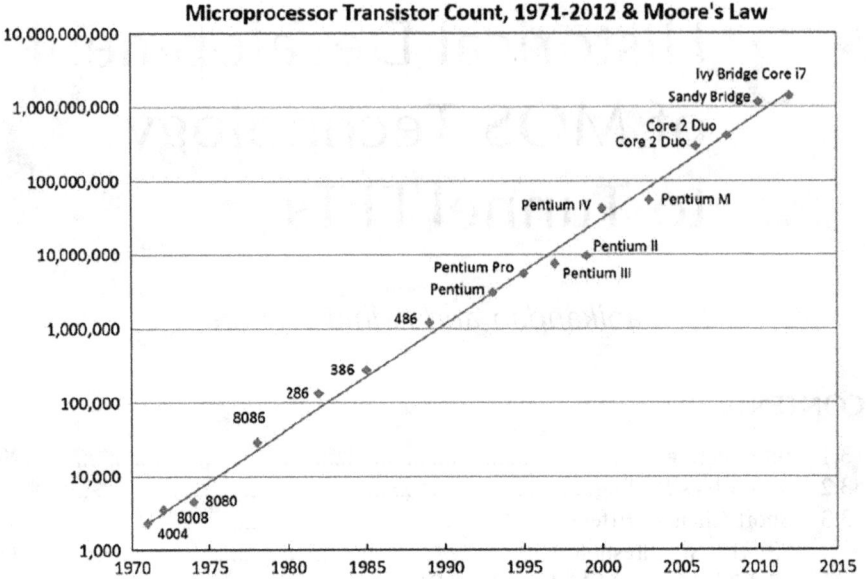

FIGURE 3.1 Transistor scaling dictates the number of transistors in the processor.

Consequently, the performance of ICs merely depends on the performance of MOSFETs. Hence, the advancement in electronics in the digital world, such as smart watches, smart TV, the Internet of things, Cloud computing, and Artificial Intelligence (AI), has become inevitable to the common man nowadays. The main success of MOSFETs is that they can be scaled down to increase the IC performance and power issues and reduce the total area. Figure 3.1 shows the plot transistor scaling that denotes the processor and its number of transistors. It is clearly visible that the number of transistors doubles in each new processor. This scaling is denoted as Moore's law by industry. However, the MOSFETs scaling leads to several critical issues such as Short Channel Effects (SCEs), Hot carrier degradation, and more leakages. Hence, the industry looked for an alternate device structure that could replace the conventional MOSFETs. The replacement is not only based on scaling but also on the device performance, I_{ON}/I_{OFF} ratio, subthreshold swing (SS), power dissipation, and so on. Several transistors have been considered to replace conventional MOSFETs, such as Silicon on Insulator (SOI) based MOSFETs, FinFETs, Junctionless FETs, Nanowires, and Tunnel FETs. Among these transistors, Tunnel FETs have better subthreshold swing and I_{ON}/I_{OFF} ratio than the conventional MOSFETs.

3.2 MOSFETS SCALING

A transistor acts as a switch in Integrated Circuits (ICs). In the last four decades, researchers worldwide have focused on reducing the size of MOSFETs, increasing the density of the transistor and the multiple functions performed in a single chip. Further, increasing the number of transistors and density in ICs significantly improved the speed and power dissipation. However, certain effects arise due to the scaling of CMOS transistors. Theoretically, current flows in a CMOS transistor when the input voltage equals $0.5V_{DD}$ and turns off when there is no input voltage. However, the subthreshold leakage current keeps flowing through the MOSFETs in CMOS circuits. This power dissipation is called static power dissipation. During the switching characteristics, power dissipation occurs due to parasitic capacitances of NMOS and PMOS transistors. This power dissipation is known as dynamic power dissipation. In order to decrease the power dissipation, the input voltage or parasitic capacitances should be reduced. The input voltage scaling is not feasible because the transistor will turn on below the threshold voltage. Although, the scaling of MOSFETs not only increased the ON current but also the OFF current, known as subthreshold current. This subthreshold current will draw current from the power source in an ideal state. Thus, another option to reduce the power dissipation is to reduce the device's parasitic capacitances by reducing the gate length to reduce the capacitances. Therefore, CMOS scaling is the best method to achieve power efficiency and increase integrated circuits' capabilities.

3.3 SHORT CHANNEL EFFECTS (SCES)

The scaling of the CMOS transistor caused some physical device degradation phenomena known as the short channel effects. This section explores the different second-order effects that degrade the device's performance.

3.3.1 SUBTHRESHOLD CURRENT

The drain current characteristics of MOSFETs in a semi-log plot are shown in Figure 3.2. The standard practical formula to define the device threshold voltage is 100 nA × W/L. Here, W is the device's width and L is the length of the device; however, we can see that the current flow in the MOSFETs is below the threshold voltage. This current is defined as the subthreshold current of the MOSFETs.

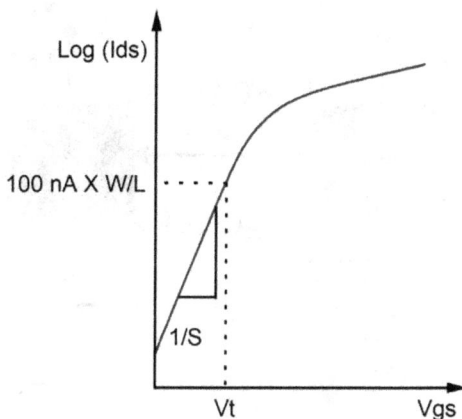

FIGURE 3.2 Illustration of subthreshold current in drain characteristics of MOSFETs.

The physics behind this is that the current flow happens because of the smaller inversion electron concentration n_s. This electron concentration proportional to the potential then rises the drain current of MOSFETs. The channel length scaling reduces the barrier distance between the source and drain region. Hence, the minority electrons available in the source region may get passed to the channel region and raise the current. The subthreshold current is in the order of $\sim 10^{-9}$; the number seems smaller for a single device. If the same current is driven for millions of transistors causes a considerable power dissipation over the processor. Hence, this current should be denoted as the OFF current (I_{OFF}), which should be as minimum as possible.

3.3.2 THRESHOLD VOLTAGE ROLL-OFF

The threshold voltage is defined as the gate voltage required to form the inversion region under the gate. After the threshold voltage only, the device gets minimum current from the source to drain. The aggressive scaling of the MOSFETs reduced the channel length and distance between the source and drain. This situation makes the minimum gate voltage reduce the barrier between the source and channel and makes the current flow out of the MOSFETs. This situation becomes even worse with the shorter channel length. This roll-off situation is known as the threshold voltage roll-off. For the larger channel length, the threshold voltage is equal to $V_{t,\,long}$; for a smaller length, the threshold voltage roll-off is a function of the channel length and larger V_{ds}. Hence, a minimum channel length is to be kept to avoid the roll-off situation in the transistor. The pictorial representation of threshold voltage roll-off is shown in Figure 3.3.

3.3.3 DRAIN INDUCED BARRIER LOWERING (DIBL)

The potential barrier height between the source and channel is higher for long-channel MOSFETs which is controlled by the gate voltage. As the channel length

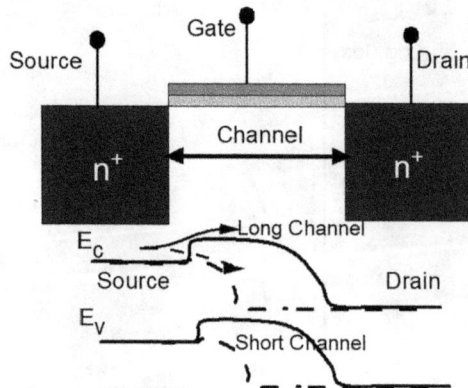

FIGURE 3.3 Comparison of threshold voltage roll-off between the long and short channel device.

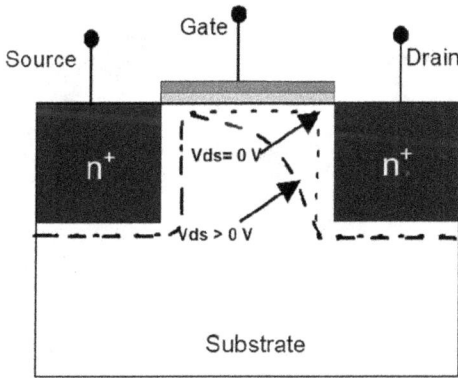

FIGURE 3.4   Schematic representation of DIBL for different drain voltages.

gets scaled, the height of the barrier at the source side is also reduced. However, when the channel length is reduced more in this regime, the source/channel and channel/drain depletion region becomes very near, as shown in Figure 3.4. This implies a lower energy requirement for charge carriers to cross the potential barrier.

The potential barrier reduction in the channel is reduced by the higher drain voltage instead of the gate voltage. In this hypothetical situation gate losses the channel control, and drain voltage control the device current. The drain region should be kept far away from the source region to avoid this effect. With the help of some engineering methods or structural modifications can reduce the DIBL by pocket implantation use of high-k spacer materials.

3.3.4   Hot Carrier Effects

Hot carrier effects are mainly caused by a high lateral electric field which creates a new electron/hole pair combination due to the Collison of atoms in the channel known as impact ionization. Short channel MOSFET has a high lateral electric field due to the higher drain voltage. The charge carriers in the channel get high momentum and kinetic energy due to the high electric field. These charge carriers collide with the atoms, exchange momentum and energy, and create a new electron-hole pair due to the impact ionization mechanism, as shown in Figure 3.5. These generated electrons are attracted by the gate electric field and get trapped in the oxide region.

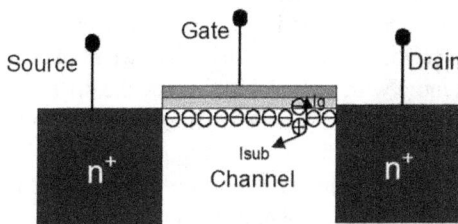

FIGURE 3.5   Hot carrier effects in conventional MOSFETs.

(a)   (b)

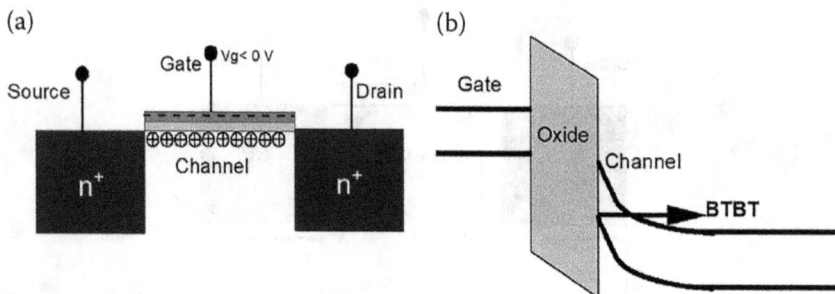

FIGURE 3.6 Schematic representation of Gate Induced Drain Leakage (GIDL) of MOSFETs (a) Cross-section view of TFETs (b) Band diagram at the oxide interface by vertical cutline.

These trapped electrons may reduce the insulator capacity of oxide, which further affects the device threshold voltage and lead current flow to the gate terminal. This charge-trapping event modifies the device threshold voltage. The main issue of this charge trapping is that this reduces the device lifetime and causes the bias instability of device characteristics. Thus, the hot carrier injected current in MOSFETs, such as gate current and substrate current, adversely affects the device's reliability.

3.3.5 GATE-INDUCED DRAIN LEAKAGE CURRENT (GIDL)

The scaling of MOSFETs leads to a high field of oxide and the gate-drain overlap region. For a high drain bias with low gate voltage, the depletion region of the drain may overlap with the gate region, as shown in Figure 3.6a. The significant overlap band bending is visible in the band diagram of Figure 3.6b. This band bending creates the tunneling of electrons from the valence band to the conduction band. This is known as band-to-band tunneling (BTBT). This tunneling of electrons is increased with the increase in gate negative bias. Since the GIDL originated at the drain side, the proper design of the drain side profile may reduce the leakage current. The lightly doped drain side profile reduces GIDL significantly in MOSFETs.

3.3.6 VELOCITY SATURATION

Another drawback of the conventional MOSFET is saturation velocity. The drift velocity is defined as the $v_d = \mu\varepsilon$, where μ is the carrier mobility, and ε is the electric field. This equation implies that the increase in the electric field can make infinite drift velocity. However, the reality is not like that; when the carrier energy exceeds the phonon, it losses more with the velocity. Hence the drift velocity can not exceed the particular value this limiting velocity is known as the saturation velocity. The most important short-channel effects of conventional MOSFETs are channel length modulation of velocity saturation, as shown in Figure 3.7. It is clearly visible that the short channel device's channel length is much less than the long channel.

On the other hand, the lateral electric field from the drain side opposes the electron transport from the channel. The reason behind this is that the drain side

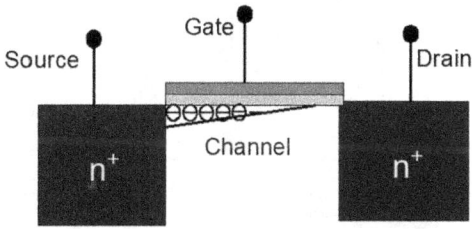

FIGURE 3.7   Schematic representation of velocity saturation of MOSFETs.

depleted field opposed the carrier transport of the device. Hence, near the drain side, the carrier transport gets saturated and allows only small carrier transport. This phenomenon of change in the device length is known as the pinch-off of the device. After the device pinches off, the device enters a region of saturation and carrier transport becomes saturated. This weird situation happens only in the short-channel device where the drain field affects carrier transport.

3.4   TRADE-OFF IN THE DEVICE DESIGN

In the earlier section, we discussed the need for device scaling and how device scaling affects device performance through short-channel effects. However, another issue in the circuit applications is the trade-offs between the I_{ON} and I_{OFF}. An ideal MOS transistor should have a higher I_{ON} for better circuit applications. But the higher the ON current causes, the higher the OFF current also. So, there is a trade-off between the ON current and the OFF current. The industry has proposed numerous design techniques to solve these trade-off problems. One is the development of multi-threshold voltage devices in the processor. In this technique, we can use the higher V_t with a lower OFF current and lower V_t with a higher ON current. This solution might also reduce the device's static and dynamic power consumption. Another suitable solution is also provided multiple V_{DD} for different circuit elements in the processor. This solution can also reduce the device's static and dynamic power consumption. However, the multi-threshold voltage creates inevitable complicated fabrication issues.

On the other hand, the oxide thickness reduction in the channel may rise issues by bias instability which causes the threshold voltage variation for different biases. Another issue is that the carrier mobility in the channel can change with respect to stress effects created during the fabrication process. The variation in mobility can change the device's current and conductance behavior. This mechanical strain can create the second-order effects of MOSFETs, such as line edge roughness, random dopant fluctuation, and fluctuations due to noise. Hence, the industry looked into the non-conventional MOSFETs that can solve these short-channel effects issues and trade-off problems.

3.5   SOI-BASED MOSFETS

Silicon on Insulator (SOI) based MOSFETs can be a better alternative to conventional MOSFETs. The cross-section view of SOI MOSFETs is shown in Figure 3.8.

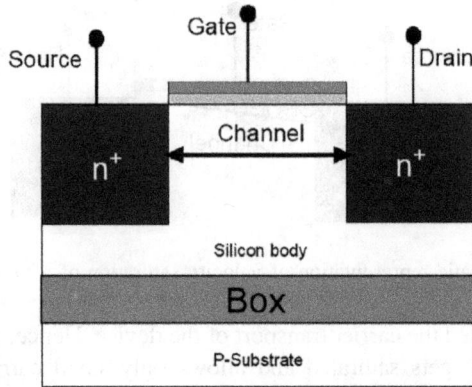

FIGURE 3.8 Cross section view of SOI-based MOSFETs.

The buried oxide under the silicon body can reduce the leakage path, and the gate can control the entire device. The fabrication of the SOI buried wafer is way cost more than the silicon-based one. However, these SOI-based MOSFETs provide better device speed and reduced parasitics than conventional MOSFETs.

3.6 MULTIGATE AND FINFETS

The Multigate MOSFETs, such as Double Gate, Trigate, and Surrounding gate MOSFETs, can be a better alternative to the conventional MOSFETs [1]. Another advantage of Multigate MOSFETs is that the gate can take control of the entire channel with multiple surroundings. In the case of Double gate (DG) MOSFETs, the top and bottom gates can be the same voltage and control the channel. The surface potential in the channel can rise and down with the change in the fewer millivolts from the gate terminal. In this working method, there is no need for higher depletion doping concentration needed in the substrate. Hence, the vertical electric field is enhanced and can improve mobility, consequently increasing the device's current. Among these structures, FinFETs have better scalability and controllability than conventional MOSFETs [2,3].

3.7 TUNNEL FETS

The tunnel FET device structure looks like a reverse-biased p-i-n diode. Interband tunneling is the primary mechanism of this device. The source region is heavily doped to achieve high tunneling efficiency to achieve the tunneling. The source/channel interface also requires abrupt junctions to create band bending and increase tunneling probability. Creating the abrupt junction requires ultra-fast annealing techniques during fabrication. The gate controls the tunneling of electrons from the source region's valence band to the channel region's conduction band during a strong inversion regime. In the OFF state, the valence band source region overlapped with the forbidden gap of the channel region. This blocks the leakage current problems associated with MOSFETs. This section will explore how the Tunnel FETs work and their challenges to the VLSI integration.

3.8   THE IMPORTANCE OF TUNNEL FETS

The scaling of the CMOS transistor leads to tremendous application in the IC industry. In order to tolerate the scaling of MOSFETs, the input power V_{DD} needs to be scaled to the reduced static power consumption of CMOS transistors. Hence, each technology node has a different VDD to maintain minimum station power consumption. Conventional MOSFETs work under the principle of thermionic emission. The increase in the gate voltage V_{GS} will create the inversion charge beneath the gate/oxide interface. This inversion charge will create a path between the source and drain terminal that causes the drain current of the device. In this working principle, the minimum voltage required to create the inversion charge density or channel charge is known as the device's threshold voltage. This working principle leads to critical issues such as subthreshold conduction and subthreshold slope degradation. The subthreshold conduction is a current flow between the source and drain terminal before the threshold voltage. The subthreshold slope is the gate voltage required to change the drain current by an order of 10. The lower subthreshold slope will lead to a higher I_{ON}/I_{OFF} ratio and lower power consumption. The subthreshold slope of transistors is denoted as follows.

$$SS = \left(\frac{d\left(\log_{10} Ids\right)}{dV_G} \right) = 2.3\frac{kT}{q}\left(1 + \frac{C_{dep}}{C_{ox}}\right) \tag{3.1}$$

Where the I_{ds} is the drain current, Vg is the gate voltage, k is the planks constant, T is the temperature, q is the channel charge, C_{dep} is the depletion capacitance and C_{ox} is the oxide capacitance.

From equation (3.1), it is visible that the minimum subthreshold slope of the transistor is 2.3 (kT/q), which is 60 mV/decade. This denotes that, to achieve the I_{ON}/I_{OFF} ratio of 10^7 the required gate voltage is $7 \times 0.06(60 \ mv) = 0.42 \ V$. Therefore, increasing the drain current by one magnitude requires the gate voltage to increase by 0.42 V. This arbitrary example denotes that the subthreshold slope calculation is very important to define the device performance by power consumption. Hence, to reduce the device's power consumption, we need to decrease the subthreshold slope less than 60 mV/decade.

The comparison of the subthreshold swing between the TFETs and MOSFETs is shown in Figure 3.9. Tunnel Field Effect Transistors (TFETs) are such devices that can provide a subthreshold slope of less than 60 mV/decade. The working principle of TFETs is much different from the conventional MOSFET. The conventional MOSFETs works on the principle of thermionic emission, where the source-to-channel potential barrier can be reduced by increasing the gate voltage. On the other

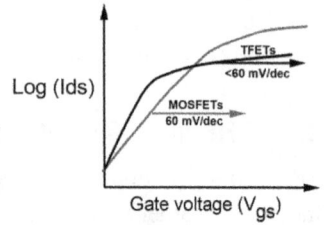

FIGURE 3.9 Comparison of subthreshold swing between MOSFETs and Tunnel FETs.

hand, the TFETs work on the Band to Band Tunneling (BTBT) principles. The TFETs have a wide bandgap during the OFF state hence less subthreshold conduction between the source and drain. The drain current of TFETs depends on the tunneling probability of the source/channel, where it can enhance the charge carriers and gate field.

3.9   TUNNEL FETS EVOLUTION

The research on Tunnel FETs began in 1928, and scientists assumed tunneling might occur in the presence of a strong electric field [4]. Zener explained interband tunneling in PN junction diodes in the form of dielectric breakdown in 1934. This is officially noted as the band –to band tunneling (BTBT) phenomenon. After Zener, Leo Esaki developed the tunnel diode in 1957, which is a heavily doped PN junction diode. The first silicon TFET was discussed by Banerjee et al. in 1987, and also they provided the analytical current model for three-terminal TFET [5]. The schematic representation of TFETs is shown in Figure 3.10. The difference between the MOSFETs and TFETs is that the source is doped with p^+ doping concentration. The other configurations, such as gate, source, and drain terminals, are similar to the MOSFETs.

3.10   TFETS DEVICE OPERATION

In this section, the device operation of TFETs will be analyzed qualitatively. During the OFF state, i.e., $V_{gs} = 0$, $V_{ds} > 0$ V, the TFETs subthreshold current is significantly lower than the MOSFETs. The physics behind this is that in the OFF state, two depletion regions have been formed between the source/channel and channel/drain. The electrons available in the conduction band of the channel can move into

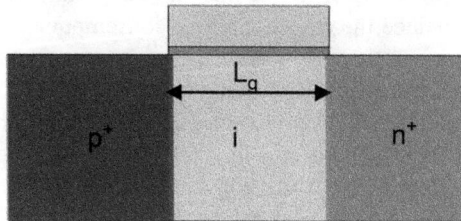

FIGURE 3.10   Device cross-section of Tunnel FETs.

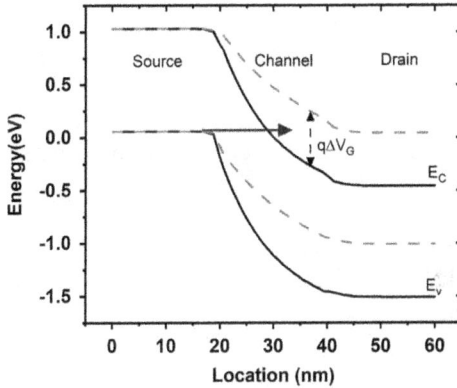

FIGURE 3.11 Band diagram of Tunnel FETs along the channel length.

the drain region. Nevertheless, the source side is doped with P$^+$ doping concentration, which makes the band gap very wide. This ensures that the few electrons available in the conduction band of the source region may not get into the conduction band of the channel. In the case of MOSFET, the source region is heavily doped with n$^+$ doping concentration, which creates minority current carriers to pass through the channel barrier and causes the subthreshold current more. The band diagram in Figure 3.11 shows that during the OFF state source/channel band is very wide, ensuring there is no current flow between the source to drain region.

In the ON state operation of TFETs, the electrons should be injected into the source region and passed through the drain region, which makes the drain current if the $V_{gs} > 0$ V, the channel conduction band may align with the source valence band. This band bending depends on the number of charge carriers available in the channel region, so it is denoted by ΔqV_g, as shown in Figure 3.11. Once the band is aligned with the source/channel and drain region, the electrons get injected into the source and pass through the drain region, denoted as the drain current of the device. The tunneling of the source to channel purely depends upon the material properties such as electron mass, band gap, and tunneling length denoted by WKB (Wentzel–Kramer–Brillouin) approximation, as shown below.

$$T_{WKB} = exp\left(-\frac{4\lambda\sqrt{2m * \sqrt{E_g^3}}}{3qh(E_g + \Delta\varnothing)}\right) \tag{3.2}$$

Where m* is the effective mass of the electron, E_g is the band gap, q is the electron charge, h is the planks constant, and λ is the tunneling length.

3.11 CHALLENGES IN TUNNEL FETS

The main objective of TFETs is to replace the conventional MOSFETs with a higher ON current (I_{ON}), lower OFF current (I_{OFF}), lower subthreshold swing (SS < 60 mv/decade), and lower V_{dd} (~0.5 V). However, the TFETs might not have a higher ON current than the conventional MOSFETs. This is because the ON current of TFETs depends on the tunneling probability between the source and channel region [6,7]. As shown in equation 3.2, the tunneling depends on the material's effective mass (m^*), band gap (E_g), and tunneling length (λ). Hence, the electron mass and band gap should be minimum to increase the tunneling probability of TFETs. Among these parameters, the tunneling length (λ) can be minimized by reducing the gate capacitance using high-k gate oxide and higher doping concentration in the intrinsic region.

The next major challenge is ambipolar current, which causes circuit failure and excessive standby leakage in CMOS-based devices. If the gate voltage $V_{gs} < 0$ V, the conduction band of the channel goes up, which consequently matches the valence band of the source region. The phenomenon is known as Band to Band Tunneling (BTBT); this BTBT causes the ambipolar current in the TFETs [8]. Because of its asymmetric doping, TFET fabrication is rather more difficult than MOSFET. Because the lowest leakage and SS values are inherently reliant on device characteristics, optimization plays an important role in enhancing performance. The delayed saturation in the output characteristics causes noise margin issues at lower supply voltage [8,9]. The structural modifications of Dual material gate (DMG) [10,11], Tri Material Gate (TMG) [12], Halo doped Tunnel FETs [13,14], underlap/overlap TFETs [15–17] are all explored to improve the TFETs device performances.

3.12 GATE-ALL-AROUND-TUNNEL FETS

The Multigate FET is a recent technological advancement in the field of semiconductor devices. The Multigate FETs can be classified as Double Gate (DG), Triple Gate (TG), and Gate -All-Around (GAA) [18,19]. Among these structure optimizations, the GAA structure provides better enhancement of gate control and reduced short-channel effects than the other. The cross-section view of GAA-based TFETs is shown in Figure 3.10. The entire structure looks like a nanowire due to the cylindrical gate structure. The I_D-V_g characteristics of GAATFETs are shown in Figure 3.12. From the transfer characteristics of GAA TFETs, we can extract the device characteristics such as I_{ON} and I_{OFF}, as shown in Figure 3.13. It is visible that the GAATFETs have a very lower I_{OFF} of 6.16 nA. The source side p$^+$ doping creates a wide band between the source and channel, making the subthreshold current much smaller than the conventional MOSFETs. However, the only issue with the TFETs, as discussed in the challenges of TFETs, is their I_{ON} current. The ON current of the higher controlled device of GAATFETs has only the order of nA. This requires a current booster to improve the device drain current that can be used for CMOS inverters.

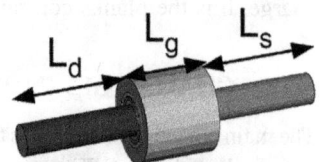

FIGURE 3.12 Device structure of GAA Tunnel FETs.

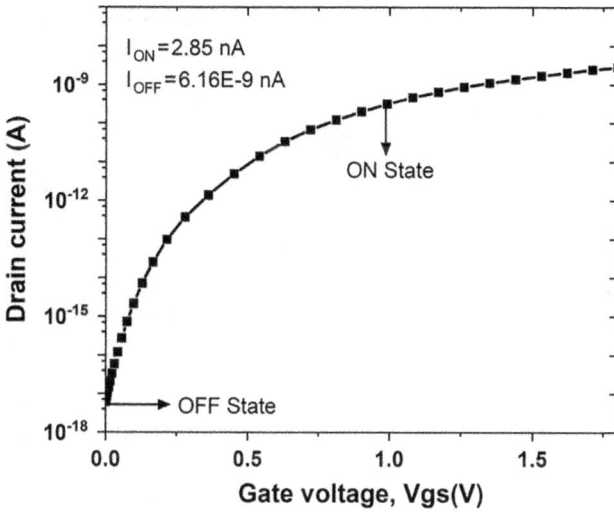

FIGURE 3.13 I_d-V_{gs} characteristics of Gate-All-Around TFETs.

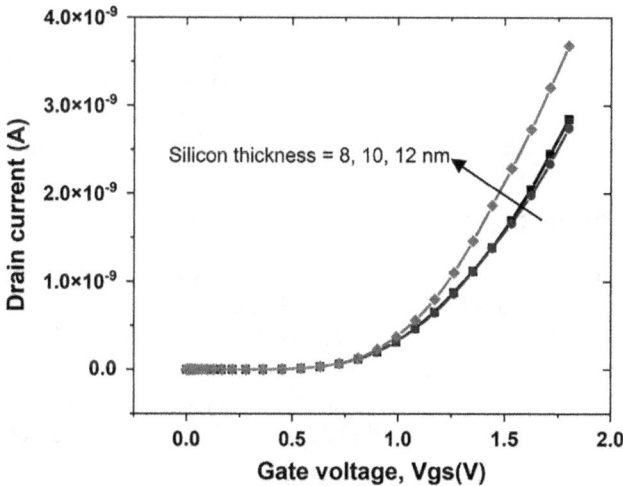

FIGURE 3.14 I_d-V_{gs} characteristics of Gate-All-Around TFETs with a variation of silicon thickness.

The drain current characteristics of GAATFETs with change in the silicon thickness is shown in Figure 3.14. Obviously, the silicon thickness of the drain current will increase due to the increase in the total area. It is calculated that the ION increases the channel thickness from 8 nm to 12 nm by 29%. This can be useful for the circuit designer to alternate the device thickness that can be used for better circuit application. Figure 3.15 shows the drain current characteristics of GAATFETs for different oxide thicknesses. As we discussed earlier, the higher permittivity of oxide can increase the higher tunneling probability of TFETs.

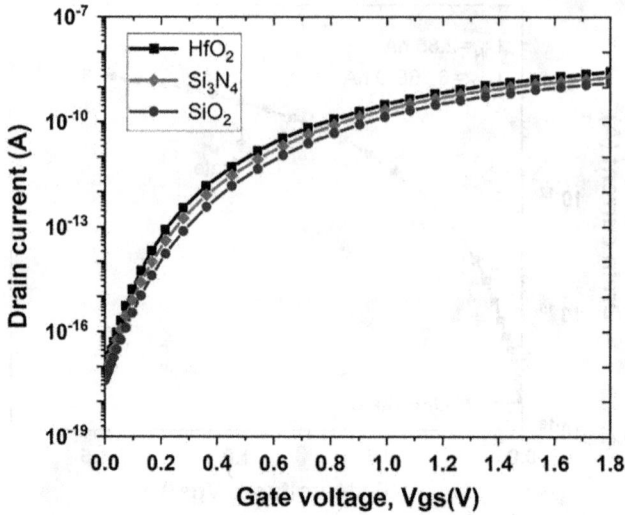

FIGURE 3.15 I_d-V_{gs} characteristics of Gate-All-Around TFETs with a variation of different high-k.

As expected, the drain current of GAATFETs with HfO_2 has a better subthreshold swing and higher I_{ON} current. The physics behind this is that higher permittivity materials allow more electric field than the lower permittivity. The tunneling current will also increase if the tunneling junction has a better electric field.

3.13 VERTICAL TUNNEL FETS

In this section, we will explore the vertical TFETs that can increase the device performance of conventional TFETs [20]. The cross-section of vertical TFETs is shown in Figure 3.16. The main difference between lateral TFETs and vertical TFETs is that the source position has been placed at the bottom of the channel. The conventional TFET work under the tunneling mechanism of point tunneling, where the tunneling probability merely depends on the device's surface charge and gate

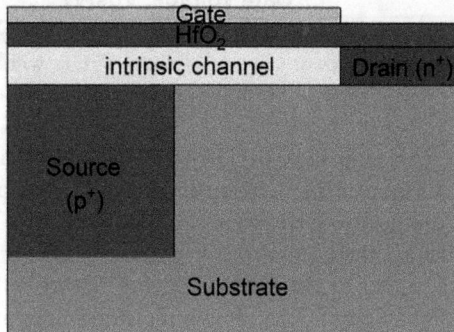

FIGURE 3.16 Device cross-section view of vertical Tunnel FETs.

electric field. Nevertheless, vertical TFETs work under line tunneling [21–23]. The key advantage of line tunneling is tunneling region is immensely that can enhance the BTBT between the source/channel region.

The contour plot of Band to Band Tunneling (BTBT) is shown in Figure 3.17. It is clearly visible that the tunneling region between the source/channel is higher than the conventional TFETs. This is because the valence and conduction bands of the source/channel region are very much aligned with the conventional TFETs. This much-matched band alignment creates the tunneling region very wide and creates the maximum possibility of tunneling current, as shown in Figure 3.17. The drain current characteristics of vertical TFETs with Ge and Si source is shown in Figure 3.18. Germanium is used as the source for higher hole mobility than silicon. As expected, the Ge-sourced vertical TFETs possess a higher I_{ON} current than the silicon. The physics behind this is that Ge has higher mobility, creating a higher tunneling region. However, Ge's subthreshold current is much higher than the

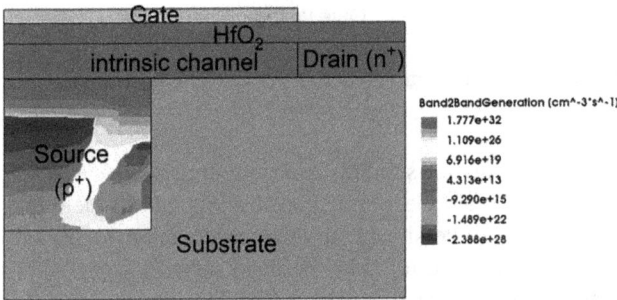

FIGURE 3.17 Band to Band Tunneling of vertical Tunnel FETs.

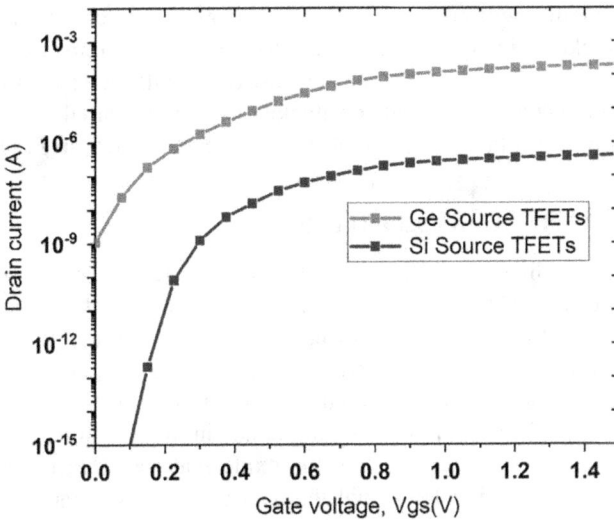

FIGURE 3.18 I_d-V_g characteristics of vertical TFETs with Ge/Si as a Source.

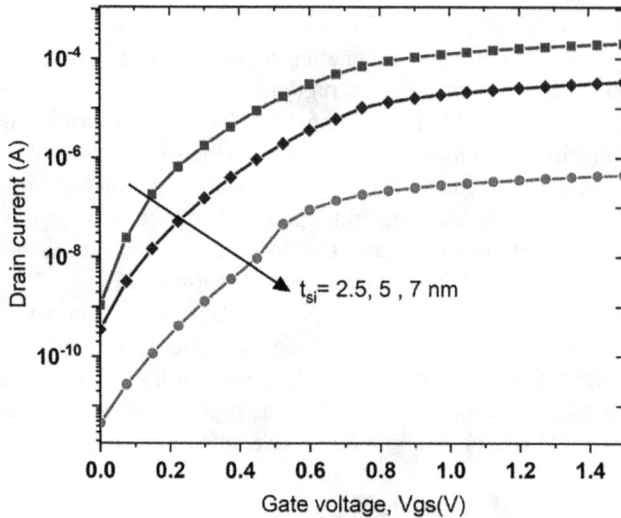

FIGURE 3.19 Drain current characteristics of vertical TFETs with the variation of silicon thickness.

Si source TFETs. This higher subthreshold current arises because of mismatched band alignment between the Ge source and the Silicon channel. The Germanium fermi level is higher than the silicon. In this consequence, higher charger carriers have to pass through the band to make the band stable.

The transfer characteristics of vertical TFETs with the variation of silicon thickness or channel thickness are shown in Figure 3.19. The drain current decreases by increasing the silicon thickness of vertical TFETs. This may look the opposite of the conventional theory. The reason behind this weird effect is that the tunneling probability will get reduced for an increase in the channel thickness. As the channel thickness increases, the source/channel interface remains the same and has little change. Due to this, the channel resistivity will increase rather than expected with the increase in the silicon thickness. This increased resistivity of the intrinsic region resists the drain current much lower, as shown in Figure 3.19.

3.14 2D- MATERIAL-BASED TFETS

The aggressive scaling of CMOS transistors causes device degradation, such as short channel effects, more power consumption, self-heating effects, and threshold voltage shift due to the Quantum mechanical effects (QME) [24]. Hence, researchers looked into different materials to replace the silicon to enable further scaling. Among those, two-dimensional material attracts and can replace silicon even at 1nm [25]. The absence of dangling bonds in the two-dimensional material can solve the interface issues between the oxide material. Hence, there is much scope in the two-dimensional material that can be used as a transistor for future technology nodes. This section will explore the basics of two-dimensional material properties and their application in the semiconductor devices field. We will give insights into how the Tunnel FETs performance can be boosted by adding two-dimensional materials.

3.14.1 Basics of 2D-Materials

A two-dimensional material is a monolayer or multilayered material in the form of AB_2 where A is the metal such as M_o, W, and B is a chalcogen like S, Se, and T. This combination can make a two-dimensional material. The fabrication process of 2D materials earlier was micro mechanically exfoliated flakes. However, the integration of these flakes cannot make the front line or back line in VLSI integration. Hence, wafer-scale synthesis is inevitable in 2D materials. The process of chemical vapor deposition (CVD) and metal-organic chemical vapor deposition (MCVD) can make thin (~2 inches) sapphire wafers for the possible integration of 2D materials such as MoS_2, MoS_{e2}, WS_2, and WS_{e2} [26]. The major challenge in the fabrication of 2D materials is that they should be free from mechanical damage because the process of VLSI integration with various etching, deposition, and cleaning may damage this ~0.6 nm of 2D materials. Hence, the different integration schemes should be processed to integrate 2D materials successfully.

3.14.2 Transistor Characteristics of 2D-Materials

As discussed earlier, a perfect transistor should have a lower OFF current (I_{ON}), higher ON current (I_{ON}), higher transconductance (g_m), and lower subthreshold slope. Apart from this, for a short channel transistor, contact resistances, mobility, and velocity saturation also matter for ultra-scaled CMOS devices. The equation shows that the velocity saturation depends on the mobility and electric field. Hence, the mobility linearly increases with the increase in the electric field. This leads to the I_{ON} depending on the velocity saturation instead of the channel mobility. For silicon, it is reported that the velocity saturation is in the order of $\sim 10^7$ cmS^{-1} at the electric field of $E > 1$ Vμm^{-1} [27]. For a MoS_2 transistor with the gate oxide of SiO_2 the velocity saturation is measured as 6×10^6 cmS^{-1} [28]. Hence, the average of this velocity of the 2D transistor is much equal to the conventional MOSFETs and can make the I_{ON} of 1μA/μm. So, carrier mobility is less important because the drain current of the 2D transistor merely depends on the velocity saturation, injection velocity, and contact resistances. However, mobility can be beneficial for identifying the second-order effects of Coulomb scattering, surface roughness, and phonon scattering. Overall, a 2D-based transistor with a monolayer thickness of ~<1nm can provide a similar device characteristic of nanometer range devices due to the 2D material properties [29,30]. In this point of view, phosphorene has higher mobility and optimum bandgap (tunable bandgap), which boosts the I_{on} as expected values. In section 2.10.4, we show the phosphorene-based TFET characteristics.

3.14.3 Reliability of 2D Material-Based Transistors

In the previous section, we briefly discussed the prerequisite of a 2D-based transistor that can provide a better possibility of replacing conventional MOSFETs for

future scaling. However, a successful transistor is not only based on the performance metrics; the reliability also evaluates during the abrupt fabrication process
[31]. As reported, the 2D-based transistors are affected mainly by the silicon-
interface issues, which create traps and defects beneath the gate. These traps in the
insulator traps create low-frequency variation known as flicker noise (1/f) [32]. This
noise can affect the device threshold voltage, and the magnitude of the spectral
density of noise can limit the device drain current. The same charge can be counted
down in the order of 10 to 100. Then, each carrier trap can cause a discrete current,
creating random telegraph noise (RTN). Apart from the charge trapping event, the
water or gaseous items present in the oxide interface can cause a hysteresis,
changing the device threshold voltage. Another important reliability issue is temperature bias instability (BTI), where the charge traps can accumulate and change
the device threshold voltage until the device operation point [33]. These insulator-
based charge trapping events mostly arise due to dangling bands of SiO_2 or HfO_2
with the 2D material such as MoS_2, and WSe_2 [34]. It is reported that the insulator
with the 2D material of hBN or $BiSeO_5$ can be the alternative for reducing the
charge trapping events in the 2D material-based transistor [35]. However, the
fabrication of different oxides also creates issues over the fabrication limits.

3.14.4 TFETs Based on the 2D Materials

In this section, we will explore the device characteristics of TFETs with the incorporation of phosphorene in the channel region. The figure shows the phosphorene DG
TFET and ADMDG-TFET structure with the below-mentioned dimension shown
in Figure 3.20.

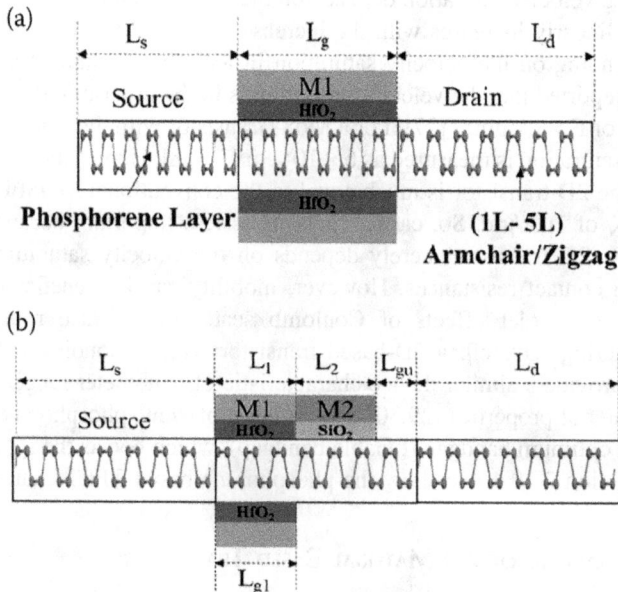

FIGURE 3.20 Schematic view of (a) Phosphorene Symmetric TFET (DG TFET)
(b) Asymmetric Phosphorene TFET (ADMDG –TFET).

The device has a 100nm overall length. L1 = 16 nm and L2 = 16 nm make up the upper gate length (Lg = L1 + L2), which is 32 nm. Underlap gate length (Lgu) is 2 nm, while the lower gate length (Lgl) is 21 nm. The ADMDG TFET device's optimized gate work functions are M1 = 2.25 eV and M2 = 4.8 eV for the source and drain sides, respectively. P++-type source, n+-type drain, and p-type channel doping concentrations are 1×10^{20} cm^{-3}, 5×10^{18} cm^{-3}, and 1×10^{20} cm^{-3}, respectively. Compare the device performance in terms of I_{on}, I_{off}, SS, and I_{on}/I_{off} ratio to demonstrate the benefit of asymmetric versus symmetric devices. Additionally, study the performance in an armchair or zigzag direction, layer by layer (1L to 5L).

The device simulation has been performed using a hybrid method in Sentaurus TCAD. A phosphorene material file (.par) containing layer- and direction-based material file (.par) was made in order to perform simulations for 1L to 5L in both directions [36]. The gate work function, fitting parameters, and BTBT tunneling are precisely tuned.

The off-state energy band diagram of monolayer/few-layer phosphorene ADMDG TFET is shown in Figure 3.20. It is realized that CB/VB is slightly shifted (with reference to armchair) due to its anisotropic effective mass nature. It is found that off-state band aliment does not favor S/C BTBT tunneling; therefore, the device is in the deep-off state; hence, Ioff is the lowest. Besides, a few layers of phosphorene ADMDG TFET has high Ioff (Figure 3.21a) due to the channel to drain (C/D) BTBT drain tunneling represented in Figure 3.21b. Hence, few-layers ADMDG TFET has a high Ioff, reducing the Ion/Ioff ratio. So this unintentional C/D BTBT process degrades off-state and as well as overall device performance.

The ON-state energy band diagram of monolayer/few-layer phosphorene ADMDG TFET is shown in Figure 3.22a. After applying a gate and drain bias, the source's valence band is aligned channel's conduction band. Hence, S/C BTBT has happened, which yields more I_{on}. It is found that armchair direction-based devices have more I_{on} than Zigzag due to their light-effective mass values shown in Figure 3.22b. Besides, In monolayer zigzag has less leakage current (I_{off}) due to heavier effective mass, which widens the tunneling window. Hence, the effective mass has more influence on the monolayer than the few-layer.

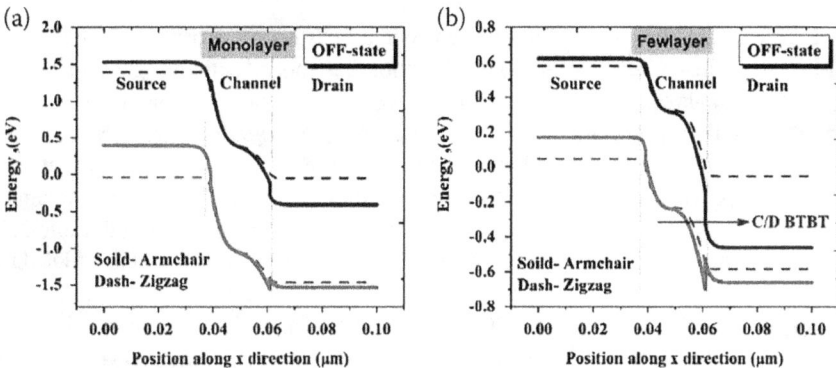

FIGURE 3.21 Off state energy band diagram of (a) monolayer phosphorene ADMDG TFET (b) few-layer phosphorene ADMDG TFET.

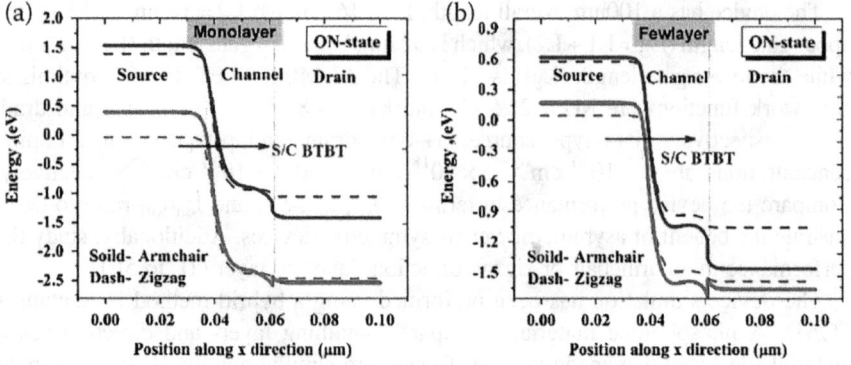

FIGURE 3.22 On-state energy band diagram of (a) monolayer phosphorene ADMDG TFET (b) few-layer phosphorene ADMDG TFET.

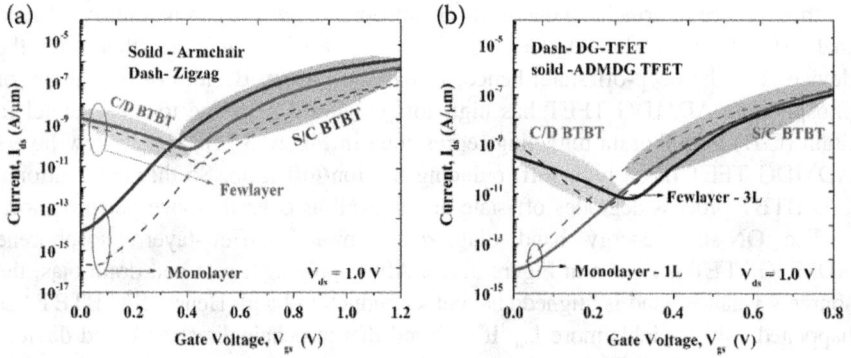

FIGURE 3.23 (a) I_d-Vgs Characteristics of phosphorene ADMDG TFET with monolayer-few-layer for both directions. (b) Comparing phosphorene DG TFET to ADMDG TFET in armchair direction. The yellow-shaded region is due to C/D BTBT tunneling. This phenomenon was more pronounced in few-layer device.

Figure 3.23a shows the drain current characteristics of phosphorene ADMDG TFET, it is clearly realized armchair direction has more I_{on} than zigzag in monolayer and few-layer due to its light-effective mass. Hence, it concluded that the armchair is the preferable direction for transport. So we compare the ADMDG TFET with DG TFET in armchair direction (Figure 3.23b). It is noticed that significant improvements are not obtained by using ADMDG TFET. The obtained FOMs of 1L-5L armchair/zigzag phosphorene are listed in Table 3.1 and Table 3.2. The maximum gained I_{on} value is 9.22 µA/µm for monolayer–armchair ADMDG TFET which is very less compared to Ge-TFET and SiGe TFET. The I_{on}/I_{off} ratio is very low in the phosphorene DG-TFET, namely in armchair 4L has only 1.14. This is due to more C/D BTBT in symmetric structure whereas ADMDG TFET has 5290 which is ~ 4.64×10^3 times high.

TABLE 3.1

Comparison of dc Parameters and Their Values for ADMDG TFET and DG TFET in Armchair Direction

Number of layer	Phosphorene ADMDG TFET				Phosphorene DG TFET			
	I_{on} (µA/µm)	SS (mV/dec)	I_{off}	I_{on}/I_{off}	I_{on} (µA/µm)	SS (mV/dec)	I_{off}	I_{on}/I_{off}
1	9.22	71.19	8.8×10^{-15}	1.0×10^9	3.95	67.895	9.07×10^{-15}	4.3×10^8
2	5.67	99.19	5.3×10^{-12}	1×10^6	0.366	56.872	1.43×10^{-12}	2.5×10^5
3	6.30	140.79	3.3×10^{-10}	18585.5	0.199	104.372	9.94×10^{-10}	2×10^2
4	6.50	209.09	1.2×10^{-9}	5290.72	0.0243	86.142	2.13×10^{-8}	1.14
5	3.21	251.31	2.1×10^{-9}	1493.84	–	–	–	–

TABLE 3.2

Comparison of dc Parameters and Their Values for DG TFET and ADMDG in Zigzag Direction

Number of layer	Phosphorene ADMDG TFET				Phosphorene DG TFET			
	I_{on} (µA/µm)	SS (mV/dec)	I_{off}	I_{on}/I_{off}	I_{on} (µA/µm)	SS (mV/dec)	I_{off}	I_{on}/I_{off}
1	0.79	87.84	4.2×10^{-16}	1.8×10^9	0.86	60	2.66×10^{-17}	32×10^9
2	0.95	83.46	1.5×10^{-12}	611286	1.06	62.6	4.45×10^{-14}	2.3×10^7
3	0.71	127.18	1.4×10^{-10}	4877.81	0.91	64.66	7.65×10^{-10}	1192.46
4	1.06	200.21	6.0×10^{-10}	1740.78	0.80	120.6	1.43×10^{-09}	561.751
5	0.94	240.5	1.1×10^{-09}	829.823	0.82	98.35	1.17×10^{-09}	710.232

The unanticipated C/D BTBT process in a few-layer phosphorene TFET prevented the expected Ion value of a few hundred A/m. This operation (between 0.3 and 0.4 V) decreased tunneling width and delayed the S/C BTBT band bending. Perhaps the question is: If this C/D BTBT process does not occur in a monolayer, why does monolayer phosphorene TFET only yield 9.22 A/m while not achieving maximum Ion? The bandgap of monolayer phosphorene is 1.45 eV, higher than silicon's, which is why it could not produce the inadequate 100 A/m currents. The requirement is armchair-based few-layer phosphorene TFET without C/B BTBT device may produce maximum Ion. To vanish the C/D BTBT in a few-layer phosphorene TFET, radical changes will be necessary to produce the maximum Ion and without affecting SS and Ion/Ioff ratio. In this regard, drain doping concentration can be reduced it can widen the C/D barrier. But these doping techniques in 2D materials are a real challenge.

3.15 SUMMARY

This chapter explored the importance of scaling MOSFETs and how physical limitations affect their performance. A brief description of short channel effects of subthreshold current, threshold voltage roll-off, Drain Induced Barrier Lowering (DIBL),

Hot carrier effects, and Gate Induced Drain Leakage (GIDL) is presented. The highlights of Tunnel FETs and their performance booster that can improve performance are analyzed. The key parameters of Tunnel FETs (TFETs) and they can improve the device's electrical characteristics are reported. It is also identified that the different configurations of TFETs, such as GAATFETs, Vertical TFETs, and 2D- material based TFETs can much better improve device performance than the conventional TFETs. The device characteristics and fabrication limitations are presented.

REFERENCES

[1] J. P. Colinge, and A. Chandrakasan, *FinFETs and other multigate transistors.* Boston, MA: Springer. 2008.

[2] S. Manikandan, N. B. Balamurugan, and T. S. Arun Samuel, "Impact of uniform and non-uniform doping variations for ultrathin body junctionless FinFETs," *Material Science in Semiconductor Processing*, vol. 104, p. 104653, December 2019, 10.1016/j.mssp.2019.104653

[3] N. B. S. Manikandan, and Balamurugan, "The improved RF/stability and linearity performance of the ultrathin - body Gaussian - doped junctionless FinFET," *Journal of Computational Electronocs*, no. 0123456789, pp. 1–9, March 2020, 10.1007/s1 0825-020-01472-y

[4] L. Esaki, "New phenomenon in narrow germanium p-n junctions," *Physical Review*, vol. 109, no. 2, p. 603, January 1958, 10.1103/PhysRev.109.603

[5] S. Banerjee, W. Richardson, J. Coleman, and A. Chatterjee, "A new three-terminal tunnel device," *IEEE Electron Device Letters*, vol. 8, no. 8, pp. 347–349, August 1987, 10.1109/EDL.1987.26655

[6] K. Boucart, and A. M. Ionescu, "Double-gate tunnel FET with high-k gate dielectric," *IEEE Transactions Electron Devices*, vol. 54, no. 7, pp. 1725–1733, 2007.

[7] A. M. Ionescu, and H. Riel, "Tunnel field-effect transistors as energy-efficient electronic switches," *Nature 2011 4797373*, vol. 479, no. 7373, pp. 329–337, November 2011, 10.1038/nature10679

[8] S. Saurabh, and M. J. Kumar, *Fundamentals of tunnel field-effect transistors.* CRC Press, 2016.

[9] J. K. Mamidala, R. Vishnoi, and P. Pandey, "Tunnel field-effect transistors (TFET): Modelling and simulation," *Tunnel Field-Effect Transistors Modelling and Simulation*, pp. 1–195, October 2016, 10.1002/9781119246312

[10] P. Vanitha, T. S. Arun Samuel, and D. Nirmal, "A new 2 D mathematical modeling of surrounding gate triple material tunnel FET using halo engineering for enhanced drain current," *AEU - International Journals of Electronics and Communications*, vol. 99, pp. 34–39, February 2019, 10.1016/J.AEUE.2018.11.013

[11] T. S. Arun Samuel, N. B. Balamurugan, S. Bhuvaneswari, D. Sharmila, and K. Padmapriya, "Analytical modelling and simulation of single-gate SOI TFET for low-power applications," *International Journal of Electronics*, vol. 101, no. 6, pp. 779–788, June 2014, 10.1080/00207217.2013.796544

[12] P. Vanitha, G. L. Priya, N. B. Balamurugan, S. T. Chandra, and S. Manikandan, "Analytical approach on the scale length model for tri-material surrounding gate tunnel field-effect transistors (TMSG-TFETs)," in *Intelligent Computing and Applications.* New Delhi: Springer, 2015, pp. 231–238.

[13] M. Venkatesh, M. Suguna, and N. B. Balamurugan, "Subthreshold performance analysis of germanium source dual halo dual dielectric triple material surrounding gate tunnel field effect transistor for ultra low power applications," *Journal of Electronic Materials*, vol. 48, no. 10, pp. 6724–6734, 2019.

[14] M. Venkatesh, M. Suguna, and N. B. Balamurugan, "Influence of germanium source dual halo dual dielectric triple material surrounding gate tunnel FET for improved analog/RF performance," *Silicon*, pp. 1–9, 2020.

[15] D. B. Abdi, and M. J. Kumar, "Controlling ambipolar current in tunneling FETs using overlapping gate-on-drain," *IEEE Journal of Electron Devices Society*, vol. 2, no. 6, pp. 187–190, November 2014, 10.1109/JEDS.2014.2327626

[16] R. Narang, M. Saxena, R. S. Gupta, and M. Gupta, "Assessment of ambipolar behavior of a tunnel FET and influence of structural modifications," *JSTS Journal Semiconductor Technology Science*, vol. 12, no. 4, pp. 482–491, December 2012, 10.5573/JSTS.2012.12.4.482

[17] J. Madan, and R. Chaujar, "Gate drain-overlapped-asymmetric gate dielectric-GAA-TFET: A solution for suppressed ambipolarity and enhanced ON state behavior," *Applied Physics A Materials Science and Processing*, vol. 122, no. 11, pp. 1–9, November 2016, 10.1007/S00339-016-0510-0/FIGURES/12

[18] S. Manikandan, N. B. Balamurugan, and D. Nirmal, "Analytical model of double gate stacked oxide junctionless transistor considering source/drain depletion effects for CMOS low power applications," *Silicon*, pp. 1–11, December 2019, 10.1007/s12633-019-00280-9

[19] M. K. Pandian, N. B. Balamurugan, and S. Manikandan, "Analytical modeling of junctionless surrounding gate silicon nanowire transitors," *Journal of Nanoelectronics and Optoelectronics*, vol. 9, no. 4, pp. 468–473, 2014.

[20] K. H. Kao *et al.*, "Optimization of gate-on-source-only tunnel FETs with counter-doped pockets," *IEEE Transactions Electron Devices*, vol. 59, no. 8, pp. 2070–2077, 2012, 10.1109/TED.2012.2200489

[21] S. W. Kim, J. H. Kim, T. J. K. Liu, W. Y. Choi, and B. G. Park, "Demonstration of L-shaped tunnel field-effect transistors," *IEEE Transactions Electron Devices*, vol. 63, no. 4, pp. 1774–1778, April 2016, 10.1109/TED.2015.2472496

[22] M. Ehteshamuddin, S. A. Loan, and M. Rafat, "A vertical-gaussian doped soi-tfet with enhanced dc and analog/rf performance," *Semiconductor Science and Technology*, vol. 33, no. 7, p. 075016, June 2018, 10.1088/1361-6641/AAC97D

[23] G. Zhou *et al.*, "Vertical InGaAs/InP tunnel FETs with tunneling normal to the gate," *IEEE Electron Device Letters*, vol. 32, no. 11, pp. 1516–1518, November 2011, 10.1109/LED.2011.2164232

[24] Y. S. Chauhan *et al.*, *FinFET modeling for IC simulation and design: Using the BSIM-CMG standard.* Elsevier Inc., 2015.

[25] W. Cao *et al.*, "2-D layered materials for next-generation electronics: Opportunities and challenges," *IEEE Transactions Electron Devices*, vol. 65, no. 10, pp. 4109–4121, 2018, 10.1109/TED.2018.2867441

[26] S. Das *et al.*, "Transistors based on two-dimensional materials for future integrated circuits," *Nature Electronics 2021 411*, vol. 4, no. 11, pp. 786–799, November 2021, 10.1038/s41928-021-00670-1

[27] J. Knoch, and J. Appenzeller, "A novel concept for field-effect transistors – The tunneling carbon nanotube FET," *Device Resolution Conference – Conf. Dig. DRC*, vol. 2005, pp. 153–156, 2005, 10.1109/DRC.2005.1553099

[28] Q. Zhang, W. Zhao, and A. Seabaugh, "Low-subthreshold-swing tunnel transistors," *IEEE Electron Device Letters*, vol. 27, no. 4, pp. 297–300, April 2006, 10.1109/LED.2006.871855

[29] A. Pon, A. Bhattacharyya, and R. Rathinam, "Recent developments in black phosphorous transistors: A review," *Journal of Electronic Materials 2021 5011*, vol. 50, no. 11, pp. 6020–6036, September 2021, 10.1007/S11664-021-09183-1

[30] R. Rathinam, A. Pon, S. Carmel, and A. Bhattacharyya, "Analysis of black phosphorus double gate MOSFET using hybrid method for analogue/RF application,"

IET Circuits, Devices System, vol. 14, no. 8, pp. 1167–1172, November 2020, 10.1049/IET-CDS.2020.0092

[31] J. Knoch, "Optimizing tunnel FET performance - Impact of device structure, transistor dimensions and choice of material," *International Symposium VLSI Technology Systems and Applications Proceedings*, pp. 45–46, 2009, 10.1109/VTSA.2009.5159285

[32] J. Na *et al.*, "Low-frequency noise in multilayer MoS 2 field-effect transistors: The effect of high- k passivation," *Nanoscale*, vol. 6, no. 1, pp. 433–441, December 2013, 10.1039/C3NR04218A

[33] S. Yang, S. Park, S. Jang, H. Kim, and J. Y. Kwon, "Electrical stability of multilayer MoS2 field-effect transistor under negative bias stress at various temperatures," *Physoca status solidi – Rapid Research Letters*, vol. 8, no. 8, pp. 714–718, August 2014, 10.1002/PSSR.201409146

[34] A. Pon, K. S. V. P. Tulasi, and R. Ramesh, "Effect of interface trap charges on the performance of asymmetric dielectric modulated dual short gate tunnel FET," *AEU – International Journal of Electronics Communication*, vol. 102, pp. 1–8, April 2019, 10.1016/J.AEUE.2019.02.007

[35] T. Li *et al.*, "A native oxide high-κ gate dielectric for two-dimensional electronics," *Nature Electrons 2020 38*, vol. 3, no. 8, pp. 473–478, Jul. 2020, 10.1038/s41928-020-0444-6

[36] A. Pon, A. Bhattacharyya, B. Padmanaban, and R. Ramesh, "Optimization of the geometry of a charge plasma double-gate junctionless transistor for improved RF stability," *Journal of Computational Electronics*, no. 0123456789, 2019, 10.1007/s10825-019-01340-4

4 Modeling and Simulation of Dual Material Double Gate Tunnel FETs

A.V. Arun and Jobymol Jacob

CONTENTS

4.1 INTRODUCTION

A tunnel field effect transistor (TFET) is an emerging low-power device that exhibits excellent subthreshold swing (SS) and low leakage current [1–15]. The problem with TFET is that they have low ON current, hence various structural modifications are being implemented to improve the ON current [4–16]. Since the carriers' transport mechanism in TFET differs from that of MOSFET, the time taken for the drain current saturation is delayed. This leads to the degradation in performance when the device is used for CMOS low-power applications. DIBL effects are also prominent in TFET, which limits the use of the device in the circuits [16]. Apart from the ON current, all these factors should be considered while designing a TFET structure. Dual material gate (DMG) TFET [17] can address all these issues, including improvement in ON current, reduction of OFF current, reduction of SS, immunity against DIBL, and better output characteristics.

The point to be noted is that DMG structure alone cannot improve the ON current. Using smaller bandgap material can enhance tunneling and improve the ON current. The coupling between the channel and gate can be improved with high-k dielectrics leading to improved ON current. Implementation of DMG technique in TFET offers reduced transistor dimensions up to 20 nm.

DOI: 10.1201/9781003327035-4

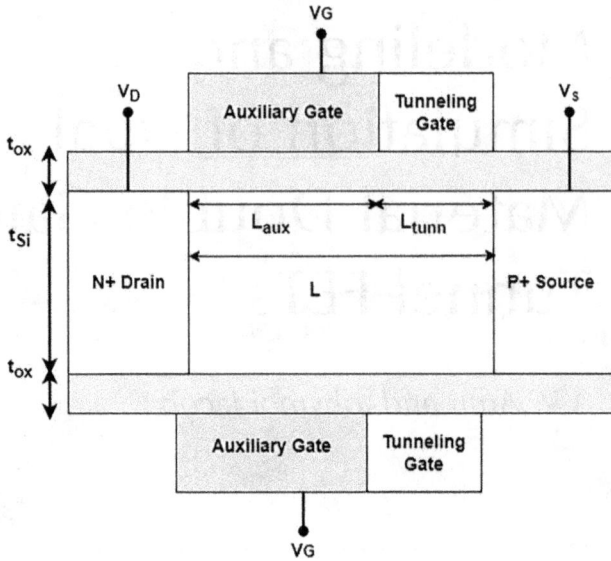

FIGURE 4.1 Cross-sectional view of DMG TFET.

TABLE 4.1
Parameters Considered for Simulation

Parameter	Value
Doping of source	$10^{20}/cm^3$
Doping of body	$10^{17}/cm^3$
Thickness of silicon body	10 nm
Thickness of gate oxide	3 nm
Channel length	50 nm
Tunneling gate length	20 nm
Auxiliary gate length	30 nm

4.2 DUAL MATERIAL GATE TFET

The structure of DMG TFET is shown in Figure 4.1. It is a double gate structure composed of gate metals with different work functions on top and bottom. The part of the gate near the source is termed the tunneling gate, and the part near the drain is termed the auxiliary gate. All simulations presented in this chapter are based on silvaco atlas. Non-local tunneling models are applied for the simulation. The parameters used for the simulation are shown in Table 4.1.

4.3 DMG TFET CHARACTERISTICS

The impact of auxiliary gate work function (Φ_{aux}) on OFF current is studied in Figure 4.2 and 4.3. In Figure 4.2, the device is under OFF condition, and when Φ_{aux}

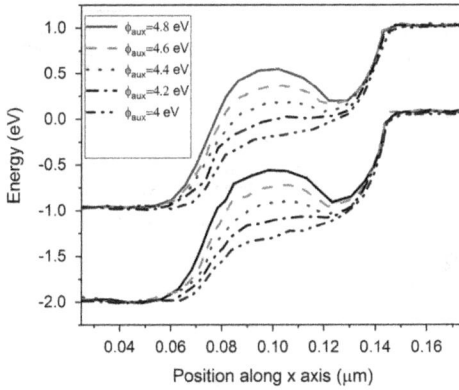

FIGURE 4.2 Simulated energy band diagram of DMG TFET under OFF condition.

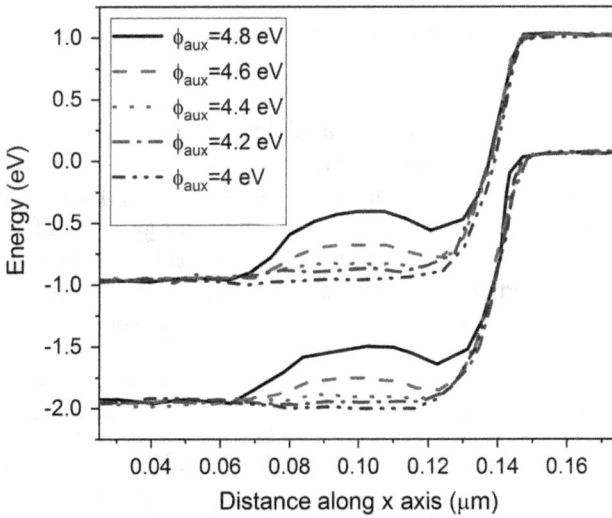

FIGURE 4.3 Simulated energy band diagram of DMG TFET under ON condition for different values of auxiliary gate work function.

increases, the overlapping of energy bands increases, leading to an increase in tunneling width. This reduces the OFF-state current of the device. In the ON condition, as shown in Figure 4.3 there is no effect of Φ_{aux} on the energy band diagram. As the auxiliary gate work function is increased beyond 4.4 eV, the band overlap begins to decrease. However, the tunneling width is already high, so tunneling the carriers is impossible.

The transfer characteristics of DMG TFET are shown in Figure 4.4 for different values of the auxiliary gate work function. It can be observed that when the auxiliary gate work function increases, the OFF current reduces, but when it reaches 4.8 eV, the tunneling of carriers in the drain side increases, leading to an increase in

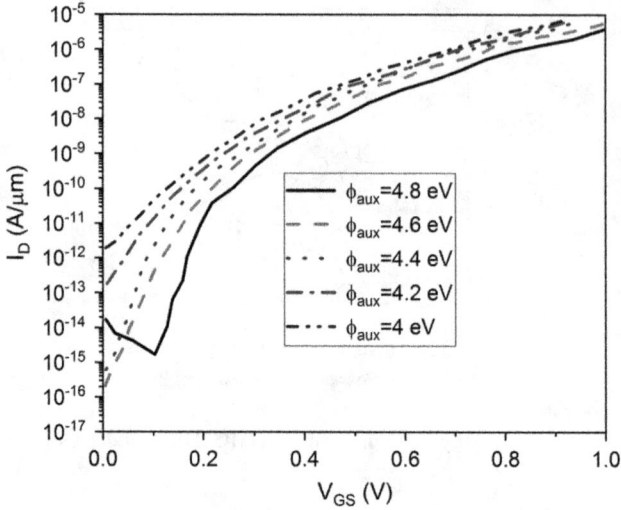

FIGURE 4.4 Transfer characteristics of DMG TFET for different values of auxiliary gate work function.

the OFF current again. There is also a slight decrease in OFF current with the increase in auxiliary gate work function.

The impact of the tunneling gate work function (Φ_{tunnel}) on the energy band diagram is shown in Figures 4.5 and 4.6. The auxiliary gate work function is kept constant at 4.4 eV, and the tunneling gate work function is decreased from 4.8 eV to 4 eV. In the OFF condition, when Φ_{tunnel} is 4 eV, the OFF current is less as the energy band overlap is not there. In the ON condition, when Φ_{tunnel} reduces, the

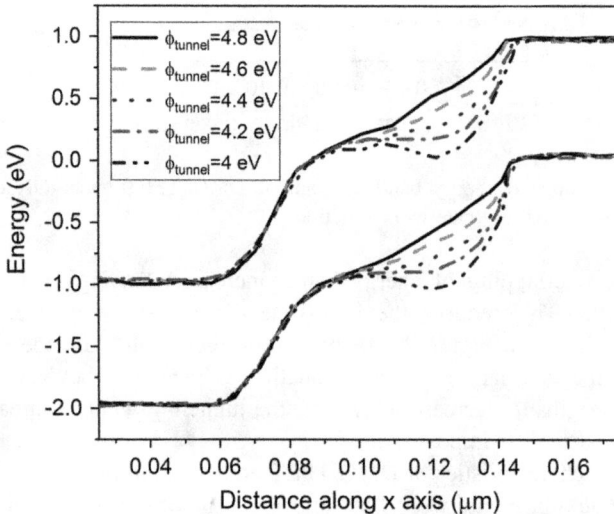

FIGURE 4.5 Simulated energy band diagram of DMG TFET under OFF condition.

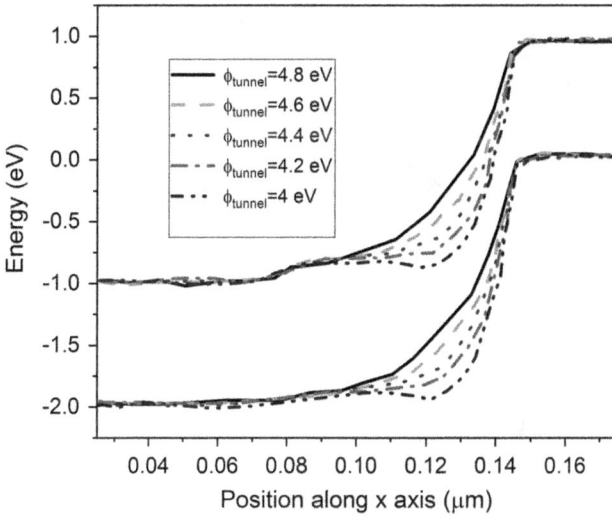

FIGURE 4.6 Simulated energy band diagram of DMG TFET under ON condition for different values of tunneling gate work function.

level of overlap between the energy band rises, and also tunneling width reduces. Thus, ON current is increased. The transfer characteristics are plotted in Figure 4.7. As the value of Φ_{tunnel} is decreased from 4.8 eV to 4 eV. ON current increases as the Φ_{tunnel} reduces.

The value of Φ_{aux} and Φ_{tunnel} can be optimized to obtain improved ON current and OFF current. A large OFF current is observed when Φ_{tunnel} is less than 4 eV and

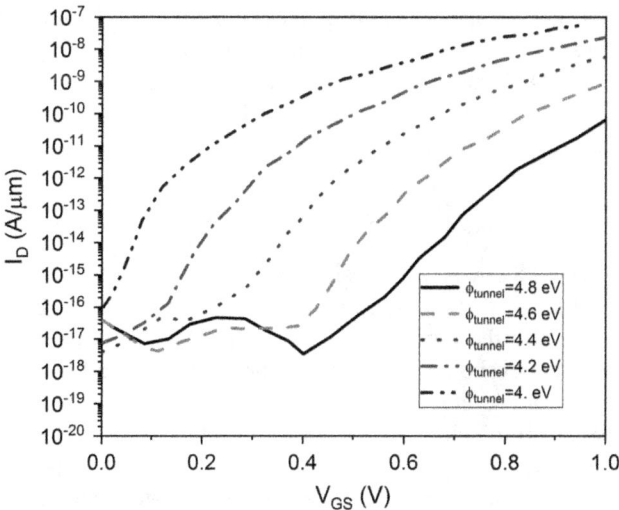

FIGURE 4.7 Transfer characteristics of DMG TFET for different values of tunneling gate work function.

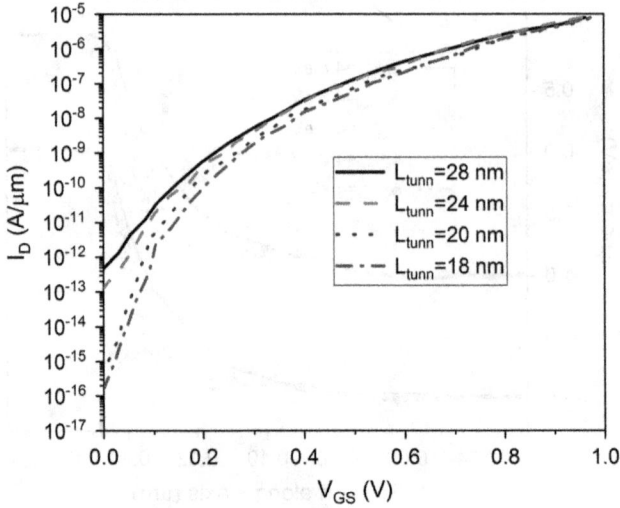

FIGURE 4.8 The transfer characteristics of DMG TFET for different values of channel length.

also when Φ_{aux} is greater than 4.8 eV. Hence, the value of Φ_{aux} and Φ_{tunnel} should lie within 4 eV to 4.8 eV. If there is an increase in Φ_{aux} beyond 4.4 eV the ON current reduces without any considerable reduction in OFF current. Hence, Φ_{aux} can be fixed at 4.4 eV. Molybdenum and tungsten offer a work function of 4.4 eV. Φ_{tunnel} is fixed at 4 eV and metals like molybdenum, nickel, etc., offer this work function. The transfer characteristics for different channel length is plotted in Figure 4.8. It is observed that the OFF current is well within the limit when the device is scaled up to 20 nm.

The comparison of transfer characteristics of DMG TFET and SMG TFET is shown in Figure 4.9. The ON to OFF current ratio is significantly high in DMG TFET. The ON to OFF current ratio can be increased in SMG TFET by changing the gate metal work function. The maximum ON current for SMG TFET is obtained at a work function of 4.2 eV.

4.4 DMG TFET DRAIN CURRENT MODELING

Traditional MOSFETS size and dimensional gauging are limited due to second-order effects that shoot up the thermionic emission carriers in the source-channel interface. Due to this, it leads to the formation of various undesirable effects like high static leakage current, subthreshold swing (SS) of the range greater than 60 mV/decade, etc. Since the thermionic emission depends upon temperature, MOSFET has only a minimum SS of 60 mV/decade at room temperature. The static leakage current increases exponentially with the supply voltage and threshold voltage in MOSFET. As a result power dissipation increases due to the mismatch between the scaling of transistors and the scaling of the power supply.

Tunnel FET(TFET) [18–20] has a SS less than MOSFET, which is steeper and is considered a favorable electron device in future upcoming technology. Tunnel FET

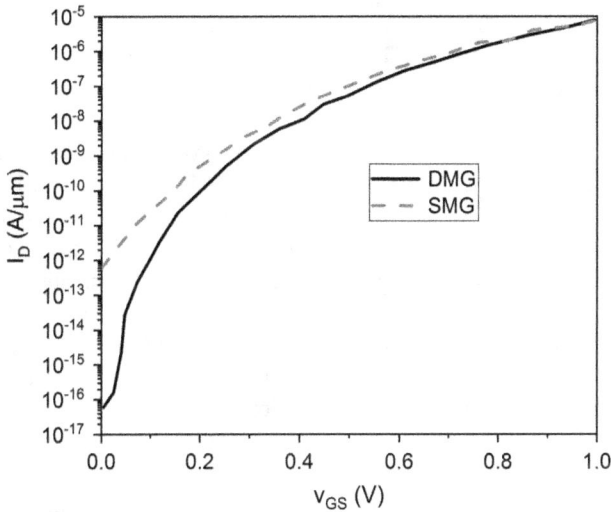

FIGURE 4.9   The transfer characteristics of DMG TFET and SMG TFET.

works on the mechanism of band-to-band tunneling (BTBT), where the transport of carriers takes place due to electron tunneling from the valence band of the source terminal to the conduction band of the channel. BTBT does not depend on temperature and due to this, it can achieve lower values of SS and hence operates at low supply voltage. It can achieve a low OFF current and lower power consumption when compared to MOSFETs [21,22].

Even though Tunnel FETs consume less power, SS below 60mV/decade and low leakage current, the ON- current (I_{ON}) is not sufficient for digital applications and there is also a chance of an increase in the OFF- current because of ambipolar behavior [23,24]. Conventional TFETs like Si-TFETs have low tunneling efficiency due to the indirect bandgap mechanism, high tunneling barriers, and heavier carriers. To overcome these problems, profound research works are explored to increase I_{ON} of Si-TFETs. Some of these include the introduction of heterojunctions [25,26] with a low bandgap in the source-channel interface [27], high k-dielectrics, etc. [5,28].

The double gate TFET structure with the installation of a dual-metal-gate gives considerable improvements in performance. Dual material gate-double gate TFETs (DMG-DGTFET) are made of two types of gates placed sideways. The gate in proximity to the drain terminal is known as the auxiliary gate and the tunneling gate is near the source terminal [29]. The tunneling gate's work function is greater than the auxiliary gate work function, which results in a higher source threshold than at the drain. The fabrication of DMG devices is a difficult task. A lot of advancements are made in the area of interest because of its advantages like steep SS and high ON-OFF current ratio [28].

Different features of dual material gate TFETs have been analyzed with TCAD simulations [28–30]. There is a need for an analytical model in a condensed manner to better understand device operation characteristics and effectively simulate these devices. There is a need for modeling electrostatics and the carrier transport in the channel for TFETs like that in the MOSFETs. The short channel effects can be neglected as TFETs work by band-to-band tunneling and not by thermionic emission [31].

The initial models of simple TFET [32–36] and DMG-DGTFET [37–39] specified in the literature have lots of drawbacks. The works related to SMG-TFETs consist of comparatively complicated and incorrectly calculated surface potential formulated using various approximations [33,34]. In DMG-DGTFET-based drain current models, the source-channel and drain-channel depletion regions are discarded, also tunneling through the drain-channel junction is not considered. There is a requirement for modeling the drain-channel depletion region to accurately model drain current.

By analyzing TCAD simulations, it is evident that there is a variation of surface potential in the drain-channel depletion region as we move from source to drain until the drain potential is achieved. This inference is made such that the drain depletion region cannot be completely discarded. The ambipolar effect in TFETs can also be detected by accurately modeling the drain depletion region to obtain an accurate drain current model.

DMG-based TFET models [37–41] take into account only the tunneling of carriers at the source-channel junction and it results in the inability to detect the carrier-induced ambipolar current at the drain terminal at OFF conditions and negative gate voltages. The doping determines the depletion region width at the drain side, controlling the tunneling current at the drain. This phenomenon was not incorporated into the existing models. So there is a need for a compact analytical drain current model for DMGTFETs that considers both the depletion regions. The analysis of tunneling at source–channel and drain–channel junctions along with the different drain doping concentrations should also be studied.

The next section focuses on the analysis of the surface potential model of DMG-DGTFETs using a pseudo-2-D solution of Poisson's equation [42]. The modeling of source and drain regions is done and a precise analysis of the tunneling dimensions is carried out with an iterative method. The drain current model is derived by assuming a non-local path tunneling approximation and the effects of variable drain doping are incorporated into the model by employing some fitting parameters. The model is framed by joining the BTBT generation rate [43] in the entire region and gives significance to source–channel and channel–drain tunneling. The junctions are assumed to have uniform electric field.

4.5   DEVELOPMENT OF THE SURFACE POTENTIAL MODEL

DMG-DGTFET structure taken for this study is shown in the Figure 6.6. It has following parameters: p type doping in source (N_S) = n type doping in drain (N_D) = 10^{20} cm^{-3}, p-type substrate doping (N_{ch}) = 10^{14} cm–3, dielectric (SiO_2) thickness (t_{ox}) = 2 nm, thickness of silicon body (t_{Si}) = 10 nm, work function of tunneling gate = 4.0 eV, length of tunneling gate (L_{tun}) = 12 nm, channel length (Ltun + Laux = L) = 30 nm work function of auxiliary gate = 4.4 eV and length of auxiliary gate (L_{aux}) = 18 nm. The ratio of L_{tun} to L_{aux} is 2:3 which is adopted from [28] gives better ON/OFF current ratio and SS. Abrupt junctions are considered for this study.

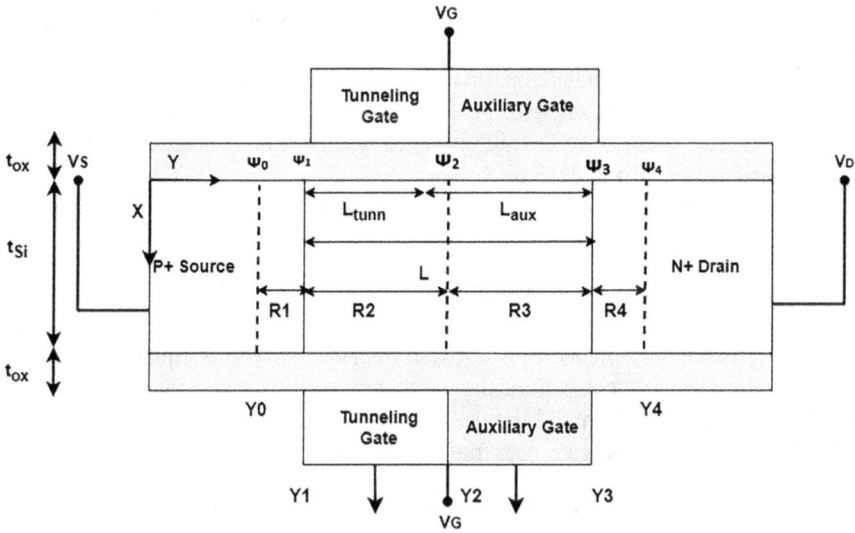

FIGURE 4.10 Cross-section of p-channel DMG-DGTFET, having regions -R1, R2, R3, and R4, with interfaces at y0, y1, y2, y3, and y4, and corresponding surface potentials ψ0, ψ1, ψ2, ψ3, and ψ4).

The voltage at the source terminal (V_S) is grounded and it is used as a reference voltage. The gate voltage is varied to do the analysis and the supply voltage is 1V. The entire device is divided into different regions and is shown in Figure 4.10. Region R1 is the depletion region at the source side and R4 represents the depletion region at the drain side. The channel is represented as two regions, R2 and R3 which are the portions under the tunneling gate and auxiliary gate. The boundary conditions are applied for all the regions and 2D Poisson's equation should be solved in the regions of interest and 2-D potential is modeled from which the electric field can be calculated. A semiempirical iterative method is used to calculate the length of depletion regions.

A. Surface Potential Model

There is not much effect in mobile charge carriers [44] when the device switches from OFF condition to the beginning of ON condition. The 2D Poisson's equation is given as

$$\partial^2\psi(x, y)/\partial x^2 + \partial^2\psi(x, y)/\partial y^2 = -q \, N_R/\varepsilon_{Si} \qquad (4.1)$$

here ψ(x, y) is the surface potential, q denotes charge of electron and ε_{Si} is the dielectric constant of Silicon, and N_R denotes the doping level of that particular portion. The parabolic approximation function for the DG devices, is given by [45]

$$\psi(x, y) = a_0(y) + a_1(y)x + a_2(y)x^2 \qquad (4.2)$$

where the coefficients $a_0(y)$, $a_1(y)$, and $a_2(y)$ can be derived by applying the boundary conditions.

The surface potential in the upper oxide semiconductor interface of the device ie, at $x = 0$ is given by $\psi_s(y) = \psi(0, y)$. Similarly at the lower oxide semiconductor interface, ie, at $x = t_{Si}$ is given as $\psi_b(y) = \psi(t_{Si}, y)$, it can be denoted as $\psi_s(y)$, due to symmetrical nature of the device.

The boundary condition related to lateral electric field ε_x at these two portions are [42]:

$$\varepsilon_x(0, y) = -\varepsilon_x(t_{Si}, y) = C'_{ox}[\psi_G - \psi_s(y)]/\varepsilon_{Si} \qquad (4.3)$$

where C'_{ox} is the capacitance at gate-oxide interface (which is equal to ε_{ox}/t, with ε_{ox} being the dielectric constant of the oxide, and $t = t_{ox}$ for regions under the gate, and for other regions, $t = \pi t_{ox}/2$ to take the fringing field effect of the caused by the gate [46]), and ψ_G is the gate potential which is given as

$$\psi_G = V_G - \varphi + \chi + Eg/(2q) + \varphi_t \ln(N_{ch}/n_i) \qquad (4.4)$$

where ψ_G can be ψ_{G2} or ψ_{G3}, in that case φ can be φ_{tun} or φ_{aux}. In equation (4.4), χ is the electron affinity E_g is the energy band gap and n_i is the intrinsic carrier and φ_t is the thermal voltage given by kT/q, where k is Boltzmann's constant and T is the absolute temperature.

Applying the boundary condition given in equation (4.3) to equation (4.2), the coefficients $a_0(y)$, $a_1(y)$, and $a_2(y)$ can be obtained as

$$\begin{aligned}
a_0(y) &= \psi_s(y) \\
a_1(y) &= C'_{ox}[\psi_s(y) - \psi_G]/\varepsilon_{Si} \\
a_2(y) &= C'_{ox}[\psi_G - \psi_s(y)]/(\varepsilon_{Si}t_{Si}).
\end{aligned} \qquad (4.5)$$

Substituting the coefficients given in equation (4.5) into equation (4.2) changes the Poisson's equation which is a two dimension differential equation into one dimensional and is given by

$$\partial^2\psi_s(y)/\partial y^2 - \beta^2\psi_s(y) = -\beta^2\psi_v \qquad (4.6)$$

where β is given by $(2C'_{ox}/(\varepsilon_{Si}t_{Si}))1/2$ and

$$\psi_v = \psi_G + q \ N_R t_{Si}/(2C'_{ox}). \qquad (4.7)$$

The factors β and ψ_V vary according to different regions due to varying changes in doping levels, work function of gate metals, and gate capacitance values. The factor $1/\beta$ is the characteristic length in different regions.

Differential equation (4.6), is solved and the generalized solution for all regions is given as

$$\psi_s(y) = bi \ exp \ [\beta_i(y - y_i - 1)] + c_i \ exp \ [-\beta_i(y - y_i - 1)] + \psi_{vi} \qquad (4.8)$$

where b_i and c_i need to be found out by applying the boundary conditions. the segments $[y_i -1, y_{i]}$ is the defined boundary for each region where i = 1 − 4. The values of ψs at extreme ends of R1 and R4 are given by

$$\psi_{s|y=y_0} = \psi_0 = -\varphi_t \, ln(N_S/n_i)$$
$$\psi_{s|y=y_4} = \psi_4 = V_D + \varphi_t \, ln(N_D/n_i)$$

$$(4.9)$$

The surface potentials in all other regions can be derived by applying continuity equations of potential and electric field

$$\psi_{si|y=yi} = \psi_{s(i+1)|y=yi}$$
$$d\psi_{si}/dy|_{y=yi} = d\psi_{s(i+1)}/dy|_{y=yi}$$

$$(4.10)$$

where i = 1–3. Solving equations (4.9) and (4.10) gives values of the constants b_i and c_i, and surface potential in each region can be solved.

B. Source and Drain Depletion Region Lengths

The existing models [42] show potential ψ_v dependent approximated depletion region lengths which do not take into account the junction influence. This is not physically admissible as it discards the source and drain junctions and gives wrong results. So to solve this issue an iterative method that takes into consideration the effects of junctions on ψ_v is detailed next

1. Step 1: To begin with, assume that ψ_v controls the depletion region lengths (L1, L4) and is given by

$$L1 = \sqrt{2\varepsilon_{Si}(\psi_{v1} - \psi0)/(q \ N_1)}$$

$$(4.11)$$

$$L4 = \sqrt{2\varepsilon_{Si}(\psi_4 - \psi_{v3})/(q \ N_4)}$$

$$(4.12)$$

where N_1 is the doping at source and N_4 is the doping at the drain. For ψ_{v3} in equation (4.12), can be obtained by substituting parameters of region 3 in equation (4.7), where ψ_{G3} can be calculated from equation (4.4); while for ψ_{v1} in equation (4.11), can be obtained by substituting parameters of region 1 in equation (4.7), with ψ_{G1} approximated as

$$\psi_{G1} = V_G - \varphi_{tun} + \chi + Eg/q$$

$$(4.13)$$

Due to high source doping.
2. Step 2: the constants b_i, c_i can be calculated for each region using these values of L1 and L4 by solving the equations (4.9) and (4.10).
3. Step 3: The surface potentials at interfaces, ψ_1 and ψ_3, are calculated by applying the constants obtained in Step 2.

4. Step 4: Recalculating L1, L4 with ψ_{V1} and ψ_{V3} calculated in Step 3 using equation (4.11) and equation (4.12)
5. Step 5: Repetition of steps from 2 to 4 until it reaches convergence is taken as L1 and L4 leading to the precise analysis of depletion region lengths.

The inference is that L1, L2 are dependent on gate voltage through the term $\psi 1$ and $\psi 3$. As gate voltage increases the potentials ψ_1, ψ_3 increase which leads to an increase in L1 but decrease in L_4. The efficient evaluation of L_1 and L_2 is implemented by an iterative method. As L_1, L_4 depend on $\psi 1$ and $\psi 3$, closed form solution is not available.

C. Model Validation

TCAD simulations [47] are used to verify the results of the surface potential model for device structure in Figure 4.10. The model is validated for a different set of parameters as shown in Figure 4.11 and Figure 4.12. Figure 4.11 compares the device simulation and the surface potential model along the lateral distance for different values of gate voltage and a fixed drain voltage of 1 V. On the other hand Figure 4.12 plots the same comparison of the surface potential model for different values of drain voltage and fixed gate voltage of 0.2 V.

The above figures show that the surface potential model has an excellent match with that of obtained TCAD simulations. Similar matches were seen for all values of gate voltage and drain voltage as well. In the region under the gate terminal as seen in Figure 4.11 from the end of the source terminal, there is a linear increase in the potential and the slope increases with an increase in the applied gate bias.

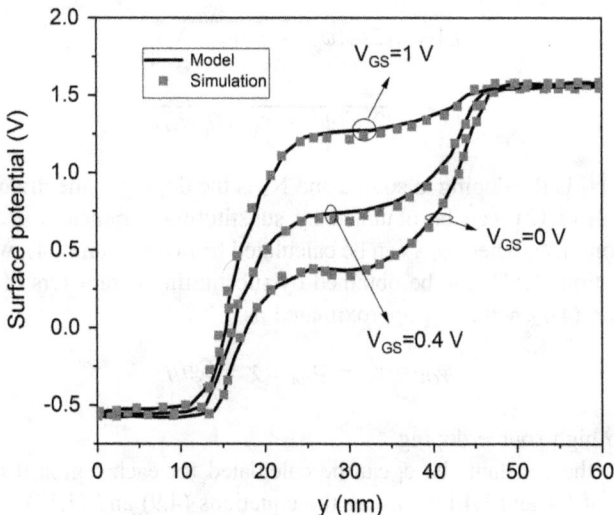

FIGURE 4.11 Comparison of device simulation and the surface potential model along the lateral distance for changing gate voltage and drain voltage kept constant at 1 V.

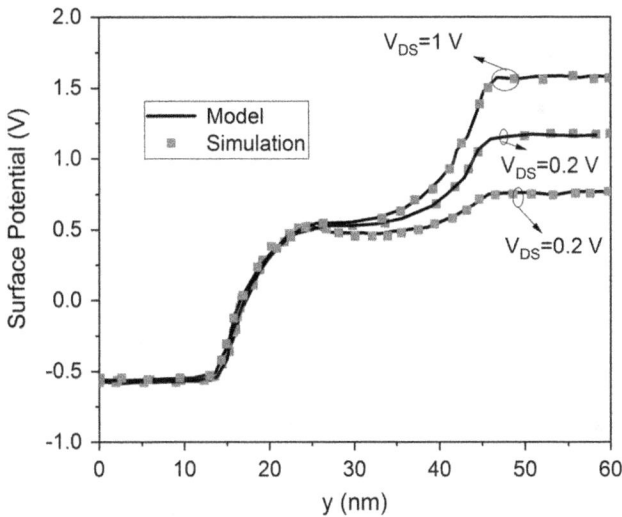

FIGURE 4.12 Comparison of device simulation and the surface potential model along the lateral distance for changing drain voltage and gate voltage kept constant at 0.2 V.

This shows that BTBT-induced electric field also increases and tunneling width decreases at the tunneling junction. At the drain side, as the electric field increases the potential rises which leads to an increase in drain current.

4.6 DRAIN CURRENT MODEL

A closed-form expression for drain current is derived here. A non-local BTBT model [47] is used to compare results obtained from TCAD simulation. Here electric field in the tunneling region is kept constant and it is obtained by taking the average of electric field or taking the maximum electric field in the tunneling region. A uniform electric field is assumed [43,48,49], the BTBT generation rate is given as [47]:

$$G_{BTBT} = A (E/E0)^P / exp (-B/E) \qquad (4.14)$$

In equation (6.14) E represents the electric field observed in the tunneling region, E0 takes the value of 1 V/cm, P is a constant that can either be 2 or 2.5 [47] depending on the tunneling process. The tunneling process can either be direct tunneling or phonon-assisted tunneling. As silicon is the material here, phonon-based tunneling is the dominating tunneling process present in this particular device. A and B are constants, that have a value of 4×10^{14} cm^{-3}s^{-1} and 1.9×10^7 V/cm.

The simulated energyband diagram of DMG-DGTFET is plotted in Figure 4.13 and the values of tunneling widths of the corresponding tunneling regions are marked. Tunneling widths vary with the surface potential under different bias conditions. When surface potential rises beyond the source potential ψ_0 by a value equal to E_g/q, the minimum tunneling width l_1 along the lateral tunneling path is

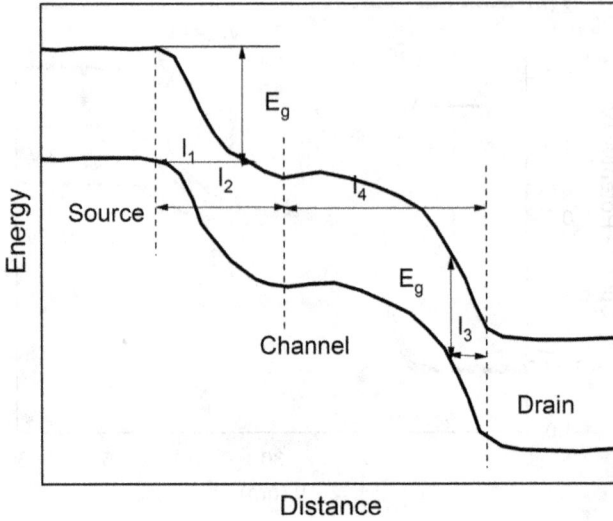

FIGURE 4.13   Schematic energyband diagram of a DMG-DGTFET, exhibiting the width at the tunneling regions.

obtained while tunneling gate length L_{tun} gives the maximum width l_2 since Φ_{tun} controls ON-state tunneling current. When surface potential goes down the potential at the drain side, ψ_4 by a value equal to E_g/q, minimum width of l_3 along the lateral tunneling path is obtained while the length of the auxiliary gate L_{aux} is equal to the maximum tunneling width l_4.

The drain current is obtained by integrating G_{BTBT} given in equation (4.14) over the entire tunneling volume. By taking the average value of an electric field in the tunneling region, the corresponding value is obtained as $E_{avg} = E_g/(ql_{path})$, here l_{path} is the corresponding width of the tunneling region, ranging from l_1 to l_2 for the source tunneling region, and from l_3 to l_4 in the tunneling region.

The drain current is given as [24]

$$I_{tun} = q \iint G_{BTBT}\, dy dx \qquad (4.15)$$

where G_{BTBT} is taken from equation (4.14) and by applying E_{avg}. Equation (4.15) is valid for source and the drain tunneling currents with limits for y given from l1 to l2 and l3 and l4.

The sum of I_{source} and I_{drain} gives the total tunneling current (I_{tun}). If the applied gate voltage is high, I_{source} dominates and I_{drain} dominates for negative values of gate voltage. It is due to higher BTBT at source–channel interface due to the increased overlap between the valence band of the source and the conduction band of the channel, at positive values of gate voltage. Thus I_{source} has higher impact on I_{tun}.

The transfer (I_D–V_G) characteristics of DMG-DGTFET are shown in Figure 4.14 with drain doping N_D as a parameter which is obtained from simulation results. Ambipolar current of the device is reduced as the doping on the drain region is

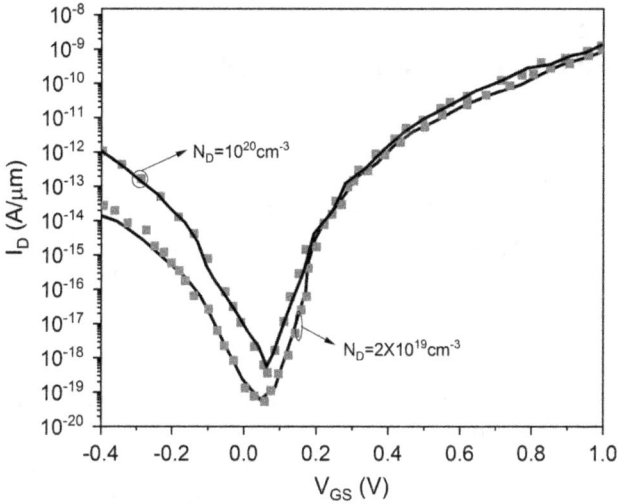

FIGURE 4.14 Transfer characteristics (I_D–V_G) of a DMG-DGTFET. Results with the model represented by lines and TCAD simulation given by symbols with varying drain doping.

reduced. As the drain doping goes down, tunneling of carriers from drain to source reduces and ambipolar current is suppressed. The drain depletion length (L_4) increases and leads to an increase in the parameter l_3 under the same V_G at the drain-channel interface resulting in less tunneling of carriers.

As gate bias voltage V_G is reduced progressively the reduction in the tunneling current can be seen due to reduced source-channel junction band overlap. As V_G is reduced consistently, the voltage V_R is approached. The VR is related to Φ_{tun} and aux, and it is chosen as 0.2V as shown in Figure 4.14. where the decrease in the overlapping of bands takes place at both junctions reducing the BTBT mechanism. As V_{GS} is again reduced further the overlapping of energy bands does not occur in the tunneling region near the source, but in contrast, there is an increase in band overlapping at the tunneling region near the drain. Here a considerable amount of current is not generated until V_G reaches 0V in this condition. With the reduction in V_G again energy band aligns at the tunneling region near the drain and hence drain current I_{drain} becomes an important parameter in the total current I_{tunn}.

For $V_{GS} < 0$ as well as for $V_{GS} > V_R$ the model results obtained gave great similarity with the TCAD simulation results but there is a mismatch when the gate bias goes below V_R. The least tunneling widths at both the junctions for these given values of voltages become high and it can be seen as direct tunneling occurs between the two terminals. Here we use the basic p-n junction theorem to determine the drain current which wrongly predicts the actual current seen through simulation. This gives the inference of other elements of current influencing the particular region which contradicts the tunneling phenomenon. Here it is assumed that the mechanism of generation of current in the reverse-biased source/drain-channel junctions leads to the formation of current in these regions which is influenced by

TABLE 4.2

Empirical Parameters Used in the Model

$N_D(cm^{-3})$	a(A/μm)	b(cm)	c(A/μm)	d(cm)
10^{20}	1.23×10^9	8.06×10^{-6}	2.71×10^{26}	1.56×10^{-5}
8×10^{19}	1.08×10^{11}	1.07×10^{-5}	6.31×10^{30}	1.72×10^{-5}
6×10^{19}	6.29×10^{12}	1.49×10^{-5}	1.18×10^{31}	1.74×10^{-5}
4×10^{19}	2.67×10^{10}	1.97×10^{-5}	2.60×10^{31}	1.75×10^{-5}
2×10^{19}	4.96×10^{11}	3.75×10^{-5}	4.73×10^{33}	1.85×10^{-5}

the widths of depletion regions in these junctions, which is again controlled by the concentration of the regions.

As V_G becomes greater, surface potential in equation (4.8) increases, therefore leading to an increase in length L_1 given by equation (4.11), but L_4 given in equation (4.12) decreases. As V_G is moved from 0 to V_{OFF} I_D reduces but an increase in I_D can occur with V_G between V_{OFF} and V_R.

Thus for obtaining a greater similarity of the model with that of device simulations, fitting using an empirical function is used with the model when V_G is less than V_R which depends on source and drain depletion lengths.

$$I_f = a \ exp(-b/L4), \quad for \ 0 \le V_G \le V_{OFF}$$
$$c \ exp(-d/L1), \quad for \ V_{OFF} < V_G \le V_R \tag{4.16}$$

the fitting parameters are given by a, b, c and d. The selected values for the parameters are given in Table 4.2 for different values of drain doping. The model is accurate [shown in Figure 4.14 (lines)] for the entire operating range with respect to different values of N_D. The drain current is non-zero when the drain voltage is zero. To compensate for this current in the model [50], a correction factor should be added and it is given by:

$$f_{Correction} = 1 - 2/[1 + exp\{V_D/(f\varphi_t)\}] \tag{4.17}$$

where f is an empirical parameter. Thus, compactness and consistency of current over entire bias regions for practical model can be used to implement the TFET simulated designs.

We can also see other approaches to model drain current when drain voltage is zero without any compensation parameters [51]. Here dual modulation effect can determine the drain current assuming that gate bias can influence the surface potential.

The modified models use the following relation corresponding to interband tunneling [52], given by

$$I_T \propto P_T(E)D1(E)[f_1(E) - f_2(E)]D_2(E)dE \tag{4.18}$$

The tunneling probability gives P_T, and Di is the density of states and fi ($i = 1, 2$) is the Fermi–Dirac distribution function corresponding to regions, where 'i' takes the value 1 or 2. This model becomes more accurate as the drain current depends not only on the tunneling probability P_T, but also on the density of states.

The tunneling probability P_T, can be calculated using the WKB approximation [41,53,54]. For TFET, the density of state function (f_1–f_2) is detached from the tunneling probability. The first method is based on drain voltage only [40] while second method depends on both the gate and drain bias. In both methods, the occupancy function is multiplied by the probability of tunneling to compensate for the zero drain voltage [53–57].

4.7 CONCLUSION

Here a 2-D DMG-DGTFET is developed to find the path of tunneling current through the devices. The source-drain depletion effect is present and as a result, better agreement of the model results for the surface potential is achieved with respect to the TCAD simulation results for different gate and drain biases. Surface potential determines the tunneling widths at both the source and drain junctions. This is again modeled for the non-local tunneling current under the assumption of an average constant electric field across the tunneling path. Consideration of BTBT is recognized by source and drain regions; hence, ambipolar behavior is limited. A fitting function with the effects of the current generation is first to be used to model the overall drain current for gate bias equal to zero. The entire model results match is analyzed as a function of gate voltage with changing drain doping concerning TCAD simulations, which provides the best results.

REFERENCES

[1] Q. Zhang, W. Zhao, and A. Seabaugh, "Low-subthreshold-swing tunnel transistors," *IEEE Electron Device Letters*, vol. 27, pp. 297–300, April 2006.

[2] S. O. Koswatta, M. S. Lundstrom, and D. E. Nikonov, "Performance comparison between p-i-n tunneling transistors and conventional MOSFETs," *IEEE Transactions on Electron Devices*, vol. 56, pp. 456–465, March 2009.

[3] T. Nirschl, P. F. Wang, C. Weber, J. Sedlmeir, R. Heinrich, R. Kakoschke, K. Schrufer, J. Holz, C. Pacha, T. Schulz, M. Ostermayr, A. Olbrich, G. Georgakos, E. Ruderer, W. Hansch, and D. Schmitt-Landsiedel, "The tunneling field effect transistor (TFET) as an add-on for ultra-low-voltage analog and digital processes," *IEDM Technical Digest*, pp. 195–198, December 2004.

[4] K. K. Bhuwalka, J. Schulze, and I. Eisele, "Scaling the vertical tunnel FET with tunnel bandgap modulation and gate work function engineering," *IEEE Transactions Electron Devices*, vol. 52, pp. 909–917, May 2005.

[5] K. Boucart, and A. M. Ionescu, "Double-gate tunnel FET with high-κ gate dielectric," *IEEE Transactions Electron Devices*, vol. 54, no. 7, pp. 1725–1733, July 2007.

[6] E. H. Toh, G. H. Wang, L. Chan, G. Samudra, and Y. C. Yeo, "Device physics and guiding principles for the design of double-gate tunneling field effect transistor with silicon-germanium source heterojunction," *Applied Physics Letters*, vol. 91, Article Number: 243505, December 2007.

[7] V. V. Nagavarapu, R. R. Jhaveri, and J. C. S. Woo, "The tunnel source (PNPN) n-MOSFET: A novel high performance transistor," *IEEE Transactions on Electron Devices*, vol. 55, pp. 1013–1019, April 2008.

[8] A. S. Verhulst, W. G. Vandenberghe, K. Maex, S. D. Gendt, M. M. Heyns, and G. Groesenrken, "Complementary silicon-based heterostructure tunnel-FETs with high tunnel rates," *IEEE Electron Device Letters*, vol. 29, pp. 1398–1401, December 2008.

[9] M. Schlosser, K. K. Bhuwalka, M. Sauter, T. Zilbauer, T. Sulima, and I. Eisele, "Fringing-induced drain current improvement in the tunnel field-effect transistor with high-k gate dielectrics," *IEEE Transactions on Electron Devices*, vol. 56, pp. 100–108, January 2009.

[10] S. Saurabh, and M. J. Kumar, "Impact of strain on drain current and threshold voltage of nanoscale double gate tunnel field effect transistor: Theoretical investigation and analysis," *Japanese Journal of Applied Physics*, vol. 48, Article Number: 064503, Part 1, June 2009.

[11] P. F. Guo, L. T. Yang, Y. Yang, L. Fan, G. Q. Han, G. S. Samudra, and Y. C. Yeo, "Tunneling field-effect transistor: Effect of strain and temperature on tunneling current," *IEEE Electron Device Letters*, vol. 30, pp. 981–983, September 2009.

[12] A. S. Verhulst, W. G. Vandenberghe, K. Maex, and G. Groeseneken, "Tunnel field-effect transistor without gate-drain overlap," *Applied Physics Letters*, vol. 91, Article Number: 053102, 2007.

[13] F. Mayer, C. L. Royer, J. F. Damlencourt, K. Romanjek, F. Andrieu, C. Tabone, B. Previtali, and S. Deleonibus, "Impact of SOI, Si1-xGexOI and GeOI substrates on CMOS compatible tunnel FET performance," *IEDM*, pp. 1–5, 2008.

[14] T. Krishnamohan, D. Kim, S. Raghunathan, and K. Saraswat, "Double-gate strained-Ge heterostructure tunneling FET (TFET) with record high drive currents and <60mV/dec subthreshold slope," *IEDM*, pp. 1–3, December 2008.

[15] S. Saurabh, and M. J. Kumar, "Estimation and compensation of process induced variations in nanoscale tunnel field effect transistors (TFETs) for improved reliability," *IEEE Transactions Devices and Materials Reliability*, vol. 10, pp. 390–395, September 2010.

[16] K. Boucart, and A. M. Ionescua, "A new definition of threshold voltage in Tunnel FETs," *Solid-State Electronics*, vol. 52, pp. 1318–1323, September 2008.

[17] S. Sneh, and M. J. Kumar, "Novel attributes of a dual material gate nanoscale tunnel field-effect transistor," *IEEE Transactions on Electron Devices*, vol. 58, no. 2, pp. 404–410, 2010.

[18] S. O. Koswatta, M. S. Lundstrom, and D. E. Nikonov, "Performance comparison between p-i-n tunneling transistors and conventional MOSFETs," *IEEE Transactions Electron Devices*, vol. 56, no. 3, pp. 456–465, March 2009.

[19] E.-H. Toh, G. H. Wang, G. Samudra, and Y.-C. Yeo, "Device physics and design of germanium tunneling field-effect transistor with source and drain engineering for low power and high performance applications," *Journal of Applied Physics*, vol. 103, no. 10, pp. 104504-1–104504-5, May 2008.

[20] A. C. Seabaugh, and Q. Zhang, "Low-voltage tunnel transistors for beyond CMOS logic," *Proceedings of IEEE*, vol. 98, no. 12, pp. 2095–2110, December 2010.

[21] M. Luisier, M. Lundstrom, D. A. Antoniadis, and J. Bokor, "Ultimate device scaling: Intrinsic performance comparisons of carbon-based, InGaAs, and Si field-effect transistors for 5 nm gate length," in IEDM Tech. Dig., Washington, DC, USA, December 2011, pp. 11.2.1–11.2.4.

[22] L. Esaki, "Long journey into tunneling," *Reviews of Modern Physics*, vol. 46, no. 2, pp. 237–244, April 1974.

[23] W. Y. Choi, B.-G. Park, J. D. Lee, and T.-J. K. Liu, "Tunneling field-effect transistors (TFETs) with subthreshold swing (SS) less than 60 mV/dec," *IEEE Electron Device Letters*, vol. 28, no. 8, pp. 743–745, August 2007.

[24] J. Appenzeller, Y. M. Lin, J. Knoch, and P. Avouris, "Band-to-band tunneling in carbon nanotube field-effect transistors," *Physical Review Letters*, vol. 93, no. 19, pp. 196805-1–196805-4, November 2004.

[25] W. Y. Choi, and W. Lee, "Hetero-gate-dielectric tunneling fieldeffect transistors," *IEEE Transactions Electron Devices*, vol. 57, no. 9, pp. 2317–2319, September 2010.

[26] A. S. Verhulst, W. G. Vandenberghe, K. Maex, S. De Gendt, M. M. Heyns, and G. Groeseneken, "Complementary silicon-based heterostructure tunnel-FETs with high tunnel rates," *IEEE Electron Device Letters*, vol. 29, no. 12, pp. 1398–1401, December 2008.

[27] T. Krishnamohan, D. Kim, C. D. Nguyen, C. Jungemann, Y. Nishi, and K. C. Saraswat, "High-mobility low band-to-band-tunneling strainedgermanium double-gate heterostructure FETs: Simulations," *IEEE Transactions Electron Devices*, vol. 53, no. 5, pp. 1000–1009, May 2006.

[28] V. Prabhat, and A. K. Dutta, "Analytical surface potential and drain current models of dual-metal-gate double-gate tunnel-FETs," *IEEE Transactions on Electron Devices*, vol. 63, no. 5, pp. 2190–2196, May 2016, 10.1109/TED.2016.2541181.

[29] W. Long, H. Ou, J.-M. Kuo, and K. K. Chin, "Dual-material gate (DMG) field effect transistor," *IEEE Transactions Electron Devices*, vol. 46, no. 5, pp. 865–870, May 1999.

[30] W. Long, and K. K. Chin, "Dual material gate field effect transistor (DMGFET)," in IEDM Tech. Dig., Washington, DC, USA, December 1997, pp. 549–552.

[31] J. Knoch, and J. Appenzeller, "Tunneling phenomena in carbon nanotube field-effect transistors," *Physica Status Solidi A*, vol. 205, no. 4, pp. 679–694, April 2008.

[32] M. J. Lee, and W. Y. Choi, "Analytical model of single-gate silicon-oninsulator (SOI) tunneling field-effect transistors (TFETs)," *Solid-State Electronics*, vol. 63, no. 1, pp. 110–114, September 2011.

[33] L. Liu, D. Mohata, and S. Datta, "Scaling length theory of double-gate interband tunnel field-effect transistors," *IEEE Transactions Electron Devices*, vol. 59, no. 4, pp. 902–908, April 2012.

[34] N. Cui, "A two-dimensional analytical model for tunnel field effect transistor and its applications," *Journal of Applied Physics*, vol. 52, no. 4R, pp. 044303-1–044303-6, April 2013.

[35] W. G. Vandenberghe, A. S. Verhulst, G. Groeseneken, B. Soree, and W. Magnus, "Analytical model for a tunnel field-effect transistor," in Proceedings 14th IEEE *Mediterranean* Electrotech. Conference, Ajaccio, France, May 2008, pp. 923–928.

[36] V. Aa, K. Mk, S. Sp et al. Drain current modeling of tunnel FET using Simpson's rule. *Silicon*, 2021, 10.1007/s12633-021-01328-5.

[37] R. Vishnoi, and M. Kumar, "Compact analytical drain current model of gate-all-around nanowire tunneling FET," *IEEE Transactions Electron Devices*, vol. 61, no. 7, pp. 2599–2603, July 2014.

[38] R. Vishnoi, and M. J. Kumar, "Compact analytical model of dual material gate tunneling field-effect transistor using interband tunneling and channel transport," *IEEE Transactions Electron Devices*, vol. 61, no. 6, pp. 1936–1942, June 2014.

[39] R. Vishnoi, and M. J. Kumar, "A pseudo-2-D-analytical model of dual material gate all-around nanowire tunneling FET," *IEEE Transactions Electron Devices*, vol. 61, no. 7, pp. 2264–2270, July 2014.

[40] A. S. Verhulst, D. Leonelli, R. Rooyackers, and G. Groeseneken, "Drain voltage dependent analytical model of tunnel field-effect transistors," *Journal of Applied Physics*, vol. 110, no. 2, pp. 024510-1–024510-10, July 2011.

[41] J. Wan, C. Le Royer, A. Zaslavsky, and S. Cristoloveanu, "A tunneling field effect transistor model combining interband tunneling with channel transport," *Journal of Applied Physics*, vol. 110, no. 10, pp. 104503-1–104503-7, 2011.

[42] M. G. Bardon, H. P. Neves, R. Puers, and C. Van Hoof, "Pseudotwo-dimensional model for double-gate tunnel FETs considering the junctions depletion regions," *IEEE Transactions Electron Devices*, vol. 57, no. 4, pp. 827–834, April 2010.

[43] E. O. Kane, "Theory of tunneling," *Journal of Applied Physics*, vol. 32, no. 1, pp. 83–91, June 1961.

[44] C. Shen, S.-L. Ong, C.-H. Heng, G. Samudra, and Y.-C. Yeo, "A variational approach to the two-dimensional nonlinear Poisson's equation for the modeling of tunneling transistors," *IEEE Electron Device Letters*, vol. 29, no. 11, pp. 1252–1255, November 2008.

[45] K. K. Young, "Short-channel effect in fully depleted SOI MOSFETs," *IEEE Transactions Electron Devices*, vol. 36, no. 2, pp. 399–402, February 1989.

[46] S. C. Lin, and J. B. Kuo, "Modeling the fringing electric field effect on the threshold voltage of FD SOI nMOS devices with the LDD/sidewall oxide spacer structure," *IEEE Transactions Electron Devices*, vol. 50, no. 12, pp. 2559–2564, December 2003.

[47] TCAD Sentaurus Manual, Synopsys, Inc., Mountain View, CA, USA, 2014.

[48] G. A. M. Hurkx, D. B. M. Klaassen, and M. P. G. Knuvers, "A new recombination model for device simulation including tunneling," *IEEE Transactions Electron Devices*, vol. 39, no. 2, pp. 331–338, February 1992.

[49] A. Schenk, "Rigorous theory and simplified model of the band-to-band tunneling in silicon," *Solid-State Electronics*, vol. 36, no. 1, pp. 19–34, January 1993.

[50] L. Zhang, J. He, and M. Chan, "A compact model for double-gate tunneling field-effect-transistors and its implications on circuit behaviors," in Proceedings of IEEE International Electron Devices Meeting (IEDM), San Francisco, CA, USA, December 2012, pp. 6.8.1–6.8.4.

[51] C. Wu, R. Huang, Q. Huang, C. Wang, J. Wang, and Y. Wang, "An analytical surface potential model accounting for the dual-modulation effects in tunnel FETs," *IEEE Transactions Electron Devices*, vol. 61, no. 8, pp. 2690–2696, June 2014.

[52] L. Esaki, "New phenomenon in narrow germanium p–n junctions," *Physical Review*, vol. 109, no. 2, pp. 603–604, January 1958.

[53] Y. Taur, J. Wu, and J. Min, "An analytic model for heterojunction tunnel FETs with exponential barrier," *IEEE Transactions Electron Devices*, vol. 62, no. 5, pp. 1399–1404, May 2015.

[54] J. Wu, J. Min, and Y. Taur, "Short-channel effects in tunnel FETs," *IEEE Transactions Electron Devices*, vol. 62, no. 9, pp. 3019–3024, September 2015.

[55] P. Wu, J. Zhang, L. Zhang, and Z. Yu, "Channel-potential based compact model of double-gate tunneling FETs considering channellength scaling," in Proceedings of International Conference SISPAD, Washington, DC, USA, September 2015, pp. 317–320, 10.1109/SISPAD.2015.7292323.

[56] L. De Michielis *et al.*, "An innovative band-to-band tunneling analytical model and implications in compact modeling of tunneling-based devices," *Applied Physics Letters*, vol. 103, pp. 123509-1–123509-5, September 2013.

[57] L. De Michielis, L. Lattanzio, and A. M. Ionescu, "Understanding the superlinear onset of tunnel-FET output characteristic," *IEEE Electron Device Letters*, vol. 33, no. 11, pp. 1523–1525, November 2012.

5 Modeling of Gate Engineered TFET: Challenges and Opportunities

N.B. Balamurugan, M. Suguna, and M. Hemalatha

CONTENTS

5.1 INTRODUCTION

In the past few years, several unconventional MOSFETs have been proposed as device topologies to facilitate the advancement of semiconductor technology toward the nanoscale. Double-Gate (D-G), Dual-Material Gate (D-MG), Circular-Gate (C-G), and Dual-Material Cylindrical Gate (D-MCG) MOSFETs are some of the device topologies that fall under this category. However, the Subthreshold Swing

DOI: 10.1201/9781003327035-5

(SS) in these non-classical devices is restricted to 60 mv/decade (Tura & Woo, 2010) [1,2]. However, the SS of a TFET is not limited to 60 mV/dec at room temperature. As a result of this, the supply voltage required to bias the transistor is lowered, resulting in reduced power consumption.

Moreover, since the TFET device is operated under reverse bias conditions, the leakage current I_{OFF} is greatly reduced. Hence TFET is considered superior CMOS devices that may be used in six-transistor SRAM memory cells and digital circuits like NAND and NOR due to its rapid speed, high density, and diminished short channel effects [3,4]. Throughout this chapter, the straight-forward and precise parabolic approximation approach was used to determine the analytical equations for lateral-electrical field surface-potential and vertical-electric field of Single Gate Silicon-On-Insulator (SG SOI) TFETs, Dual-Material Gate TFET, Dual Material Double Gate TFETs and Surrounding Gate TFETs. The parameters are modeled and simulated by various channel lengths and applied biases.

5.2   DEVICE STRUCTURE

Figure 5.1 represents the cross-section structure of single gate SOI TFET. The doping in the p+ type source and n+ drain areas is kept at 1×10^{20} cm^{-3} and 1×10^{20} cm^3, respectively. The buried-oxide thickness and the oxide layer thickness typically have values of 2 nm. In the case of the electron as a majority carrier device, the channel area is made up of intrinsic material, whereas the source is made of holes as a majority carrier and the drain is made of n-type. Gate metal used in this device has the work function of 4.5eV [5].

It is assumed that the device is operating in its sub-threshold zone. As a result, the source and drain channel intersections are guaranteed to be abrupt, and the existence of mobile carriers is minimal. The buried oxide layer is connected to the ground. Since the B_{OX} layer thickness is supposed to be extremely thin, the potential in the B_{OX} layer is expected to be zero. An accurate tunneling model is essential to properly investigate the effect of tunneling on FET devices since it is the

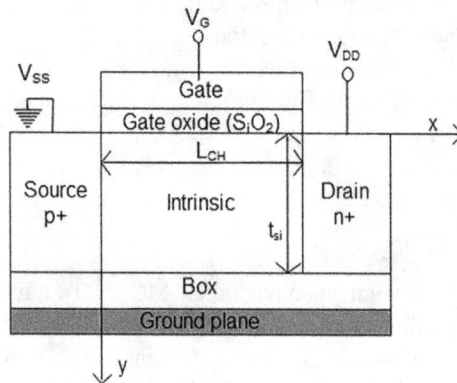

FIGURE 5.1   Cross-sectional view of Single-Gate SOI TFET.

TABLE 5.1

Abbreviations and Their Corresponding Explanations

S.No	Abbreviation	Explanation
1.	TFET	Tunnel Field Effect Transistor
2.	CMOS	Complementary Metal Oxide Semiconductor
3.	SOI	Silicon On Insulator
4.	SG TFET	Single Gate TFET
5.	DMG TFET	Dual Material Gate TFET
6.	DMDG TFET	Dual Material Double Gate TFET
7.	CG TFET	Circular Gate TFET
8.	DM-CG TFET	Dual Material- Cylindrical Gate TFET
9.	SS	Subthreshold Swing
10.	SRAM	Static Random Access Memory
11.	DRAM	Dynamic Random Access Memory
12.	B_{OX}	Buried Oxide
13.	DIBL	Drain Induced Barrier Lowering

primary and fundamental physical process that captures the device physics of the tunneling FET [6]. The mathematical modeling for band-to-band tunneling was derived by (Kane 1960). The list of abbreviations is given in Table 5.1.

5.3 ANALYTICAL MODELING OF SINGLE GATE SOI TFETs

5.3.1 MODELING OF SURFACE POTENTIAL

The 2-D Poisson's equation gives the potential distribution for the TFET's gate oxide and channel region.

$$\frac{\partial^2 \phi(x, y)}{\partial x^2} + \frac{\partial^2 \phi(x, y)}{\partial y^2} = 0 \tag{5.1}$$

The parabolic approximation is used to solve eq. 5.1. using this method, the potential distribution $\phi(x, y)$ throughout the 2 D space is estimated and as well as the potential profile along the region is estimated by

$$\phi(x, y) = C_0(x) + C_1(x)y + C_2(x)y^3 \tag{5.2}$$

Where $C_0(x)$, $C_1(x)$ and $C_2(x)$ are arbitrary constants, which depend on the x direction only.

Below are the boundary conditions needed to solve eq. 5.1 is given below

1. The surface potential is assumed to be $\phi_s(x)$

$$\phi(x, y)|_{y=0} = \phi_s(x) \tag{5.3}$$

2. The source end potential is expressed as

$$\phi(0 \cdot y) = \phi_{bi} \tag{5.4}$$

3. The drain end potential is defined as

$$\phi(L_{ch}, y) = \phi_{bi} + V_{DS} \tag{5.5}$$

4. For Single-Gate TFETs, the flux at a top gate oxide contact is continuous.

$$\left.\frac{\partial \phi(x, y)}{\partial y}\right|_{y=0} = \frac{\phi_{s(x)} - V_{GS} + V_{FB}}{t_{ox}} \cdot \frac{\varepsilon_{ox}}{\varepsilon_{si}} \tag{5.6}$$

5. The electric flux in the BOX layer is zero.

$$\left.\frac{\partial \phi(x, y)}{\partial y}\right|_{y=tsi} = 0 \tag{5.7}$$

Where, $\phi_{bi} = \frac{E_G}{q}$, E_G is Band-gap, V_{GS} is known as Gate -Source voltage, V_{FB} is known as the flat band voltage assumed to be zero, V_{DS} is known as Drain-Source Voltage, ϕ_{bi} is known as Built-in potential, q is the elementary charge, ε_{si} is silicon permittivity and ε_{ox} is silicon-dioxide relative permittivity.

The coefficients $C_0(x)$, $C_1(x)$, and $C_2(x)$ may be reformulated as a function of surface potential $\phi_s(x)$ by solving eq. 5.2. The above boundary conditions are utilized to solve the equation.

$$C_0(x) = \phi_s(x) \tag{5.8}$$

$$C_1(x) = \frac{\phi_s(x) - V_{Gs} + V_{FB}}{t_{ox}} \cdot \frac{\varepsilon_{ox}}{\varepsilon_{si}} \tag{5.9}$$

$$C_2(x) = \frac{1}{2t_{si}} \frac{\varepsilon_{ox}}{\varepsilon_{si}} \frac{\phi_s(x) - V_{Gs} + V_{FB}}{t_{ox}} \tag{5.10}$$

Using the above eq. (5.8), (5.9), and (5.10), we may express the 2D potential-distribution $\phi(x, y)$

$$\phi(x, y) = \phi_s(x) + \frac{\phi_s(x) - V_{Gs} + V_{FB}}{t_{ox}} \cdot \frac{\varepsilon_{ox}}{\varepsilon_{si}} y + \frac{1}{2t_{si}} \frac{\varepsilon_{ox}}{\varepsilon_{si}} \frac{\phi_s(x) - V_{Gs} + V_{FB}}{t_{ox}} y^2 \tag{5.11}$$

The 2D Poisson's equation (5.1) may be written as,

$$\phi''_s(x) + \frac{\phi''_s(x)}{t_{ox}} \frac{\varepsilon_{ox}}{\varepsilon_{si}} y - \frac{1}{2t_{si}} \frac{\varepsilon_{ox}}{\varepsilon_{si}} \frac{\phi''_s(x)}{t_{ox}} y^2 - \frac{\varepsilon_{ox}}{t_{si}\varepsilon_{si}} \left(\frac{\phi_s(x) - V_{Gs} + V_{FB}}{t_{ox}} \right) = 0 \tag{5.12}$$

The following 1D differential equation is created from the 2D Poisson's equation by using eq. (5.12)

$$\phi''_s(x) - k^2\phi_s(x) = k^2\phi_d \tag{5.13}$$

Where, $\phi_d = V_{FB} - V_{GS}$ and $k^2 = \frac{\varepsilon_{ox}}{t_{ox}t_{si}\varepsilon_{si}}$

Taking the partial derivatives of eq. (5.13) and using the boundary conditions, we obtain the surface potential $\phi_s(x)$ can be written as

$$\phi_s(x) = Ae^{kx} + Be^{kx} - \phi_d \tag{5.14}$$

A and B are arbitrary coefficients we can express the values as,

$$A = \frac{1}{2\sinh(kl)}[\phi_{bi}(1 - e^{-ki}) + \phi_d(1 - e^{-ki}) + V_{GS}] \tag{5.15}$$

$$B = \frac{-1}{2\sinh(kl)}[\phi_{bi}(1 - e^{ki}) + \phi_d(1 - e^{ki}) + V_{GS}] \tag{5.16}$$

5.3.2 ELECTRIC FIELD OF SG SOI TFET

By analytically calculating the potential as shown below, the electric field in the vertical direction (E_y) and electric field in the horizontal direction (E_x) are founded.

$$E_y(x, y) = -\frac{\partial \phi(x, y)}{\partial y} = -\frac{\varepsilon_{ox}}{\varepsilon_{si}}\frac{\phi_s(x) - V_{Gs} + V_{FB}}{t_{ox}} + \frac{1}{t_{si}}\frac{\varepsilon_{ox}}{\varepsilon_{si}}\frac{\phi_s(x) - V_{Gs} + V_{FB}}{t_{ox}}y \tag{5.17}$$

$$E_x(x, y) = -\frac{\partial \phi(x,y)}{\partial y} = \frac{1}{2\sinh(kl)}[\phi_{bi}(1 - e^{-ki}) + \phi_d(1 - e^{-ki}) + V_{GS}]ke^{kx}$$
$$+ \frac{-1}{2\sinh(kl)}[\phi_{bi}(1 - e^{ki}) + \phi_d(1 - e^{ki}) + V_{GS}]ke^{-kx} \tag{5.18}$$

5.3.3 DRAIN CURRENT OF SG SOI TFET

Carrier generation tunneling rate (G) throughout the device's area is numerically integrated to calculate the overall drain current. Numerical integration is done by using Mathematica software. Therefore,

$$I_{DS} = qW_{CH}\int Gdxdv \tag{5.19}$$

Where, W_{CH} – channel width.

Here Kane's model, applied to, compute the generation rate G. (Kane 1961) [7].

$$G(E) = a \cdot E^2 \exp\left(-\frac{b}{E}\right) \tag{5.20}$$

Using equations (5.17) and (5.18), the generation rate $G(E)$ can be written as

$$G(E) = a\left[\left[\left(-\frac{\varepsilon_{ox}}{\varepsilon_{si}}\frac{\phi_s(x)-V_{Gs}+V_{FB}}{t_{ox}} + \frac{1}{t_{si}}\frac{\varepsilon_{ox}}{\varepsilon_{si}}\frac{\phi_s(x)-V_{Gs}+V_{FB}}{t_{ox}}y\right)^2 + \left(\begin{array}{c}-\frac{1}{2\sinh(kl)}[\phi_{bi}(1-e^{-ki}) + \phi_d(1-e^{-ki})\\ + V_{GS}]ke^{kx} + \frac{-1}{2\sinh(kl)}[\phi_{bi}(1-e^{ki})\\ + \phi_d(1-e^{ki}) + V_{GS}]ke^{-kx}\end{array}\right)^2\right]^2\right]$$

$$\times \exp\left(-\frac{b}{\sqrt{\left(-\frac{\varepsilon_{ox}}{\varepsilon_{si}}\frac{\phi_s(x)-V_{Gs}+V_{FB}}{t_{ox}} + \frac{1}{t_{si}}\frac{\varepsilon_{ox}}{\varepsilon_{si}}\frac{\phi_s(x)-V_{Gs}+V_{FB}}{t_{ox}}y\right)^2 + \left(-\frac{1}{2\sinh(kl)}[\phi_{bi}(1-e^{-ki}) + \phi_d(1-e^{-ki}) + V_{GS}]ke^{kx} + \frac{-1}{2\sinh(kl)}[\phi_{bi}(1-e^{ki}) + \phi_d(1-e^{ki}) + V_{GS}]ke^{-kx}\right)^2}}\right)$$

(5.21)

The parameters utilized for the model are a and b. The values are 3.0×10^7 V/cm and 9.6615×10^{18} $cm^{-1}s^{-1}V^{-2}$ respectively. We can express the field intensity by adding both vertical and horizontal applied fields.

$E = \sqrt{E_x^2 + E_y^2}$. Finally, the drain current can be calculated after substituting eq. (5.20) and (5.21).

$$I_{DS} = qW_{CH}\int_0^L\int_0^{W_{CH}} a\left[\left[\left(-\frac{\varepsilon_{ox}}{\varepsilon_{si}}\frac{\phi_s(x)-V_{Gs}+V_{FB}}{t_{ox}} + \frac{1}{t_{si}}\frac{\varepsilon_{ox}}{\varepsilon_{si}}\frac{\phi_s(x)-V_{Gs}+V_{FB}}{t_{ox}}y\right)^2 + \left(\begin{array}{c}-\frac{1}{2\sinh(kl)}[\phi_{bi}(1-e^{-ki}) + \phi_d(1-e^{-ki}) + V_{GS}]ke^{kx}\\ + \frac{-1}{2\sinh(kl)}[\phi_{bi}(1-e^{ki}) + \phi_d(1-e^{ki}) + V_{GS}]ke^{-kx}\end{array}\right)^2\right]^2\right]$$

$$\exp\left(-\frac{b}{\sqrt{\left(-\frac{\varepsilon_{ox}}{\varepsilon_{si}}\frac{\phi_s(x)-V_{Gs}+V_{FB}}{t_{ox}} + \frac{1}{t_{si}}\frac{\varepsilon_{ox}}{\varepsilon_{si}}\frac{\phi_s(x)-V_{Gs}+V_{FB}}{t_{ox}}y\right)^2 + \left(-\frac{1}{2\sinh(kl)}[\phi_{bi}(1-e^{-ki}) + \phi_d(1-e^{-ki}) + V_{GS}]ke^{kx} + \frac{-1}{2\sinh(kl)}[\phi_{bi}(1-e^{ki}) + \phi_d(1-e^{ki}) + V_{GS}]ke^{-kx}\right)^2}}\right) dxdy$$

(5.22)

Figure 5.2 shows the variation in the electric field and is compared with the values obtained from the simulation for various $V_{GS} = 0.3$ V and $V_{DS} = 0.1$ V over the entire length of the channel ($L_{Channel} = 18$ nm) for single-gate TFET. The electric field in the lateral direction of the channel is much less than that in the vertical-electric field of the channel. Also, for various gate biases, the entire channel is covered by the electric field. The minimum field near the source will help us reduce the channel's hot carrier.

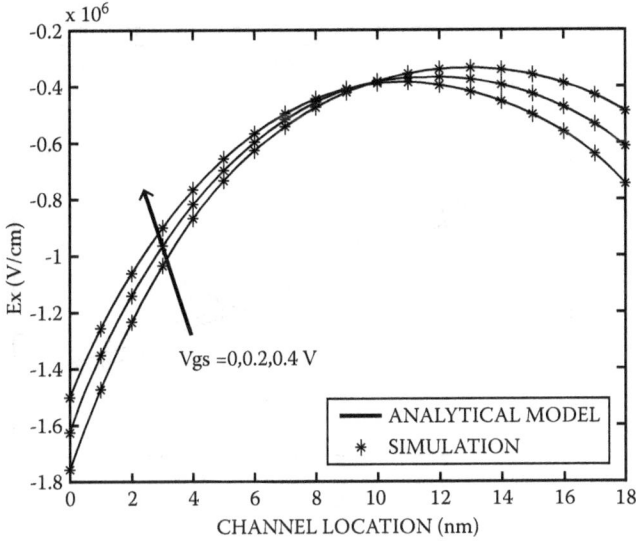

FIGURE 5.2   Effect of gate to source voltage on Ex and $V_{DS} = 0.1$ V (LChannel = 18 nm).

The gate-source voltage is increased from 0V to 0.4 V. It can be inferred from Figure 5.3 that the electric field peak is obtained at the right ordinate of the graph. In the middle of the channel, the reduced field indicates the reduction in carrier scattering. The vertical electric field distribution has become larger near the source side, causing more electron tunneling. It may be observed from the figure that the short channel TFET's vertical electric field is higher than those long channel ($L_{CH} = 100$ nm) TFETs which leads to enhanced performance by the SG TFET.

Figure 5.4 shows the surface potential versus channel length for SG SOI TFETs with various Silicon film thickness t_{si} values. The figure shows how significantly lessening the thickness of the active Si film for SG SOI TFET may diminish the reliance of the surface potential on the channel length. As the film thickness increases, the potential profile along the channel, because higher thickness gives more tunneling path, more electrons will start to flow in the channel region. As well as the small value of t_{si} gives a small amount of carriers getting flow into the channel.

Figure 5.5 depicts the drain current comparison obtained by the model and the simulation for SG SOI TFETs. The drain current rises with increasing gate-source bias, as indicated in the figure. In the figure, As the positive supply of V_{GS} rise, electrons get tunneled into the heavily doped p-type source region's valence band to the conduction band, increasing the tunneling current. Hence the device function as an N-type SG SOI TFET [8]. Similarly, when the negative values of VGS rise, more electrons tunnel from the N+ drain region's valence band to the channel region's conduction band, increasing the tunneling current. As a result, the device functions as a P-type SG SOI TFET.

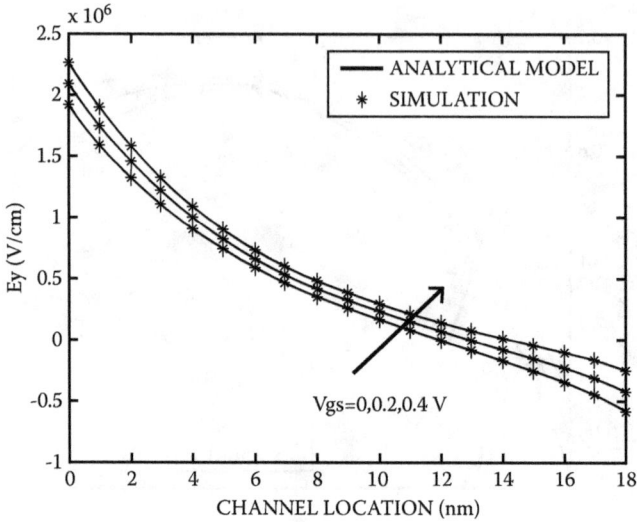

FIGURE 5.3   Vertical electric field at $L_{CH} = 18$ nm and $V_{DS} = 0.1$ V.

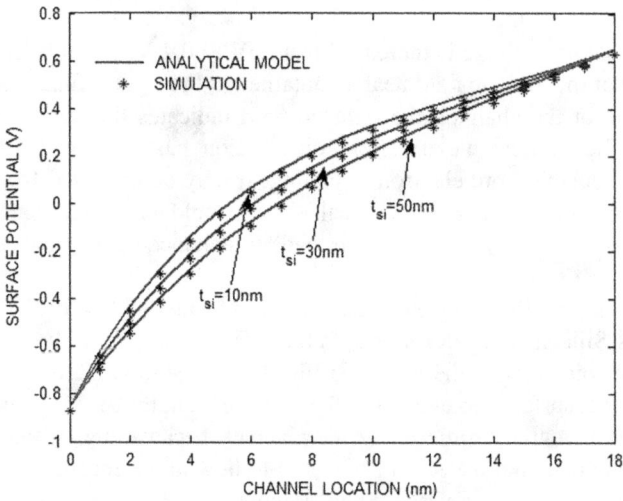

FIGURE 5.4   The effect of silicon-thickness on the surface potential profile of a SG SOI TFET (Lchannel = 18 nm).

5.4   DMG TFET-ANALYTICAL FORMULATION

Dual Material Gate (DMG) SOI TFETs are promising devices to continue CMOS technology scaling in the sub-22-nm regime because they 85 provide better control of short-channel effects (SCE) as compared with SG TFETs. The DMG TFET device is fabricated using Silicon on insulator (SOI), which offers higher

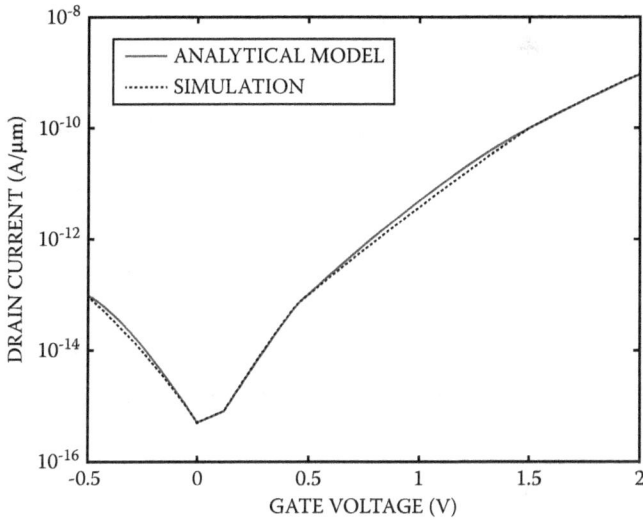

FIGURE 5.5 I_{DS}-V_{GS} characteristics of short channel SG TFETs with channel length.

electrical characteristics over SG SOI TFET, such as reduced leakage current, excellent latch-up immunity, and enhanced output current. With the aid of the parabolic approximation approach, the prospective advantages of the DMG TFETs are to be investigated., which provides simple and accurate predictions.

5.4.1 DEVICE SCHEMATIC OF DUAL MATERIAL GATE TFET

The schematic structure of dual material gate TFET is illustrated in Figure 5.6. Strongly doped p-area and n-areas serve as the source and drain, respectively. An n-type layer with considerable doping makes up the intermediate channel region. The gate dielectric employed in this system is silicon dioxide (SiO2). The gate is made up of two different materials, named M_1 and M_2, and has lengths, L_1 and L_2, as well as two distinct work functions, ϕ_{m1} and ϕ_{m2}. The gate material close to the source has a lower work function, while the gate material closer to the drain has a higher work function.

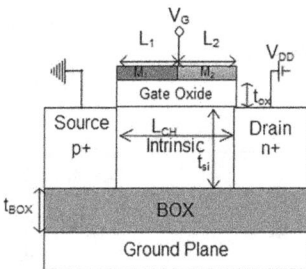

FIGURE 5.6 Schematic diagram of DMG SOI.

Due to the presence of two materials in the gate, it is possible to write the potential under Material (M_1) and Material (M_2) as

$$\phi_1(x, y) = \phi_{s1}(x) + C_{11}(x)y + C_{12}(x)y^2 \quad 0 \leq x \leq L_1 \qquad (5.23)$$

$$\phi_2(x, y) = \phi_{s2}(x) + C_{21}(x)y + C_{22}(x)y^2 \quad L_1 \leq x \leq L_1 + L_2 \qquad (5.24)$$

Similarly for Dual material gate the surface potential is obtained by solving eq. (5.23), (5.24).

$$\phi_{s1}(x) = Ae^{\lambda x} + Be^{-\lambda x} - \psi_{g1} \quad 0 \leq x \leq L_1 \qquad (5.25)$$

$$\phi_{s1}(x) = Ce^{\lambda x} + De^{-\lambda x} - \psi_{g2} \quad L_1 \leq x \leq L_1 + L_2 \qquad (5.26)$$

Where $\lambda = \sqrt{\frac{\varepsilon_{ox}}{\varepsilon_{si} t_{ox} t_{si}}}$, $\psi_{g1} = V_{gs} - \phi_{m1} + \chi + E_g/2$, $\psi_{g2} = V_{gs} - \phi_{m2} + \chi + E_g/2$. The expression for the coefficients A, B, C and D is

$$A = \frac{[(V_{bi} - \psi_{g1})e^{\lambda(L_1 - L_2)}] - [V_{bi} + V_{DS} - \psi_{g2}] + [(\psi_{g1} - \psi_{g2})\cosh(\lambda L_2)]}{e^{\lambda(L_1 - L_2)} - e^{\lambda(L_1 + L_2)}}$$

$$(5.27)$$

$$B = \frac{[V_{bi} + V_{DS} - \psi_{g2}] - [(V_{bi} - \psi_{g1})e^{\lambda(L_1 - L_2)}] - [(\psi_{g1} - \psi_{g2})\cosh(\lambda L_2)]}{e^{\lambda(L_1 - L_2)} - e^{\lambda(L_1 + L_2)}}$$

$$(5.28)$$

$$C = Ae^{\lambda L_1} + \frac{\psi_{g1} - \psi_{g2}}{2} \qquad (5.29)$$

$$D = Be^{-\lambda L_1} + \frac{\psi_{g1} - \psi_{g2}}{2} \qquad (5.30)$$

The lateral and vertical electric field for DMG TFET is given by

$$E_1(x) = -\frac{d\phi_{s1}(x)}{dx} = -\left[\begin{array}{l} \left(\frac{[(V_{bi} - \psi_{g1})e^{\lambda(L_1 - L_2)}] - [V_{bi} + V_{DS} - \psi_{g2}] + [(\psi_{g1} - \psi_{g2})\cosh(\lambda L_2)]}{e^{\lambda(L_1 - L_2)} - e^{\lambda(L_1 + L_2)}} \right)\lambda e^{\lambda x} \\ -\left(\frac{[V_{bi} + V_{DS} - \psi_{g2}] - [(V_{bi} - \psi_{g1})e^{\lambda(L_1 - L_2)}] - [(\psi_{g1} - \psi_{g2})\cosh(\lambda L_2)]}{e^{\lambda(L_1 - L_2)} - e^{\lambda(L_1 + L_2)}} \right)\lambda e^{-\lambda x} \end{array} \right]$$

$$0 \leq x \leq L_1 \qquad (5.31)$$

$$E_2(x) = -\frac{d\phi_{s2}(x)}{dx} = -\left[\left(Ae^{\lambda L_1} + \frac{\psi_{g1} - \psi_{g2}}{2}\right)\lambda e^{\lambda x}\right.$$

$$\left. -\left(Be^{-\lambda L_1} + \frac{\psi_{g1} - \psi_{g2}}{2}\right)\lambda e^{-\lambda x}\right] \quad L_1 \le x \le L_1 + L_2 \quad (5.32)$$

$$E_1(y) = -\frac{d\phi_{s1}(x, y)}{dy} = \left[-\left(\frac{\varepsilon_{ox}}{\varepsilon_{si}}\frac{\phi_{s1}(x) - \psi_{g1}}{t_{ox}}\right) - 2y\left(-\frac{1}{2t_{si}}\frac{\varepsilon_{ox}}{\varepsilon_{si}}\frac{\phi_{s1}(x) - \psi_{g1}}{t_{ox}}\right)\right]$$

$$0 \le x \le L_1 \quad\quad\quad\quad\quad\quad\quad\quad\quad\quad\quad\quad\quad\quad\quad\quad (5.33)$$

$$E_2(y) = -\frac{d\phi_{s2}(x, y)}{dy} = \left[-\left(\frac{\varepsilon_{ox}}{\varepsilon_{si}}\frac{\phi_{s2}(x) - \psi_{g2}}{t_{ox}}\right)\right.$$

$$\left. - 2y\left(-\frac{1}{2t_{si}}\frac{\varepsilon_{ox}}{\varepsilon_{si}}\frac{\phi_{s2}(x) - \psi_{g2}}{t_{ox}}\right)\right] \quad L_1 \le x \le L_1 + L_2 \quad (5.34)$$

In a DMG TFET, BTBT of electrons from the source's valence band to the channel region's conduction band is generated by current IDS. Kane's Model may determine the tunneling generation rate (G). The total drain current is calculated by numerically integrating the tunneling probability throughout the device's area. Therefore,

$$I_{Ds} = q \int_0^L \int_0^{W_{CH}} G dx dy \quad\quad\quad\quad (5.35)$$

By using eq. (5.20), the drain current is computed.

Figure 5.7 demonstrates the estimated values that are obtained from the simulation of the vertically applied electric field throughout the channel location for various gate voltage levels. The vertical electric field's peak may be seen in the graph, and it is visible to be close to the source side. i.e., region under metal M_1. This causes an increase in the rate of tunneling. This finding results in a rise in the tunneling current. With the modification of the work function, the electric field close to the drain diminishes. i.e., the area beneath metal M_2. It is evident from the findings that the analytical model's computed values closely match the simulated ones.

For DMG TFETs with various gate oxide thickness tox values, the vertical-electric field versus the channel length was displayed in Figure 5.8. The increase in the gate oxide thickness results in worse gate electrostatic control. The graphic illustrates how decreasing the gate oxide's thickness can significantly increase the vertical electric field's dependence on channel length. The electric field within the oxide might become quite strong if the oxide layer is too thin. Tunneling leakage current becomes the most significant limiting factor for SiO_2 layers thinner than 1.5 nm.

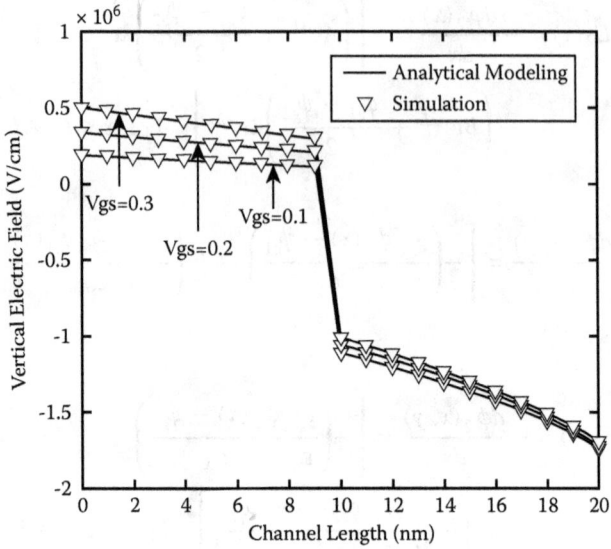

FIGURE 5.7 Electric field comparison of DMDG-TFET along the length L = 20 nm for various gate biases and $V_{DS} = 0.1$ V.

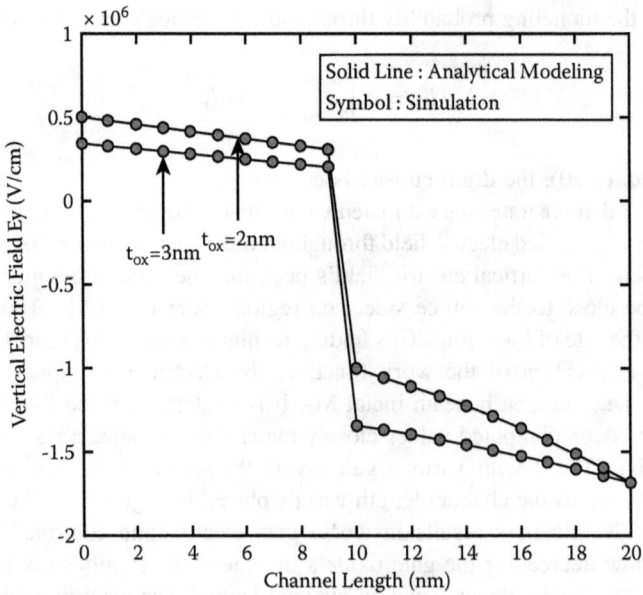

FIGURE 5.8 Effect of gate oxide thickness on vertically generated electric field of DMG TFET.

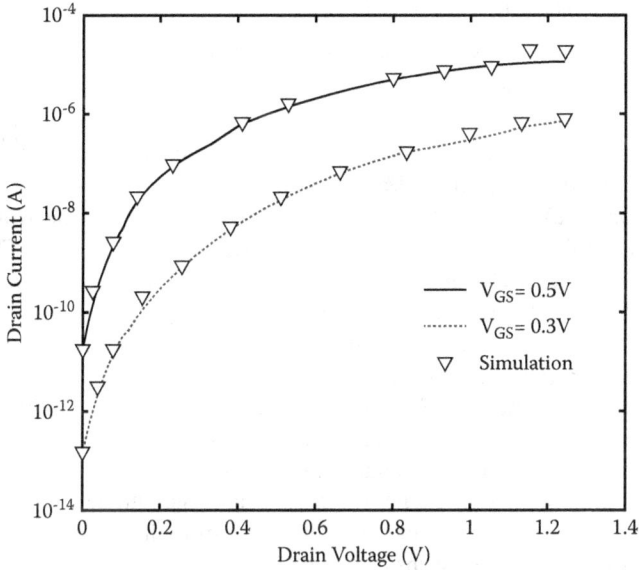

FIGURE 5.9 I_D-V_{DS} variation of Dual-Material gate TFET.

Figure 5.9 displays the drain current variation of DMG TFET with a channel length of 20 nm. As the applied drain voltage is greater than the gate voltage, the electron concentration in the channel region is not influenced by the gate voltage, and electrons are depleted, leading to high channel resistance. Due to this feature, most of the applied drain voltage decreases at the junction of the channel and drain, and the drain voltage likewise has no impact on the tunneling width at the source and channel junction.

5.5 ANALYTICAL MODELING OF DUAL MATERIAL DOUBLE GATE TUNNEL FIELD EFFECT TRANSISTOR

5.5.1 INTRODUCTION

Balestra et al. (1987) proposed a new Double-Gate (DG) control Silicon On Insulator (SOI) transistors for the first time, which is operated under a strong inversion region. This device offers excellent performance in terms of an increase in subthreshold slope, transconductance, and drain current. However, this new device model analysis is limited to the strong inversion region. The DMG principle is implemented in the Double Gate TFET device to obtain better ON current characteristics of the device. Two-dimensional analytical modeling is necessary to derive the surface-potential and electric field distribution models that properly account for the channel dependence of the DMDG TFET device.

5.5.2 DEVICE STRUCTURE OF DMDG TFET

Figure 5.10 depicts the schematic construction of the DMDG TFET. That contains two gate metals of lengths L_1 and L_2. Strongly doped n-type and p-type areas make up the drain and source, which have corresponding carrier concentrations of 10^{20} cm^{-3} and 10^{20} cm^{-3}. A p-type material has a concentration of 10^{17} cm^{-3} used to create the intermediate channel area. Gate material near the source has a lower work function (4 eV) than gate material at the drain (4.6eV). The work function variation shows the advantages of the extremely capable DM-DG TFET device. Contrary to normal MOS devices, tunnel field effect transistors operate fundamentally differently. In order to obtain better I_{OFF} and I_{ON} current, the study of DM-DG TFET operation is studied by taking into account two distinct situations of variable work functions, ϕ_{m1} and ϕ_{m2} alternately.

Case (i): Consider the work function ϕ_{m1} is decreased to 4.0 eV in the OFF state, there is no band-overlap on the source side, therefore, the I_{OFF} is anticipated to be relatively low. The band overlap and tunneling width grow in the ON-state as ϕ_{m1} declines, resulting in a considerable rise in the source-side tunneling probability. As a result, the electrons travel through drift diffusion from the majority hole source to the conduction band of the intrinsic region.

Case (ii): As ϕ_{m2} is raised, the OFF-state tunneling width grows and overlapping bands on the source region drop, resulting in a significant fall in the OFF-state tunneling probability. However, in the ON-state, the band diagram is not greatly altered by the rise in ϕ_{m2}.

The surface potential comparison for a single gate, dual material double gate, and double gate is shown in Figure 5.11. The graphic shows that there is a relationship between surface potential and channel length. To analyze the result, the gate-source voltage is fixed at 0V. By comparison, the surrounding gate gives a better surface potential of 0.541V.

Figure 5.12 demonstrates the estimated and modeled outputs of a DM-DG TFET surface potential over the channel for various gate lengths corresponding M1 and M2 combinations. Because when gate length L_1 is shortened, the peak-electric field changes to the source side. This peak electric potential results in a more uniform

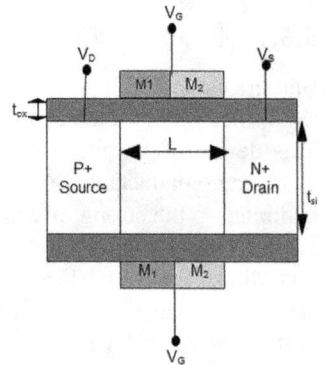

FIGURE 5.10 Schematic Diagram of Dual Material Double Gate Structure.

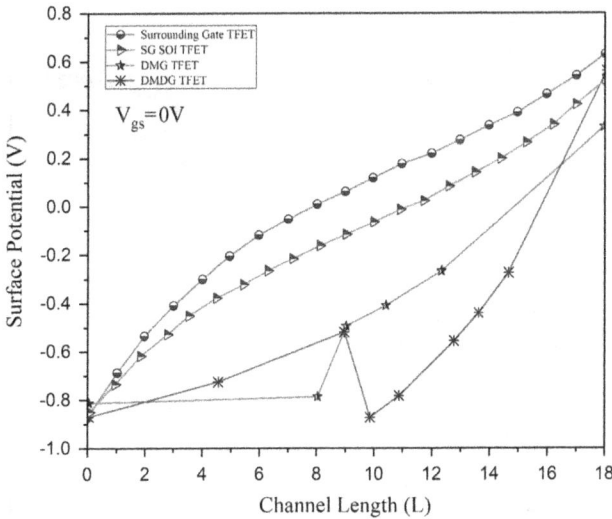

FIGURE 5.11   Surface Potential comparison at fixed gate-source voltage at 0 V.

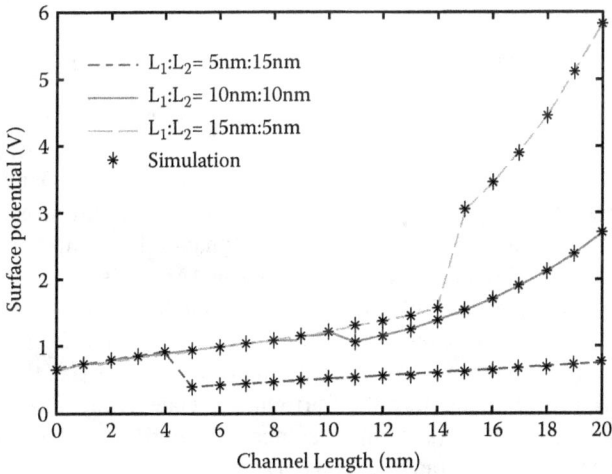

FIGURE 5.12   Different gate length dependency on surface potential profile.

field in the channel, which allows the device to function at a greater carrier drift velocity. A higher value of L_2 produces a significant rise in the potential at the left ordinate of the graph, indicating the reduction of SCEs.

Figure 5.13 demonstrate the drain current properties of Dual Material DG-TFET for change in metal two work-function-ϕ_{m2}. At the source side, metal one work-function of-ϕ_{m1} is kept constant at 4 eV. The figure makes it clear that the increased work function of M2 does not greatly impact the tunneling current at the drain end.

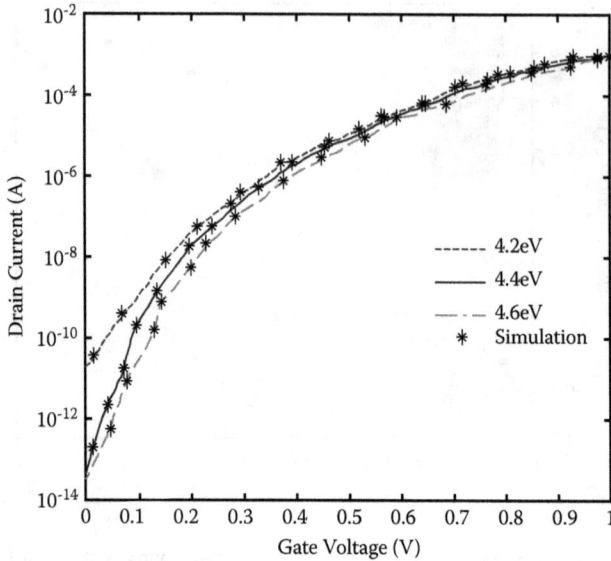

FIGURE 5.13 Effect of gate work function on the device drain-current with fixed $\phi_{m1} = 4$eV.

5.6 SURROUNDING GATE TFET- ANALYTICAL FORMULATION

Hence, the surrounding gate principle concept was implemented in the TFET structure by Verhulst et al. (2010). A novel model-based framework was created to show how a Double-Gate TFET, single-gate TFET, and Gate-All-Around TFET configuration may be directly compared. When employed as a switching device in arrayed structures like high-density Dynamic Random Access Memory (DRAM) and Static Random Access Memory (SRAM) cells, where a tiny device shape and low leakage current are crucial, surrounding-gate MOS devices are very attractive. While moving to the sub-22 nm region, the Surrounding Gate MOS produces a deleterious leakage current. In order to avoid this problem, a novel approach to implementing the Surrounding Gate in the TFET structure is developed in this thesis work. Simple, compact, analytical models are needed to apply the devices to integrated circuits. Device models for circuit simulation should be precise, process-based, and computationally light. It is generally acknowledged that using surface potential may give a decent representation of the physical behavior of transistors and produces more accurate forecasts of the functionality of integrated circuits than other approaches, especially in nanostructures and high-frequency operations.

5.6.1 DEVICE SCHEMATIC OF SURROUNDING GATE TUNNEL FET

Figure 5.14 depicts a Surrounding-Gate TFET in cross-section. The coordinate system is made up of an angular direction (θ), a vertical direction (z), and a radial direction (r). Strongly doped n-region and p-areas make up the source and drain. An

(a)   (b)

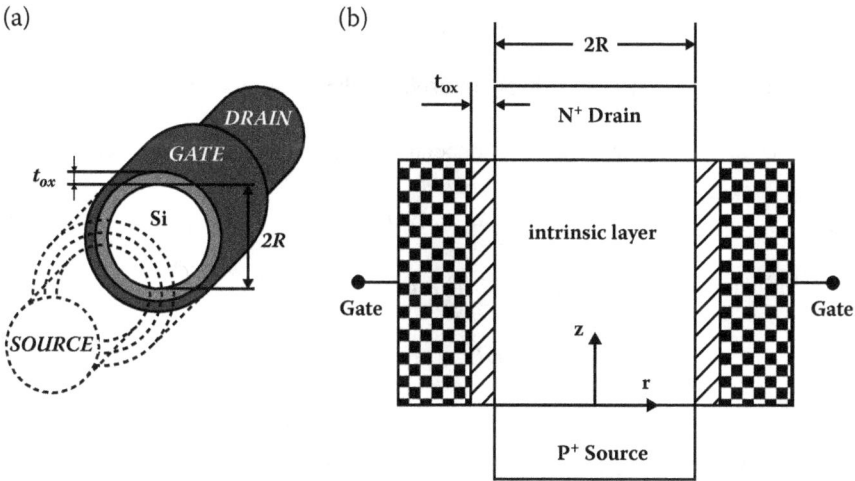

FIGURE 5.14 (a) Schematic diagram of surrounding gate TFETs, (b) Cross-sectional view of the sectional diagram.

n-type layer with considerable doping makes up the intermediate channel region. The gate dielectric employed in this system is silicon dioxide (SiO2). A positive gate voltage causes the transistor to act as an n-TFET, while a negative gate voltage causes it to behave as a p-TFET. The energy barrier between the source and the intrinsic body is reduced by raising the positive gate voltage. Then, electrons flow via drift diffusion toward the n-doped drain after tunneling from the p-doped source to the intrinsic body.

5.6.2 MODELING OF SURFACE POTENTIAL

For the sake of simplicity, it has been assumed that the junction between the interface of source/ channel and interface of drain/channel is very sharp and the TFET operates at the subthreshold regime, where the charge carriers are minimal. The source/drain region has not been presumed to really be depleted. The following is the 2D Poisson's/Laplace equation that was used to determine the potential across the device and the channel-potential at the SiO_2/Si interface:

$$\frac{1}{r}\frac{\partial}{\partial r}\left(\frac{r\partial\phi(r, z)}{\partial r}\right) + \frac{\partial^2\phi(r, z)}{\partial^2 z} = 0 \qquad (5.36)$$

The parabolic estimation is used to approximate the 2D Poisson's Equation for the n-channel FET. The used equation is presented as:

$$\phi(r, z) = C_0(z) + C_1(z)r + C_2(z)r^2 \qquad (5.37)$$

Where, $C_0(z)$, $C_1(z)$, and $C_2(z)$ are arbitrary constants which are the function of vertical direction.

The below boundary conditions are used for the solution of equation (5.37) are,

1. The surface-potential $\phi_s(z)$ is depends on the vertical direction (z)

$$\phi(R, z) = S_1(z) = \phi_s(z) \tag{5.38}$$

2. Silicon pillar has electric field of zero in the middle.

$$\left.\frac{\partial\phi(r, z)}{\partial r}\right|_{r=0} = 0 \tag{5.39}$$

3. Silicon/oxide interface experiences a continuous electric field.

$$\left.\frac{\partial\phi(r, z)}{\partial r}\right|_{r=R} = \frac{\varepsilon_{ox}}{\varepsilon_{si}R}\left(\frac{\psi_G - \phi_s(z)}{\ln\left(1 + \frac{t_{ox}}{R}\right)}\right) \tag{5.40}$$

4. Source end potential is defined as

$$\phi(0, 0) = \phi_s(0) = V_{bi} \tag{5.41}$$

5. Drain end potential is expressed as

$$\phi(r, L) = \phi_s(L) = V_{bi} + V_{ds} \tag{5.42}$$

The coefficients $C_0(z)$, $C_1(z)$, and $C_2(z)$ could be reformulated in terms of surface potential $\phi_s(z)$ by imposing all boundary conditions to eq. (5.37).

$$C_0(z) = \phi_s(z) - \frac{R}{2}\frac{C_f}{\varepsilon_{si}}(\psi_G - \phi_s(z)) \tag{5.43}$$

$$C_1(z) = 0 \tag{5.44}$$

$$C_2(z) = \frac{1}{2R}\frac{C_f}{\varepsilon_{si}}(\psi_G - \phi_s(z)) \tag{5.45}$$

Substituting the eq. (5.43)–(5.45) in (5.37)

$$\phi(r, z) = \phi_s(z) - \frac{R}{2}\frac{C_f}{\varepsilon_{si}}(\psi_G - \phi_s(z)) + \frac{1}{2R}\frac{C_f}{\varepsilon_{si}}(\psi_G - \phi_s(z))r^2 \tag{5.46}$$

Using eq. (5.46), the 2-D Poisson's equation (5.36) can be expressed as

$$\frac{2}{R}\frac{C_f}{\varepsilon_{si}}(\psi_G - \phi_s(z)) + \frac{d^2\phi_s(z)}{dz^2} + \frac{R}{2}\frac{C_f}{\varepsilon_{si}}\frac{d^2\phi_s(z)}{dz^2} - \frac{r^2}{2R}\frac{C_f}{\varepsilon_{si}}\frac{d^2\phi_s(z)}{dz^2} = 0 \quad (5.47)$$

Equation (5.47) can be rearranged as

$$\frac{d^2\phi_s(z)}{dz^2} - k^2\phi_s(z) = -k^2\psi_G \quad (5.48)$$

By solving the second order differential equation

$$\phi_s(z) = Ae^{kz} + Be^{-kz} + \psi_G \quad (5.49)$$

The coefficients A and B can be expressed as

$$A = V_{bi} - \psi_G - \left\{ \frac{\{e^{Lk}V_{biS} - (V_{biS} + V_{DS})\} - \left\{ e^{\frac{Lk}{2}}\psi_G\left(2jsinh\left(\frac{Lk}{2}\right)\right)\right\}}{2jsinh(Lk)} \right\} \quad (5.50)$$

$$B = \frac{\{e^{Lk}V_{biS} - (V_{biD} + V_{DS})\} - \left\{ e^{\frac{Lk}{2}}\psi_G\left(2jsinh\left(\frac{Lk}{2}\right)\right)\right\}}{2jsinh(Lk)} \quad (5.51)$$

5.6.3 Modelling of Electric Field

By differentiating the surface potential, it is possible to derive the distribution of the electric field over the channel length. The equation for lateral electric field is

$$E_z = -\frac{\partial\phi(r, z)}{\partial z} = \frac{\partial\phi_s(z)}{\partial z} \quad (5.52)$$

$$E_z = k\left(\left[\left[V_{bi} - \psi_G - \left\{ \frac{\{e^{Lk}V_{biS} - (V_{biS} + V_{DS})\} - \left\{ e^{\frac{Lk}{2}}\psi_G\left(2jsinh\left(\frac{Lk}{2}\right)\right)\right\}}{2jsinh(Lk)} \right\} \right]\right]e^{kz} - \left[\frac{\{e^{Lk}V_{biS} - (V_{biD} + V_{DS})\} - \left\{ e^{\frac{Lk}{2}}\psi_G\left(2jsinh\left(\frac{Lk}{2}\right)\right)\right\}}{2jsinh(Lk)} \right]e^{-kz}\right) \quad (5.53)$$

The formula for vertical-electric field is,

$$E_r = \frac{\partial \phi(r, z)}{\partial r} \tag{5.54}$$

$$E_r = \frac{\partial \phi(r, z)}{\partial r} = \frac{1}{R}\frac{C_f}{\varepsilon_{si}}(\psi_G - \phi_s(z))r \tag{5.55}$$

5.6.4 DERIVATION OF DRAIN CURRENT

After that, the tunneling rate throughout the device's area is numerically integrated to get the overall drain current.

$$I_{DS} = q\pi r^2 \iint G(r, z)drdz \tag{5.56}$$

The tunneling generation rate is given by,

$$G(r, z) = A_{kane}\frac{|E|^2}{\sqrt{E_g}}e^{\left[-B_{kane}\frac{E_g^{\frac{3}{2}}}{|E|}\right]} \tag{5.57}$$

Where, $A_{kane} = \frac{q^2\sqrt{2m_{tunnel}}}{h^2\sqrt{E_g}}$, $B_{kane} = \frac{\pi^2 E_g^{3/2}\sqrt{m_{tunnel}}/2}{qh}$, $\frac{1}{m_{tunnel}} = \frac{1}{m_h m_0} + \frac{1}{m_e m_0}$

m_0- a mass of the free electron, m_e, m_h- the effective electron mass and effective hole mass.

Besides the parameters, the film thickness can also be used to control the surface potential of the device. Figure 5.15 illustrates the overall surface- potential versus channel length for surrounding gate TFETs with various Si film-thickness 2R values. As seen in the picture, increasing the thickness of an active Si film for the Surrounding gate TFET may significantly increase the dependency of the surface potential on the channel length.

The effect of the oxide-thickness t_{ox} in the surface potential has been shown for the Surrounding gate TFET in Figure 5.16. The surface potential has been plotted against the L for different values of t_{ox}. Decreased oxide thickness has been proven to improve the surface potential's dependence significantly. Therefore, downscaling of the oxide thickness reduces the DIBL effect. When the drain bias rises, the gate never loses control over the channel. However, to prevent tunneling through thin-oxide and hot carrier effects, oxide-thickness cannot be reduced to extremely low levels. Table 5.2 lists the potential values at the surface for various device parameters.

Figure 5.17 shows estimated values of the lateral electric field for various gate voltages of the Surrounding-gate TFET structure, based on an equation. The device's Drain/Source bias is the major contributor to the lateral-electric field.

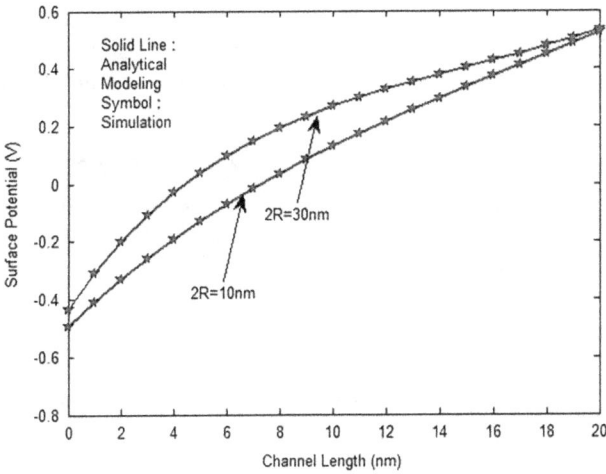

FIGURE 5.15 Effect of Si film thickness on the Potential at the surface.

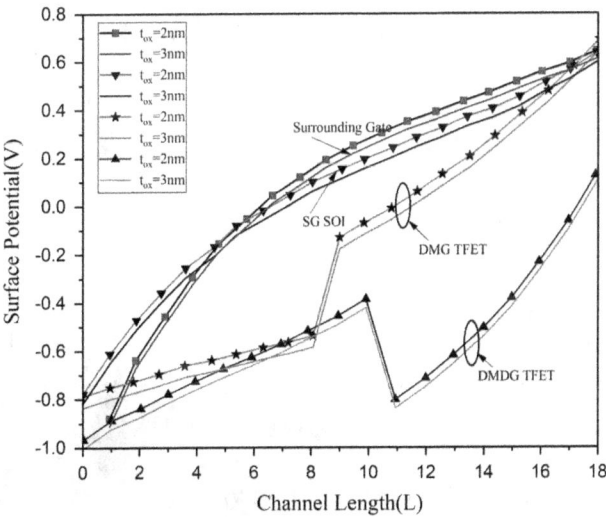

FIGURE 5.16 Surface Potential comparison for various oxide-thickness.

As the lateral-electric field increases and the source-drain bias becomes more pronounced, the gate influence through the channel reduces. Whenever gate voltage is changed, the vertical electric field inside the channel area changes more dramatically, whereas the lateral electric-field changes less. The horizontal electric field is much more prominent in the device configuration, as evident from the figure.

Figure 5.18 demonstrates the surrounding-gate TFET's drain current value at drain voltages of 0.4 V and 0.8 V and the channel length is about 20 nm for both modeled and simulated values. When VGS is positive, electrons travel from the

TABLE 5.2

Surface Potential of Different Device Structures for Various Device Parameters

Device		SG SOI TFET	DMG TFET	DMDG TFET	Surrounding Gate TFET
Parameter		Surface Potential (V)	Surface Potential (V)	Surface Potential (V)	Surface Potential (V)
V_{GS}	0.1 V	0.61	1.35	1.45	0.61
	0.2 V	0.71	1.82	1.99	0.69
	0.3 V	0.81	2.24	2.54	0.767
t_{ox}	3 nm	–	1.23	1.53	–
	2 nm	–	1.34	1.78	–

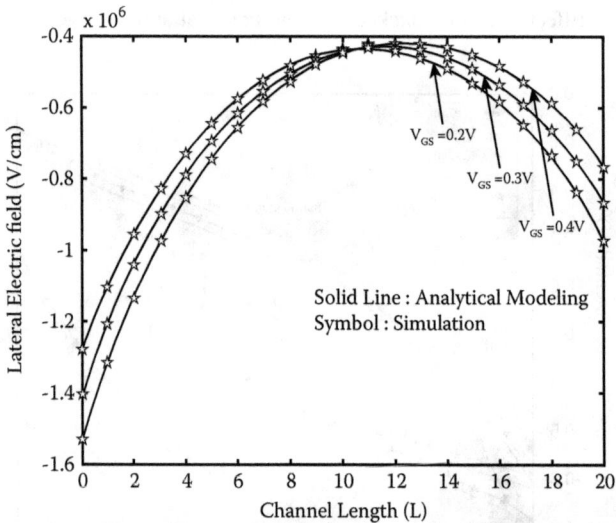

FIGURE 5.17 Lateral electric-field for L = 20 nm, V_{DS} = 0.5 V with various gate voltages.

p+ source area to the channel region, increasing the tunneling current. The device acts as an n-type surrounding-gate TFET under these circumstances.

Figure 5.19 depicts the simulated potential profile and computed surface-potential profile for several TFET structures with a channel length of L = 18 nm for various drain voltages. The potential in the vicinity of the drain area rises as the drain voltage rises. However, the potential has not significantly changed close to the source location. Consequently, it is determined that the drain voltage seems to have no effect on tunneling at the source side. It can be seen that the surrounding gate offers improved surface-potential results in relation to drain-source voltage. The drain current for the different devices is shown in Table 5.3.

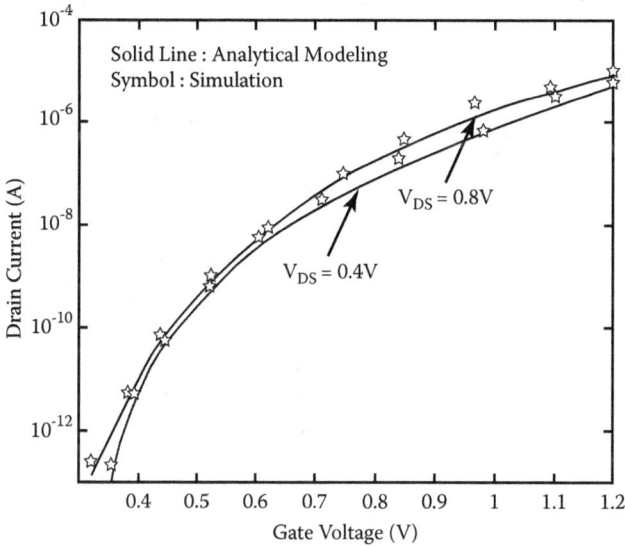

FIGURE 5.18   Effect of drain-source voltage on the drain current of surrounding-gate TFET.

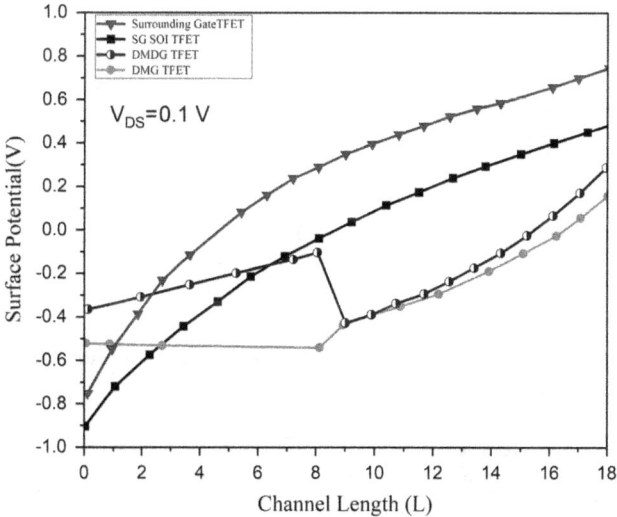

FIGURE 5.19   Surface-potential comparison of DMG TFET, SG SOI, DMDG TFET, and surrounding gate TFET.

Figure 5.20 demonstrate the modeled results obtained from the simulation of vertical electric-field distribution through the distance from the source end for various values of gate biases. The channel length is considered 20 nm, which is analytically obtained by Equation (5.54). The graph shows that the vertical electric

TABLE 5.3

Drain Current Comparison of Advanced TFET Structures

Device Parameter (V_{Gs})	SG SOITFET I_{DS} (A)	DMG TFET I_{DS} (A)	DMDG TFET I_{DS} (A)	Surrounding Gate TFET I_{DS} (A)
0 V	0.5×10^{-15}	2.9×10^{-15}	2.12×10^{-11}	1.08×10^{-11}
0.5 V	1×10^{-13}	1.9×10^{-13}	2.6×10^{-5}	5.6×10^{-10}
1 V	0.05×10^{-10}	6.4×10^{-9}	8.9×10^{-4}	2.1×10^{-6}

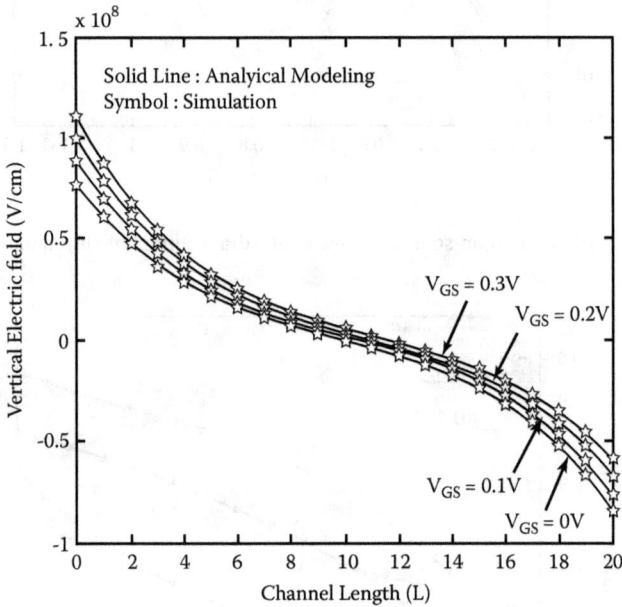

FIGURE 5.20   Effect of gate bias on vertical-electric field of surrounding-gate TFET.

field's peak is close to the source side. The pace of tunneling is increased as a result—the tunneling current increases as a result of this phenomenon.

5.7 SUMMARY

The high speed, ultra-dense integration, power consumption, and matching will be the principal issues for future MOS Technology with few nanometers of channel length. Persistent downscaling of MOS transistors results in severe short-channel effects and a large leakage current, preventing MOSFETs from becoming more miniature. TFET is believed to be the ultimate field effect transistor that is scalable to the nanometer regime. Future technology can overcome formidable obstacles with the help of modeling and simulation. This chapter describes physical device difficulties with modern TFET architectures. A device model for circuit design should be precise and computationally effective. It is commonly acknowledged that the use of surface potential can produce better forecasts of integrated circuit performance and can offer a decent approximation to the actual behavior of transistors.

REFERENCES

[1] A. Tura and J. C. S. Woo, "Performance comparison of silicon steep subthreshold FETs," *IEEE Transactions on Electron Devices*, vol. 57, no. 6, pp. 1362–1368, 2010.

[2] G. E. Moore, "Cramming more components onto integrated circuits Electronics Magazine," vol. 38, no. 8, pp. 33–35, 1965.

[3] P. F. Wang, *Complementary tunneling-fets (CTFET) in CMOS technology*, Ph.D. Thesis, Technical University of Munich, 2003.

[4] Q. Zhang, W. Zhao, and A. Seabaugh, "Low-subthreshold-swing tunnel transistors," *IEEE Electron Device Letters*, vol. 27, no. 4, pp. 297–300, 2004.

[5] T. S. Arun Samuel and N. B. Balamurugan, "Analytical Modeling and Simulation of Germanium Single Gate Silicon on Insulator TFET," *Journal of Semiconductors*, vol. 3, 2014. (ISSN: 1674-4926, Impact Factor: 0.378).

[6] W. E. Zhang, F. C. Wang, and C. H. Yang, "Design and Modeling of a New Silicon-Based Tunneling Field-Effect Transistor," *IEEE Transactions on Electron Devices*, vol. 43, no. 9, pp. 1441–1447, 1996.

[7] E. O. Kane, "Zener tunneling in semiconductors," *Journal of Physics and Chemistry of Solids*, vol. 12, no. 2, pp. 181–188, 1960.

[8] T. S. Arun Samuel, N. B. Balamurugan, S. Sibitha, R. Saranya, and D. Vanisri, "Analytical Modeling and Simulation of Dual Material Gate Tunnel Field Effect Transistors," *Journal of Electrical Engineering & Technology*, vol. 8, no. 6, pp. 1481–1486, 2013. (ISSN: 1975-0102, Impact Factor: 0.579)

6 Evolution of Heterojunction Tunnel Field Effect Transistor and its Advantages

C. Usha and P. Vimala

CONTENTS

6.1 INTRODUCTION

In 1963, complementary metal oxide semiconductor technology was introduced by Frank Wanlass. CMOS technology is an organization of two types of MOSFET: P-type and N-type. CMOS technology has dominated the silicon industry due to various advantages. In 1966 the first ion-implanted MOSFET was introduced [1]. The transistor packing density of integrated circuits has progressively increased from small-scale integration to very-large-scale integration

DOI: 10.1201/9781003327035-6

to reduce the size and lower device cost. The performance of the MOSFET is increased by reducing some critical parameters known as scaling.

The scaling of the device decreases the channel length, due to which the device behavior deviates from the long channel effects to the short channel effects. The potential distribution and high electric field across reduced channels cause short-channel effects. The decrease in channel length or increase of drain bias results in decreased punch through voltage, an increase in subthreshold current, an increased subthreshold gate swing, and a fall in the threshold voltage. Due to the high electric field, the channel mobility becomes field sensitive, leading to velocity saturation. On further increase in the field across the drain region, carrier multiplication ensues, leading to the parasitic bipolar transistor action and substrate current. High electric fields also promote hot-carrier injection hooked on the oxide, resulting in oxide charge and transconductance loss.

To overcome these short channel effects and subthreshold swing reduction, the Tunnel Field Effect Transistor (TFET) is introduced. The Single Gate (SG)TFET was initially introduced using a band-to-band tunneling mechanism. The current performance of the SGTFET device is low. To increase the current performance of SGTFET, various TFET devices with double gate, tri-gate, and nanowire were introduced [2]. Also furthermore, to increase the performance of the TFET device, heterojunction using different materials across the source and channel is introduced.

6.2 DOUBLE GATE HOMOJUNCTION TFET

This chapter will describe double gate homojunction TFET first, and then double gate heterojunction TFET will be explained. Figure 6.1a briefly illustrates the classification of modern TFETs. Even though there are many kinds of TFET, in this chapter, we will cover heterojunction TFET for on-current improvement [Figure 6.1a]. Especially various materials (Ge, InAs, GaAs, GaSb, GaAsSb, InGaAs) would be introduced for designing heterojunction TFET.

Figures 6.1b and 6.1c show the schematic diagram of DG TFET and its energy band diagram. The DG TFET can also be called a gate-controlled PIN diode, as the electrons tunnel from the source valence band to the channel conduction band based on the gate voltage control. The tunneling barrier width can be varied based on gate voltage changes [3].

6.3 HETEROJUNCTION TFET

Homojunction TFET operates as a reverse-biased p-i-n structure. For example, considering nTFET working principle, the device is generally in OFF-state condition with a very modest IOFF current involvement when the valence band at the source region and conduction band at the channel region are not allied at zero gate voltage. As positive voltage is applied at the gate terminal, the channel regime's conduction band is dragged down below the valence band of the source regime, and the energy window $\Delta\Phi$ overlays two bands. As a result, only carriers with energies inside this window can tunnel into the conductive channel.

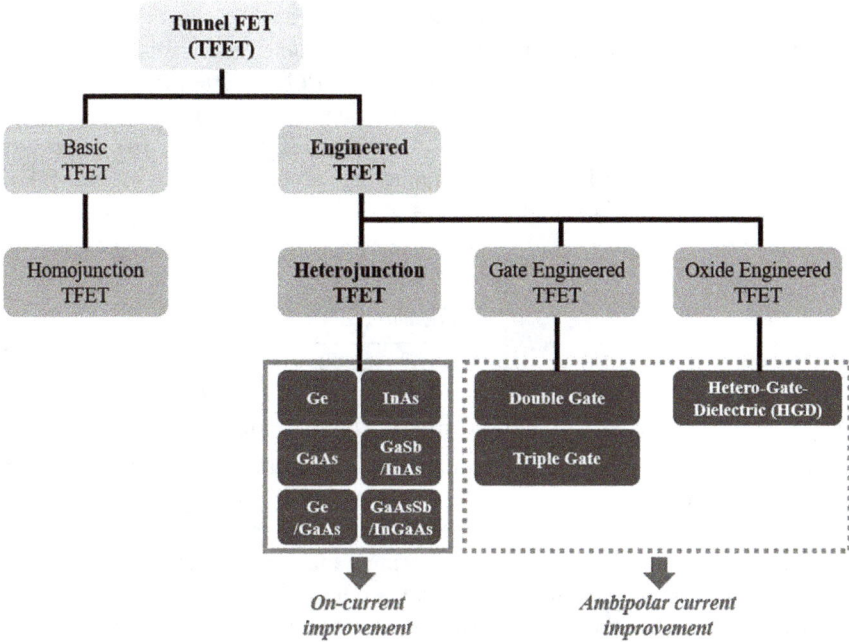

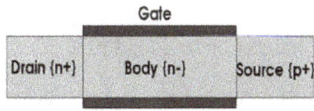

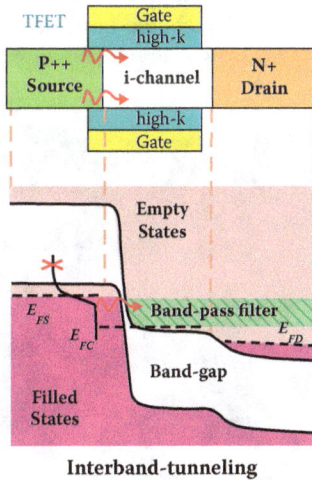

FIGURE 6.1 (a) Schematic summary explaining the classification of modern TFETs. (b) Schematic view of double gate (DG)TFET. (c) DG TFET band diagram and its operation.

This is an essential aspect of TFET functioning because it allows for the active filtering of electrons with high energy in the source Fermi distribution. It's also worth noting that certain TFET devices use a PN-junction instead of a pin-junction to operate appropriately, which necessitates using a gate bias to deplete the regime adjacent to the source and switch off tunneling [4]. The ON-state current is proportional to the tunneling probability (T) in the BTBT mechanism, which can be calculated using the Wentzel-Kramer-Brillouin (WKB) approximation.

$$T \approx -\frac{4\lambda\sqrt{2m^*}\sqrt{E_g^3}}{3q\hbar(E_g + \Delta\Phi)} \tag{6.1}$$

where m* is the effective carrier tunneling mass, Eg is the energy band gap, λ is the tunneling screening length. The tunneling probability can be achieved close to 1 by optimizing the Eg, m*, and λ with the proper selection of the materials using heterojunction. In heterostructures, outstanding transport qualities can be combined with band alignment engineering, and an optimal blend of band alignment and tunneling interface can be produced. Different material combinations can be used for the formation of heterojunction structures, such as Germanium (Ge), Indium Arsenide (InAS), Gallium Arsenide (GaAs), Gallium Nitride (GaN), etc.

6.3.1 GERMANIUM HETEROJUNCTION DGTFET

Germanium material usage in heterojunction provides a powerful solution for the low ION current performance and subthreshold swing.

Ge material can be used for all transistor regions, source regions, or drain regions for the TFET structure. Figure 6.2 depicts the basic structure of the TFET device. The combination of different materials study is shown in Table 6.1 [5].

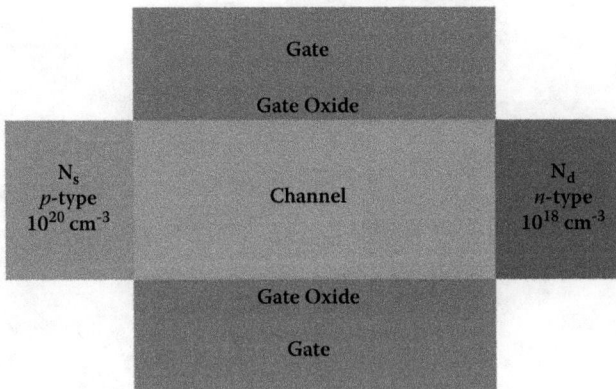

FIGURE 6.2 Basic tunnel field effect transistor schematic view [5].

TABLE 6.1

Germanium Material Combinations of the Devices Analyzed

Case	Source (p-type 10^{20}cm^{-3})	Channel (Undoped)	Drain (n-type 10^{18}cm^{-3})
Case 1	Si	Si	Si
Case 2	Ge	Ge	Ge
Case 3	Ge	Si	Ge
Case 4	Si	Ge	Si

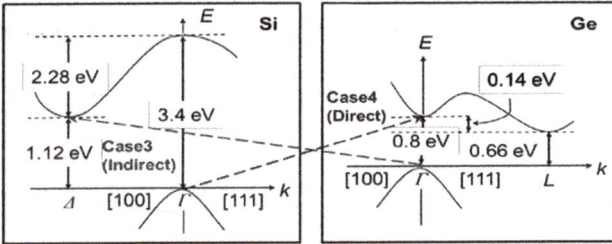

FIGURE 6.3 Schematic energy band diagram of silicon and germanium material.

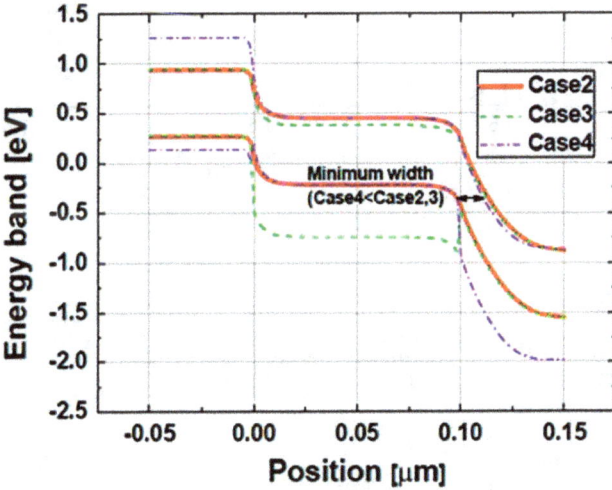

FIGURE 6.4 Comparison graph of the energy band of different combinations of material from case 2 to case 4.

The schematic energy band diagram of germanium and silicon material is shown in Figure 6.3. It depicts that the tunneling width of the Ge material is smaller than the Si material. Thus, the leakage in the Ge material is negligible, due to which the ambipolarity of the device reduces. Figure 6.4 compares the energy band diagrams for different material combinations from case 2 to case 4; out this case 4, when the channel material is Ge provides the lower energy bandwidth due to which tunneling of charges increases from the source valence band to channel conduction band.

Figure 6.5a shows the comparison plot of case 1 to case 4, it represents that the Si homojunction TFET device resulted in very low I_{ON} and Sub-threshold Swing characteristics, ascribed large tunneling resistance because of large silicon bandgap(-1.12 eV) and in part to the low band to band tunneling efficiency due to indirect bandgap. Ge homojunction TFET from table case 2 showed significant improvement in ION and SS but suffered from the ambipolar current because BTBT resistance at drain and channel junction was also reduced. Heterojunction (Ge/Si/Ge) structures (case 3 and 4) has shown higher

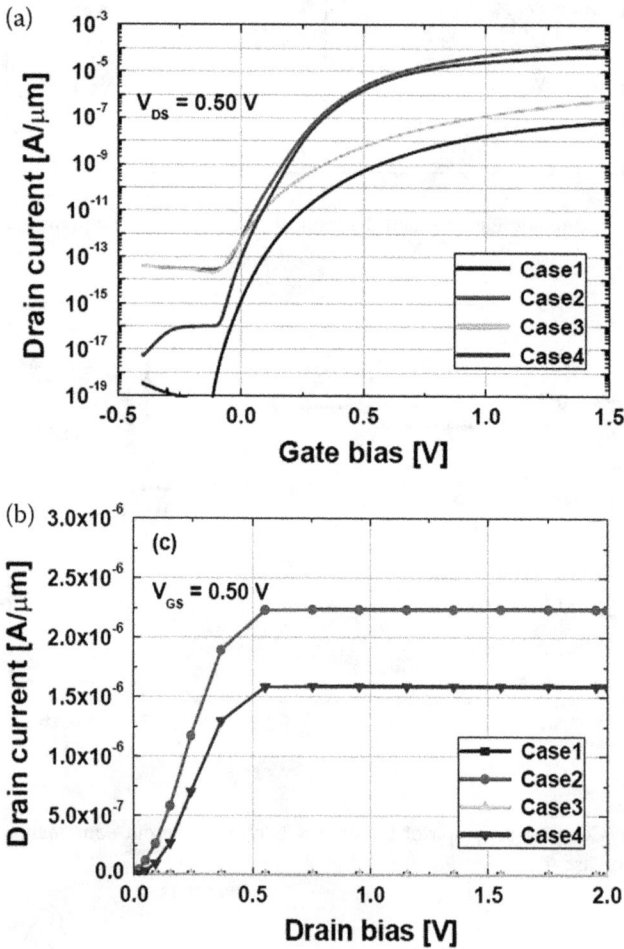

FIGURE 6.5   (a) Drain current versus Gate voltage comparison plot for V_{DS} = 0.50V (b) Drain current versus Drain voltage comparison plot for V_{GS} = 0.50V.

current performance, such I_{ON} and I_{OFF} which also significantly reduces the ambipolar current.

6.3.2 GERMANIUM HETEROJUNCTION DGTFET

Figure 6.6 presents the schematic energy band diagram of heterojunction (InAs/Si) TFET device across the source-channel interface and gate oxide semiconductor interface. It is studied that charges outside the narrow band of tunneling may interject to flow of current; this might potentially improve current levels but simultaneously has an impact on variations of subthreshold swing. Figure 6.7 represents the two-dimensional view of TFET with either Si/InAs/Ge source material for the formation of heterojunction.

From Figure 6.8a, it is studied that in the OFF state condition, the energy band gap at the source region corresponds to silicon (1.1 eV), germanium (0.67 eV), and indium arsenide (0.36 eV) and all three regions fermi level are aligned. Figure 6.8b shows the ON state condition; when the gate bias is applied, the fermi level displacement occurs, leading to the steep band bending. As the voltage

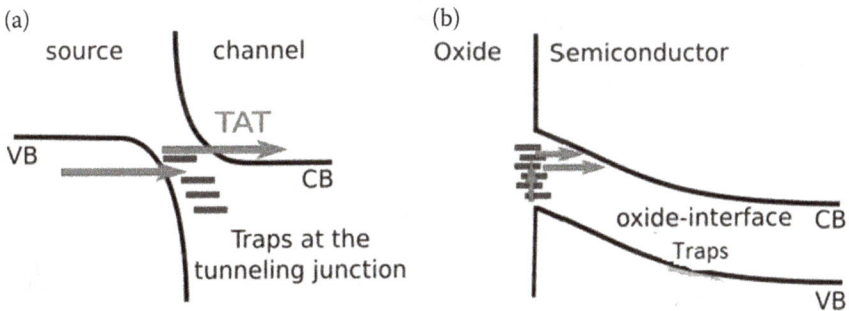

FIGURE 6.6 (a) Energy band diagram of InAs/Si TFET (a) source channel junction (b) gate oxide semiconductor interface [6].

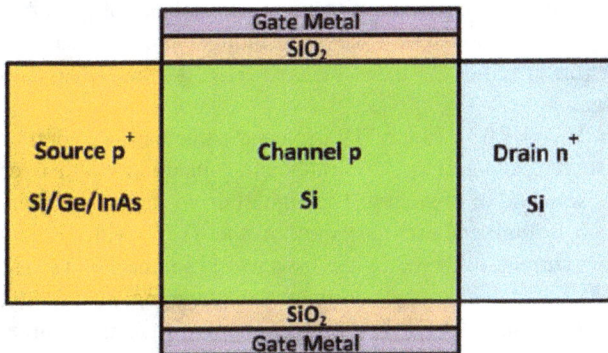

FIGURE 6.7 2D schematic view of the Heterojunction TFET.

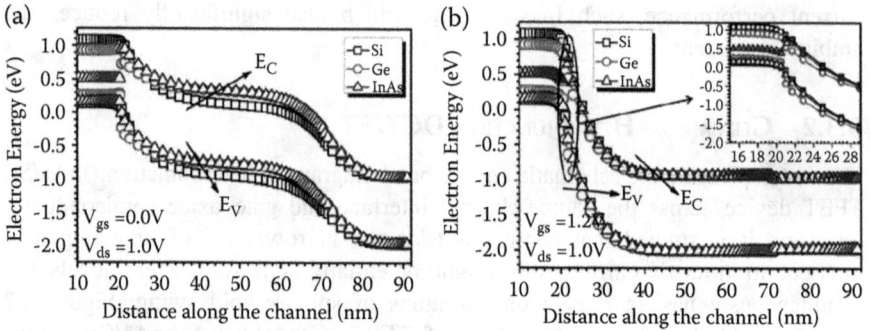

FIGURE 6.8 Comparison of energy band diagrams for silicon, germanium, and indium arsenide at (a) OFF State (b) ON state.

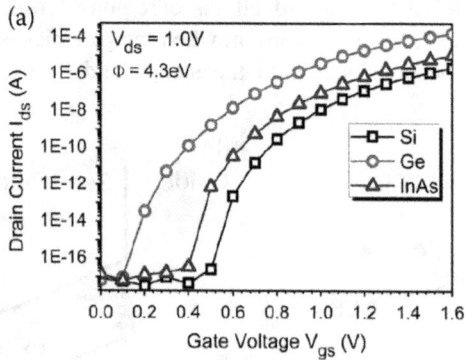

FIGURE 6.9 I_D versus V_{GS} comparison plot for three different materials Si, Ge, and InAs.

is applied to the channel, the fermi level lowers at the width of the barrier across the tunneling junction and around the source channel interface. Herein, the Ge source exhibits the most band bending, followed by materials with InAs and Si as sources. The occurrence of a bend in the conduction band of an InAs source DG-TFET in both the OFF and ON states is a thought-provoking observation. The different electron affinities of InAs and Si across the heterojunction cause this kink to emerge.

At constant $V_{DS} = 1.0$ V, Figure 6.9 depicts the fluctuation of drain current (Ids) with gate-to-source voltage (V_{GS}). The tunneling junction width is critical in influencing the likelihood of tunneling. The highest drive current is obtained when the band gap width is the smallest. Furthermore, the TFET with Ge source has the maximum drain current, followed by the InAs and Si source TFETs. The interaction of several material parameters determines the tunneling probability. The drain current of the Ge source TFET is more progressive than the source InAs, even though the InAs band gap is lower than the Ge band gap. As a result, the DG-drive TFET's current is determined not only by the source material's bandgap but also by the parameters of the entire heterojunction.

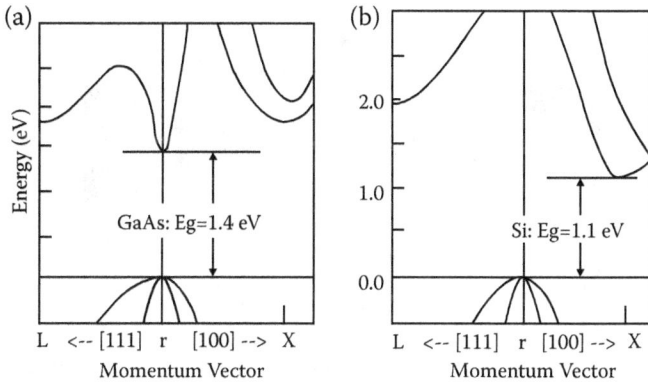

FIGURE 6.10 A comparison of the schematic energy band diagrams for (a) GaAs and (b) Si [8].

6.3.3 GALLIUM ARSENIDE HETEROJUNCTION DG TFET

The schematic band diagrams of GaAs and Si are shown in Figure 6.10. The lowermost point in the conduction band and the uppermost point in the valence band in GaAs coincide at r, at the center of the Brillouin zone, this alienation is 1.4 eV, as noted, and it is dubbed a direct gap because of the coincidence at r. Si is classified as an indirect gap material, with a 1.1 eV indirect gap. GaAs have electrostatic properties that are exceptional to this kind of semiconductor due to the presence of the satellite valley property. The direct nature of the energy band gap in GaAs is the source of exceptionally auspicious optical transition probabilities. The features of electron and hole conduction in GaAs and Si semiconductors are also explained by their energy band structures. The bending of valence and conduction bands of the energy band diagrams is inversely related to the effective mass of holes and electrons [7,8].

Figure 6.11 shows that the ambipolar current is suppressed using a GaAs/Si TFET construction without impacting the ON state performance. Figure 6.12 represents the energy band diagram of GaAs/Si heterojunction TFET, and it is studied that the ambipolar state is reduced due to the high tunneling of the electrons from the valence band to the conduction band. Figure 6.13 shows the comparison plot of silicon doping-less TFET and GaAs/Si heterojunction TFET. It is observed that the GaAs/Si interface impacts OFF current, which is lower than the homojunction and has no impact on the ON current performance.

6.3.4 HETEROJUNCTION GaN/InN/GaN DG-TFET

Gallium Nitride is a direct bandgap binary III/V semiconductor compatible with high-power transistors that can operate at high temperatures. Gallium nitride (GaN) is a direct bandgap semiconductor with a binary III/V structure that is extensively utilized in blue light-emitting diodes. The substance has a Wurtzite

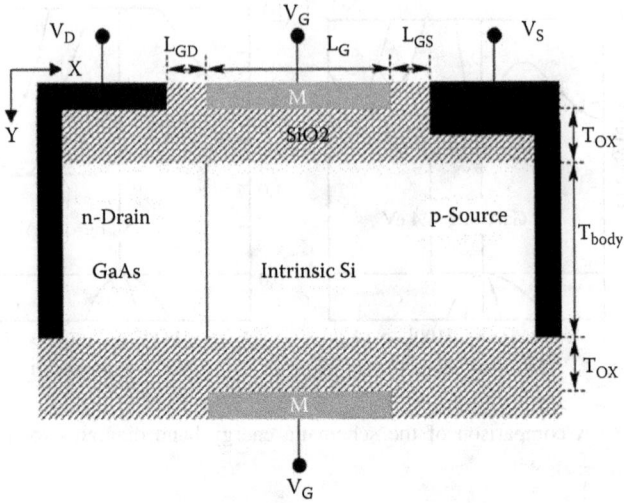

FIGURE 6.11 2D schematic view of the GaAs/Si heterojunction TFET [9].

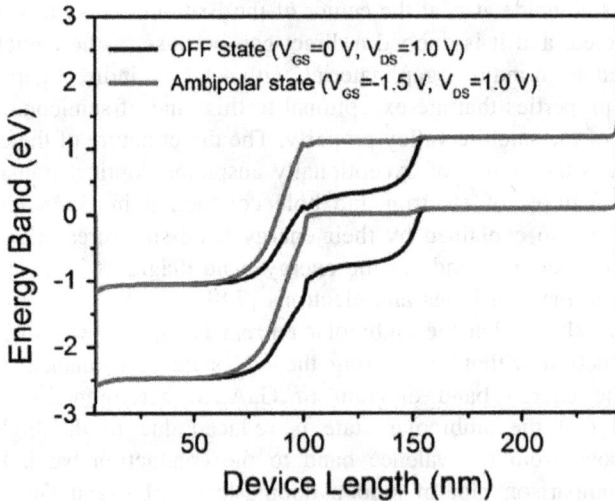

FIGURE 6.12 Energy band diagram of GaAs/Si TFET at OFF state and ambipolar state [9].

crystal structure and is extremely hard. Its wide band gap of 3.4 eV is applicable for high-frequency devices such as high optoelectronic power applications. The bandgap of gallium nitride is 3.4 eV, whereas the band gap of silicon is 1.12 eV. Because gallium nitride has a wider band gap than silicon MOSFETs, it can withstand higher voltages and temperatures.

The GaN/InN/GaN DGTFET is a heterojunction of three III-V binary compounds: GaN for the drain and source and InN for the channel, as shown in

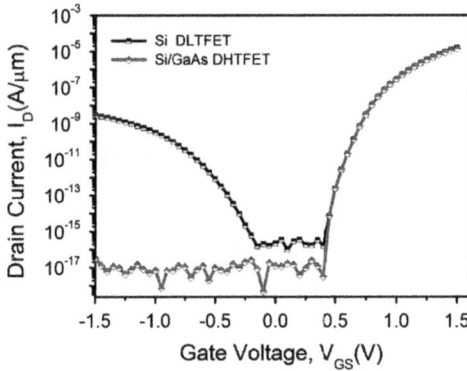

FIGURE 6.13 I_D versus V_{GS} comparison plot of Si DLTFET and Si/GaAs DHTFET [9].

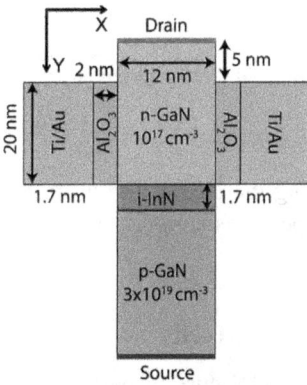

FIGURE 6.14 GaN/InN/GaN heterojunction DG TFET [10].

Figure 6.14. GaN's 3.2-eV bandgap allows for a low OFF-state current, I_{OFF}, while removing the ambipolar current. The usage of InN, which has a bandgap of 0.7 eV and a high polarisation discontinuity compared to GaN, results in a 20 MV/cm built-in electric field and a tunneling distance of about 1 nm within the p-GaN/i-InN/n-GaN heterojunction. The energy gap between the InN conduction and GaN valence bands, combined with the InN's narrow bandgap, allows for an increase in TWKB and a rise in ON-state current by lowering the Eg parameter and increasing $\Delta\Phi$

$$T_{WKB} \approx \exp\left(-\frac{4\lambda\sqrt{2m^*}\sqrt{E_g^3}}{3q\hbar(E_g + \Delta\Phi)}\right) \tag{6.2}$$

Figure 6.15 shows the drain current characteristics, it is studied that TFETs' drain current is caused by Band-To-Band-Tunneling rather than thermionic emission adds another quirk to these devices. Because the BTBT phenomenon

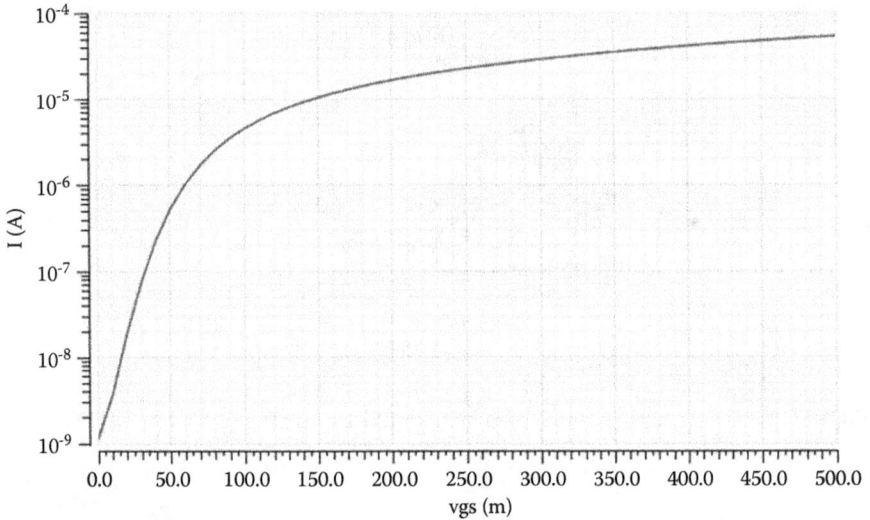

FIGURE 6.15 I_D versus V_{GS} plot of GaN/InN/GaN heterojunction DG TFET at $V_{DD} = 0.5V$.

is practically independent of carrier mobility, changing the channel width to get a specular characteristic between n-type and p-type transistors is no longer necessary [11].

6.4 NANOWIRE TUNNEL FIELD EFFECT TRANSISTOR

To attain better electrostatic coupling and increase in ON current performance is obtained by modifying gate channel geometry using nanowire structure, which can be produced either by bulk or SOI technology. The Process incorporation of TFET device with nanowire structure will intensify on-chip model density while also showing virtuous gate controllability. Figures 6.16a and 6.16b show a cylindrical gate structure allowing wider channel width per unit area of the material, resulting in a higher current drive per unit area. Furthermore, this device uses a very reedy body to eliminate a sub-surface leakage path between the source region and drain region, resulting in enhanced gate control, deferment of short channel effects, highly superior short channel immunity with sub-threshold swing less than 60 mV/decade of ON current I_{ON}, and a high I_{ON}/I_{OFF} ratio [12,13].

6.5 GATE ALL AROUND (GAA) HETEROJUNCTION TFET

An interface between two regions of different semiconductors is acknowledged as a heterojunction. In difference from a homojunction, these semiconducting materials demonstrate asymmetrical energy band gaps. A heterostructure is a device that combines many heterojunctions. The condition that each material is a

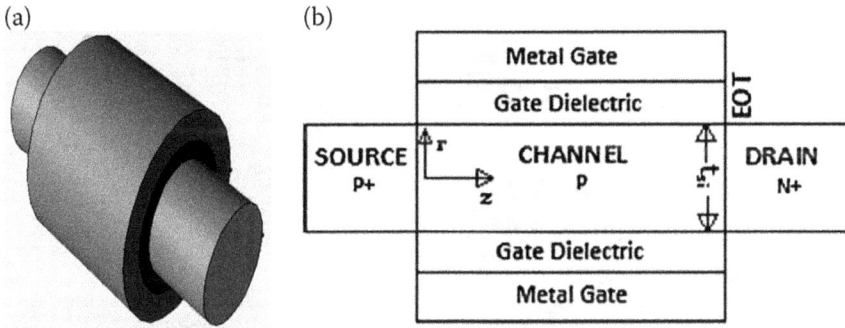

FIGURE 6.16   (a) Three-dimensional view of nanowire TFET. (b) Cross-sectional schematic view of nanowire TFET.

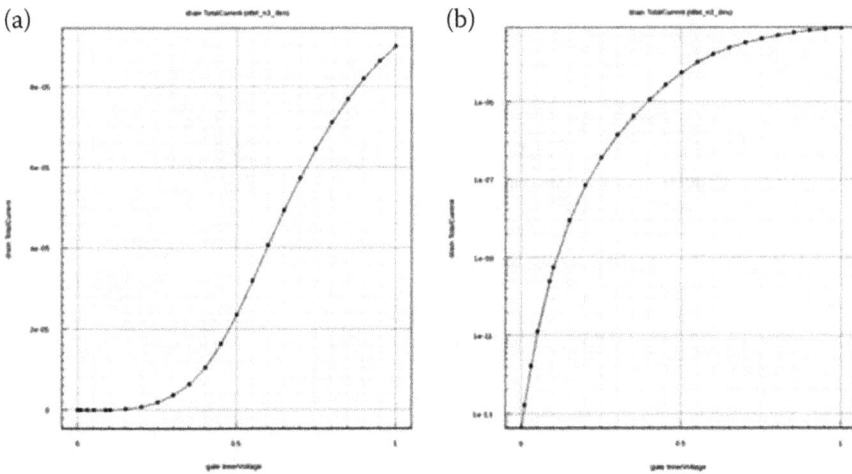

FIGURE 6.17 I_D versus V_{GS} characteristics of nanowire TFET (a) linear scale (b) logarithmic scale plot.

semiconductor with unequal band gaps is a little sloppy, especially at nanoscales, when spatial attributes influence electrical properties. The different combinations of the materials from the heterojunction for GAA TFET.

Figure 6.17 presents the drain current varied for the gate to source voltage for the nanowire structure. It depicts that the ON current is high and significantly less OFF current. It also represents that the subthreshold swing is less than the conventional MOSFET device.

6.5.1   GERMANIUM/SILICON GAA HETEROJUNCTION (HJ) TFET

Figure 6.18a and b represent the three-dimensional and cross-sectional view of Ge/Si GAA HJ TFET. In a Si-Ge heterojunction, the device's energy band moves downwards to the intrinsic area, improving I_{ON} current while suppressing

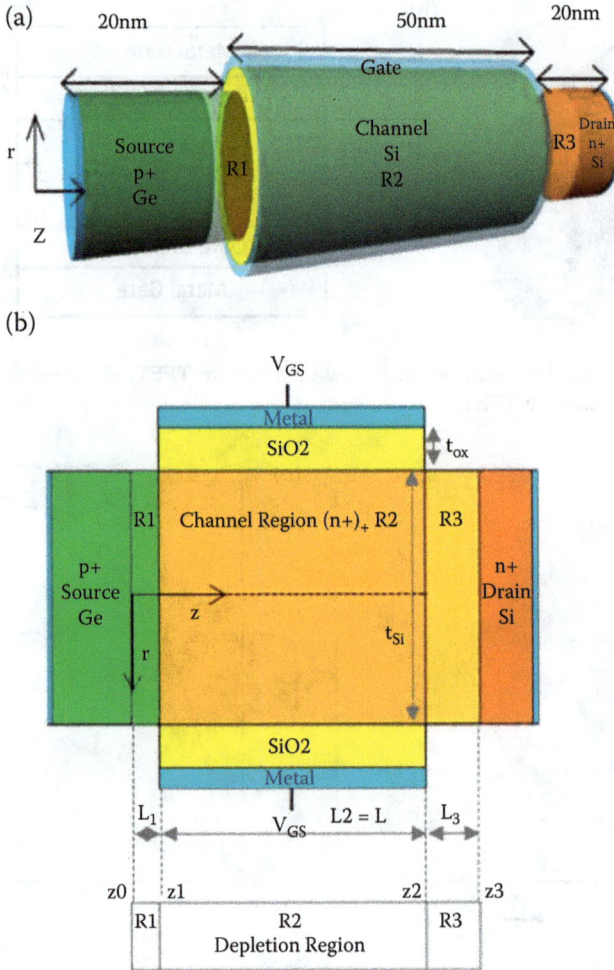

FIGURE 6.18 (a) Three-dimensional view of Ge/Si GAA heterojunction TFET (b) Two-dimensional cross-sectional view of Ge/Si GAA HJ TFET.

ambipolar behavior and lowering subthreshold swing. Figure 6.19 represents the comparison plot of homojunction and heterojunction GAA TFET. It is observed that the I_{ON} current is high for heterojunction compared to homojunction GAA TFET, but the I_{OFF} is slightly higher than the homojunction [14].

6.5.2   GASB/INAS HETEROJUNCTION GATE ALL AROUND TUNNEL FET

Figure 6.20 depicts a three-dimensional view and its cross-sectional view of GaSb/InAs heterojunction GAA TFET, the GaSb material is used for the source regime and InAs for the channel and drain regime. The heterojunction structure with a broken gap is used to attain the high ON current at the source-channel

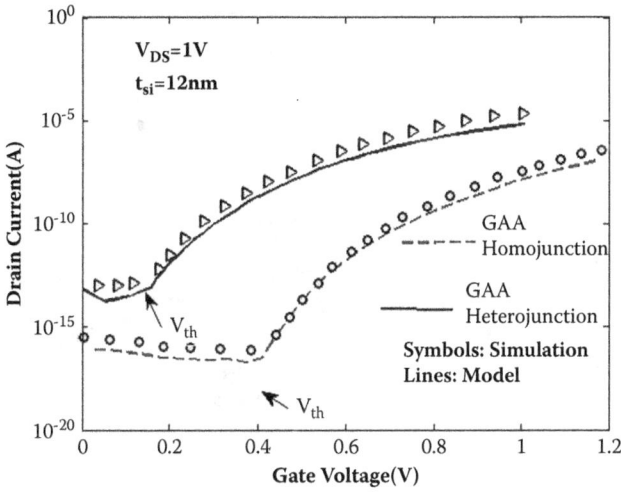

FIGURE 6.19 I_D versus V_{GS} comparison plot of GAA homojunction and GAA heterojunction TFET [14].

(a)

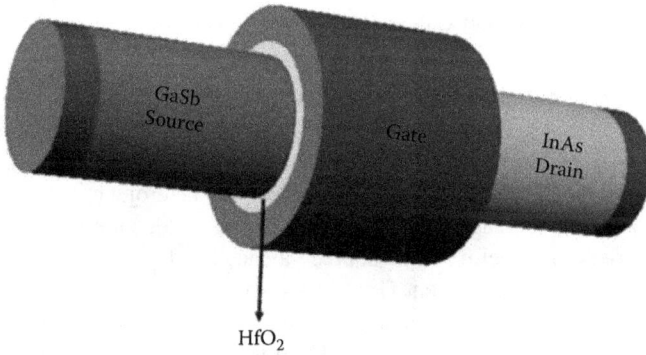

(b)

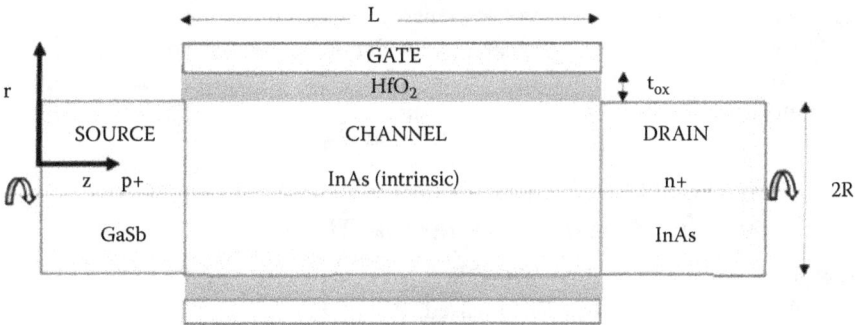

FIGURE 6.20 GaSb/InAs heterojunction GAA TFET (a) three-dimensional view. (b) Two-dimensional cross-sectional view [15].

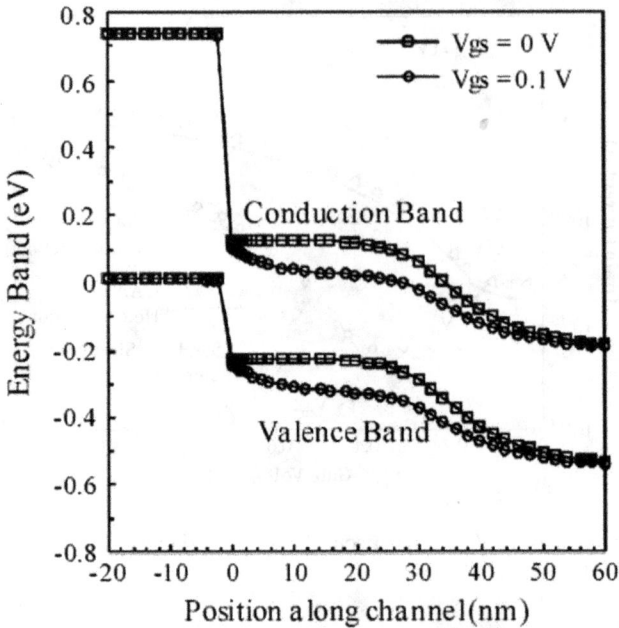

FIGURE 6.21   Energy band diagram of GaSb/InAs heterojunction GAA TFET [14].

junction. The broken-gap alignment of hetero junction structures would aid in BTBT tunneling as opposed to any other alignment. Thus, the current obtained would be larger than other hetero junction alignments.

Figure 6.21 presents the energy band gap of a GaSb/InAs heterojunction with a broken gap aligned. As seen, increasing gate voltage roots the conduction band to be reduced, allowing carriers to tunnel from the source valence band into the channel's conduction band and drain. The alignment GaSb/InAs was designed to produce the energy band diagram shown, allowing for a high drain current value [16]. The benefit of this alignment is that the forbidden energy band gap is large enough to provide a low value of current when the device is in the OFF state and small enough to produce a suitable high current when the device is in the ON state. The band diagram can also be used to show the choice of low-energy band gap material.

The variation of the drive current along with a gate to source voltage is shown in Figure 6.22, and it is studied that the slope of drain current in the sub-threshold regime changes as the channel length upsurges.

6.5.3   GE/GAAS-BASED HETEROJUNCTION GATE-ALL-AROUND (GAA) TUNNELING FIELD-EFFECT TRANSISTOR (TFET)

The germanium with gallium arsenide (Ge/GaAs)-based heterojunction gate-all-around (GAA) TFET (Ge/GaAs-based A-TFET) is optimally designed and studied among III–V compound material-based TFETs shown in the Figure 6.23.

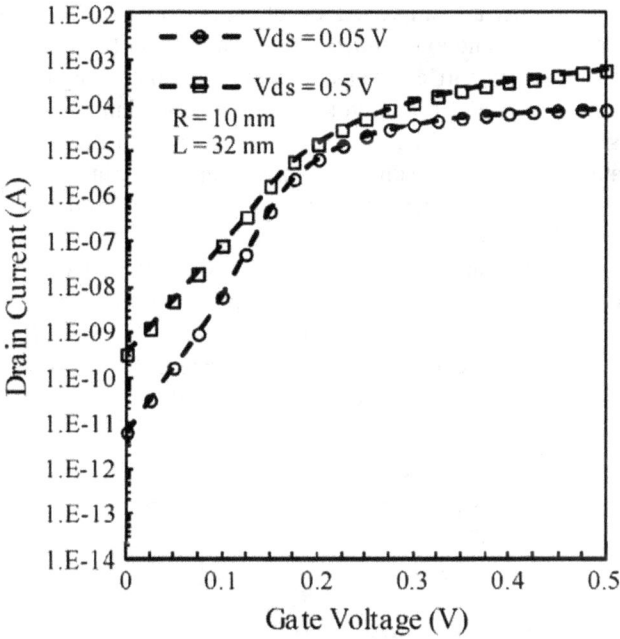

FIGURE 6.22 I_D versus V_{GS} plot for GaSb/InAs heterojunction GAA TFET [14].

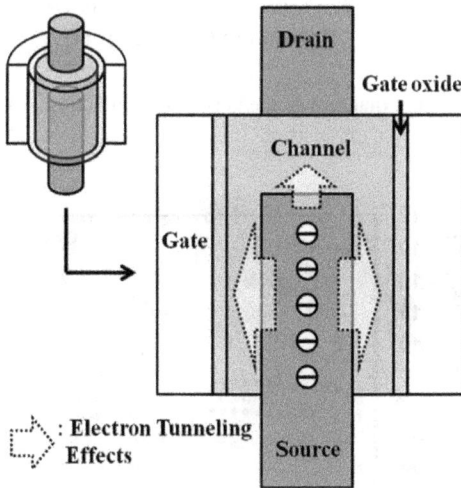

FIGURE 6.23   Three-dimensional (3D) and schematic cross-sectional view of Ge/GaAs heterojunction GAA TFET [17].

High current drivability and improved switching performance are achieved using Ge/GaAs-based heterojunction semiconductor materials. In Figure 6.24, lateral and vertical routes transport electrons from the Ge source regime to the GaAs channel. The main current of an arch-shaped TFET is a lateral tunneling current because most electron tunneling occurs at the source and drain boundary areas under the gate region. As a result, the source regime height (Hsource) and epitaxially grown channel thickness (tepi) were considered key design variables in the following optimization processes.

The I_D–V_{GS} of the Ge/GaAs-based A-TFET for different doping concentrations in the Ge source area are shown in Figure 6.25. The OFF-state current (I_{OFF}) diminishes as the Ge source concentration increases, but the ON-state current (I_{ON} and S) exhibits excellent features. The energy barriers of the source and channel regions are reduced under low doped with OFF-state operation

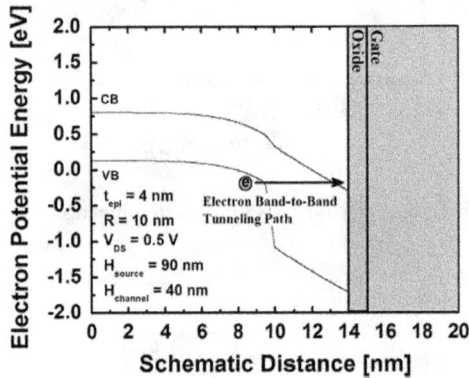

FIGURE 6.24   Energy band diagram of Ge/GaAs heterojunction GAA TFET [15].

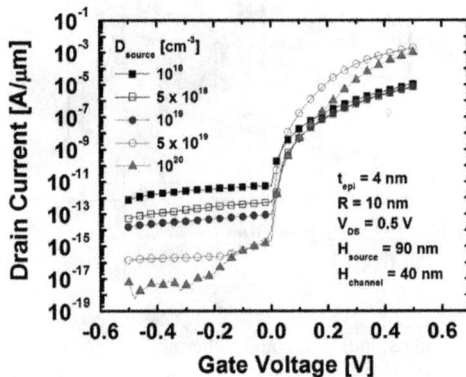

FIGURE 6.25 I_D versus V_{GS} plot for different Ge concentrations of Ge/GaAs Heterojunction GAA TFET [17].

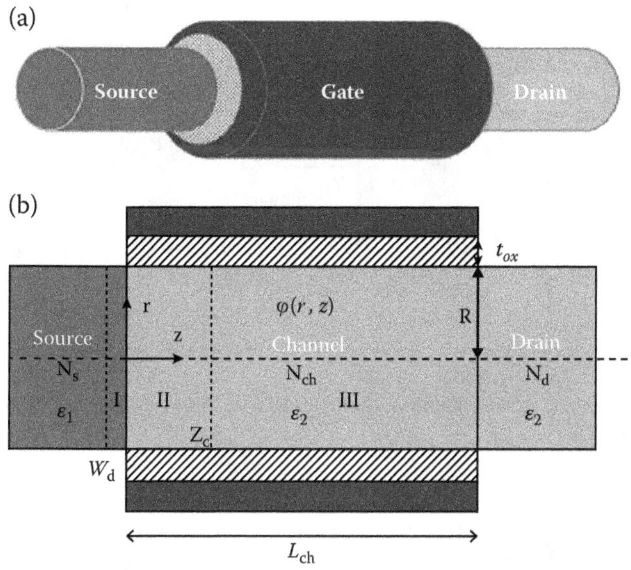

FIGURE 6.26 (a) 3D view of the GaAs$_{0.5}$Sb$_{0.5}$/In$_{0.53}$Ga$_{0.47}$As heterojunction GAA TFET. (b) 2D cross-sectional view [18].

circumstances, and the diffusion current increases I$_{OFF}$. The heavily doped source area lowers the tunneling barrier between the source and channel regions, resulting in higher lateral and vertical currents.

6.5.4 GaAs$_{0.5}$Sb$_{0.5}$/In$_{0.53}$Ga$_{0.47}$As Heterojunction GAA TFET

GaAs$_{0.5}$Sb$_{0.5}$/In$_{0.53}$Ga$_{0.47}$As heterojunction GAA TFET structure three-dimensional view is shown in Figure 6.26a and its cross-sectional view in Figure 6.26b. GaAsSb/InGaAs compound material falls under the category of the III-V compound group of the periodic table. It is analyzed that the III-V compound materials are the promising materials for the GAA heterojunction TFET which produce the highest performance than double gate heterojunction TFET. To improve the device performance the staggered gap heterojunction TFET structure is used. From Figure 6.27 it is observed that effective tunneling length is negative in the broken gap and operates normally.

Figure 6.28 shows that the I$_{DS}$ is improved and tends to be constant over the channel and drain region. The drain current is high for increasing channel voltages owing to the increased electric field at the junction and tends to be constant due to screening of the inversion charge.

6.6 GATE ENGINEERING HETEROJUNCTION TFET

Homojunction TFETs endure the low ON-state current and ambipolar condition across the drain region. The usage of different materials for different work

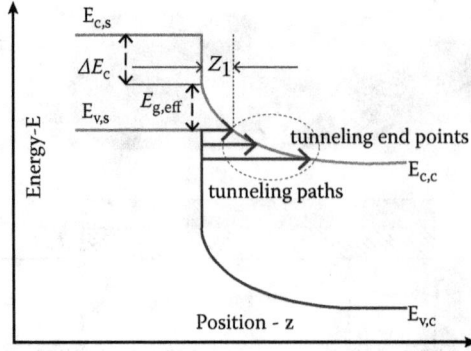

FIGURE 6.27 Energy band diagram of staggered gap $GaAs_{0.5}Sb_{0.5}/In_{0.53}Ga_{0.47}As$ heterojunction GAA TFET [18].

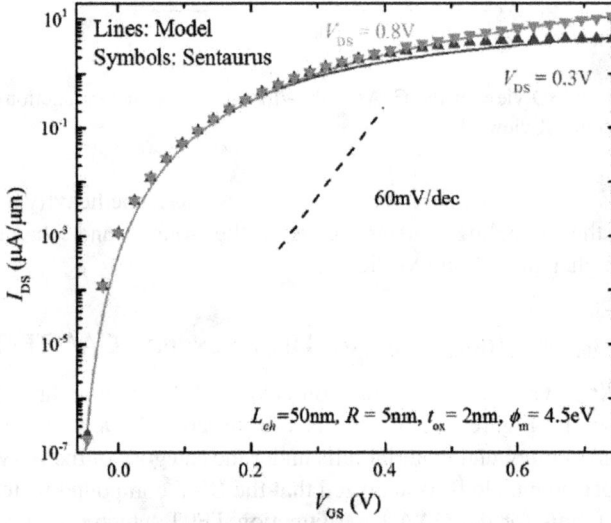

FIGURE 6.28 I_{DS} versus V_{GS} of the $GaAs_{0.5}Sb_{0.5}/In_{0.53}Ga_{0.47}As$ heterojunction GAA TFET [18].

functions across the gate terminal. A careful selection of gate material work functions at the source end will aid in tuning through the barrier width, allowing carriers to tunnel from source to channel. The choice of different materials, especially near the drain region with higher work function material, has an impact on the reduction of the ambipolar condition and lowers the OFF-state current, which further leads to the rise in the I_{ON}/I_{OFF} ratio. The two materials and three materials can be chosen for heterojunction TFET.

6.6.1   DUAL MATERIAL (DM) ELLIPTICAL GATE ALL AROUND (GAA) HETEROJUNCTION TFET

Figure 6.29(a) and (b) show the three-dimensional and cross-sectional view of DM elliptical GAA heterojunction TFET. A staggered band alignment hetero material system (GaAs0.5Sb0.5/In0.53Ga0.47As) is used for the formation of a hetero-junction structure [19]. Two different materials with different work functions are used to form Dual materials across the gate terminal.

Figure 6.30 shows the energy band diagram in comparison of the single material and Dual material elliptical GAA heterojunction TFET for the analysis of the band diagram, which represents the operation of the device. The selection of a lower work function gate material increases the potential at the source regime, sinking the conduction band energy, as seen in Figure 6.30. The device switches on when the suitable gate voltage is applied, allowing carriers to tunnel across the source/channel tunneling junction. According to the energy band diagram, the essence of 'work function engineering is tempering the band diagram and subsequent tuning of the barrier width at both the source-channel and drain-channel interfaces, which affects the tunneling rate and leakage conduction of the device simultaneously.

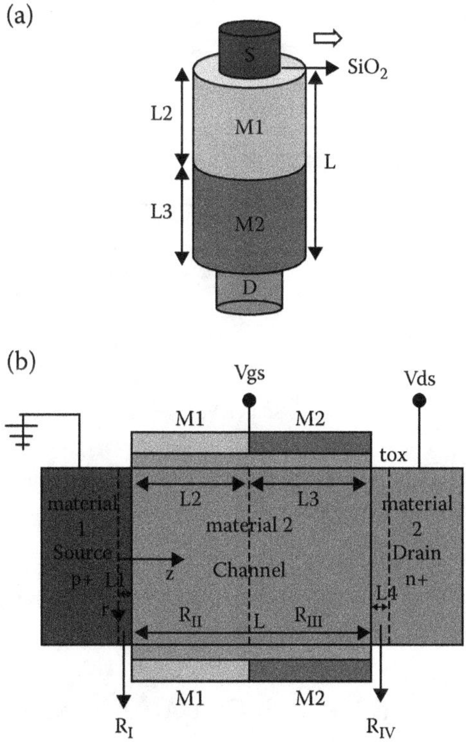

FIGURE 6.29   (a) 3D structure of the DM elliptical heterojunction GAA TFET. (b) Cross-sectional view of DM elliptical heterojunction GAA TFET [19].

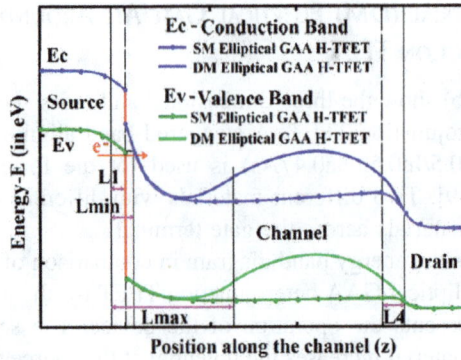

FIGURE 6.30 Energy band diagram of DM eliptical GAA heterojunction TFET (Φ_{m1} = 4.2eV and Φ_{m2} = 4.6eV) and single material (SM) (Φ_m = 4.2eV) elliptical GAA heterojunction TFET during ON-state ($V_{GS} = V_{DS} = 0.5$V) [19].

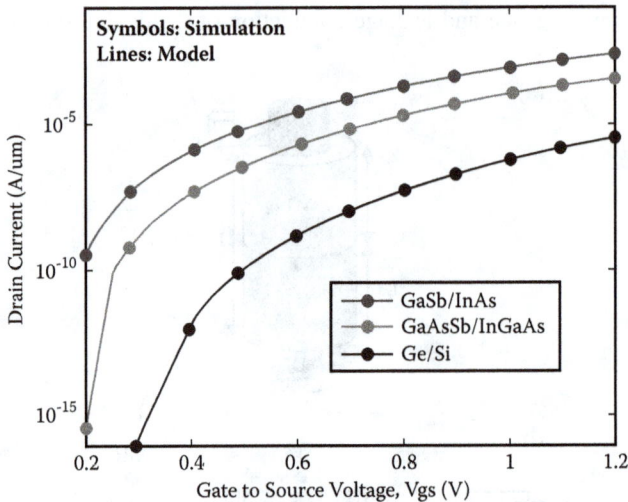

FIGURE 6.31 I_{DS} versus V_{GS} of the DM Elliptical GAA heterojunction TFET for different heterojunction material combinations [19].

Figure 6.31 represents the qualitative comparison of the drain current characteristics for different material combinations to form the heterojunction.

6.6.2 TRIPLE MATERIAL (TM) GE/SI GATE ALL AROUND(GAA) HETEROJUNCTION TFET

Figure 6.32(a) and (b) represent the 3D and 2D view of the TM Ge/Si GAA Heterojunction TFET. Germanium is used across the source region and silicon

(a)

(b)

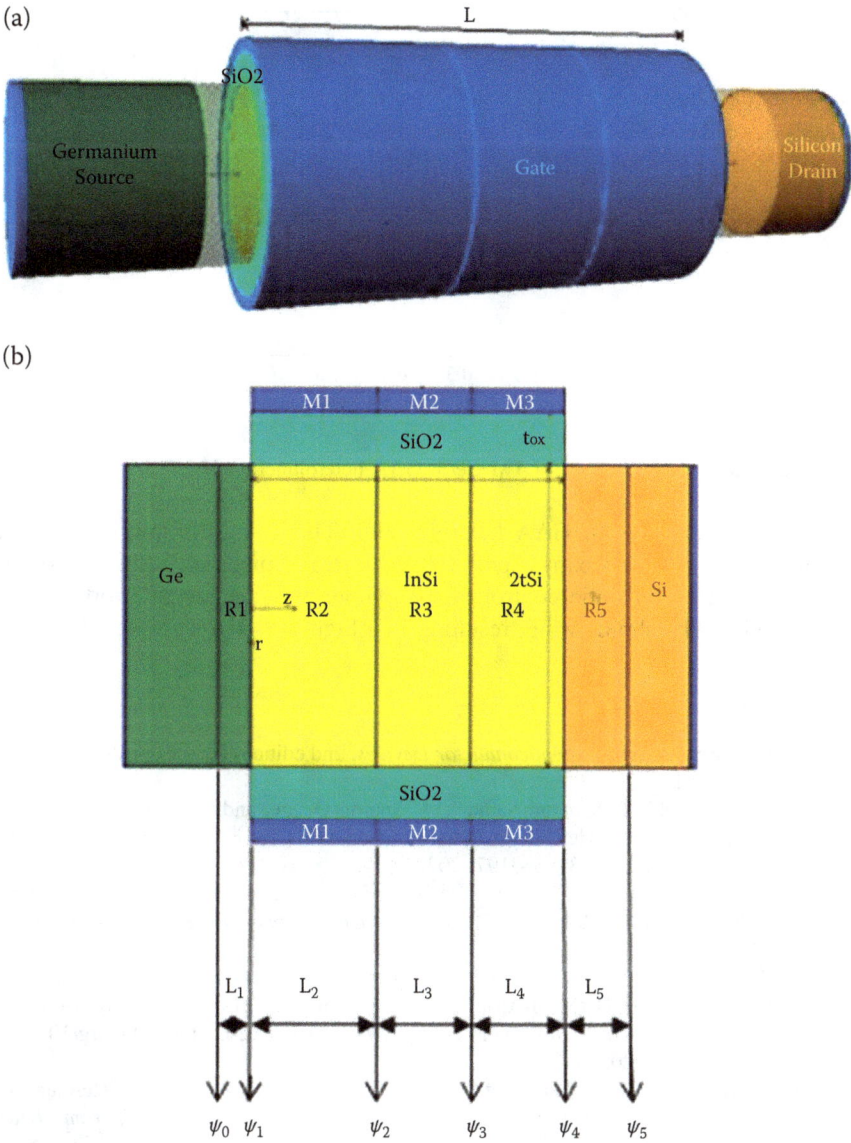

FIGURE 6.32 (a) Three-dimensional view of TM Ge/Si GAA heterojunction TFET. (b) Two-dimensional cross-sectional schematic view of TM Ge/Si GAA heterojunction TFET [20].

at the channel and drain region, leading to the heterojunction formation across the source-channel region [20]. Three materials with dissimilar work functions for the gate terminal. Figure 6.33 depicts the drain current variation with respect to gate-to-source voltage. It presents the comparison plot of the single material

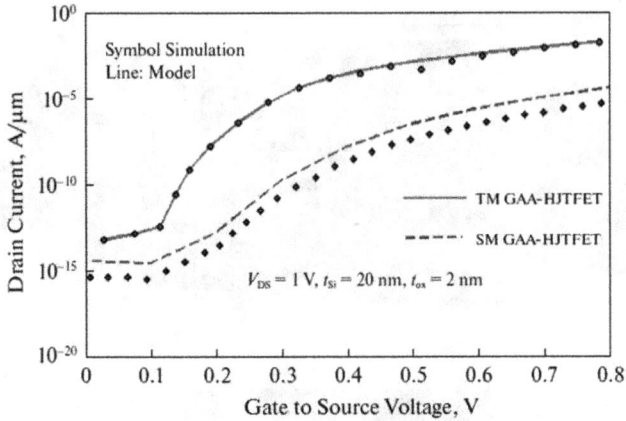

FIGURE 6.33 I_{DS} versus V_{GS} of TM Ge/Si GAA heterojunction TFET.

and triple material Ge/Si GAA heterojunction TFET. I_{OFF} is low because the forbidden gap creates a barrier prohibiting electron tunneling in the OFF-state. The increased electrostatic control of the cylindrical gate lowers short channel effects and subthreshold swing, resulting in a high ON-state current [21,22].

REFERENCES

[1] S. M. Sze, *Physics of Semiconductor Devices*, 2nd edition. New York: John Wiley & Sons, 1981.

[2] J. E. Jeyanthi, T. S. Arun Samuel, A. Sharon Geege, and P. Vimala, "A Detailed Roadmap from Single Gate to Heterojunction TFET for Next Generation Devices," *Silicon*, vol. 14, pp. 3185–3197, 2022.

[3] S. Datta, "Interband Tunnel Transistor: Oppurtunities and Challenges," *Thesis University Lecture:Electrical Engineering, University of Notre Dame, Notre Dame*, 2015.

[4] C. Usha, & P. Vimala, "Analytical Drain Current Modeling and Simulation of Triple Material Gate-All-Around Heterojunction TFETs Considering Depletion Regions," *Semiconductors*, vol. 54, pp. 1634–1640, 2020. https://doi.org/10.1134/S1063782620120398

[5] G. Kim, J. Lee, J. H. Kim, and S. Ki, "High On-Current Ge-Channel Heterojunction Tunnel Field-Effect Transistor Using Direct Band-to-Band Tunneling," *Micromachines*, vol. 10, p. 77, 2019.

[6] C. Convertinoa, C. Zotaa, H. Schmida, A. M. Ionescub, and K. Moselund, "III-V Heterostructure Tunnel Field-Effect Transistor," *Journal of Physics: Condensed*, vol. 30, p. 264005, 2018.

[7] J. Madan, S. Shekhar, and R. Chaujar, "Source Material Assessment of Heterojunction DG-TFET for Improved Analog Performance," International conference on Microelectronic Devices, Circuits and Systems (ICMDCS), 1–5, 2017.

[8] L. Canham, "Silicon quantum wire array fabricaiton by electrochemical dissolution of wafers," *Appl. Phys. Lett*, vol. 57, pp. 1046–1048, 1990.

[9] Y. Li, and W. H. Chen. "Numerical simulation of electrical characteristics in nanoscale si/gaas mosfets," *Journal of Computational Electronics*, vol. 5, pp. 255–258, 2006, 2006.

[10] Tesi di Laurea Magistrale, 2018/2019. "Evaluation of TFETs perfromances for low power applications," *Thesis, Anno Accademico*.

[11] A. Seabaugh, S. Fathipour, W. Li, H. Lu, J. H. Park, A. C. Kummel, D. Jena, S. K. Fullerton-Shirey, and P. Fay, "Steep subthreshold swing tunnel fets: Gan/inn/ gan and transition metal dichalcogenide channels," In *2015 IEEE International Electron Devices Meeting (IEDM)*, 35–36, 2015.

[12] A. Ravindran, A. George, C. S. Praveen, and N. Kuruvilla. Gate All Around Nanowire TFET with High ON/OFF Current Ratio. *Materials Today: Proceedings*, vol. 4, pp. 10637–10642, 2017.

[13] C. Usha and P. Vimala, "An Analytical Modeling of Conical Gate-All-Around Tunnel Field Effect Transistor," *Silicon*, vol. 13, pp. 2563–2568, 2020.

[14] C. Usha, P. Vimala, T. S. Arun Samuel, and M. Karthigai Pandian. "A novel 2-D analytical model for the electrical characteristics of a gate-all-around heterojunction tunnel field-effect transistor including depletion regions," *J Comput Electron*, vol. 19, pp. 1144–1153, 2020.

[15] M. Ajay, S. Gupta, and S. Pandey Bhattacharya, "Analysis of GaSb/InAs heterojunction Gate All Around Tunnel FET (HGAATFET)," 2015 Annual IEEE India Conference (INDICON), 1–5, 2015.

[16] S. O. Koswatta, S. J. Koester, and W. Haensch, "On the Possibility of Obtaining MOSFET-Like Performance and Sub-60-mV/dec Swing in l-D Broken-Gap Tunnel Transistors," *Electron Devices*, vol. 57, pp. 3222–3230, 2010.

[17] J. H. Seo, Y. J. Yoon, and I. M. Kang, "Design Optimization of Ge/GaAs-Based Heterojunction Gate-All-Around (GAA) Arch-Shaped Tunneling Field-Effect Transistor (A-TFET)," *Journal of Nanoscience and Nanotechnology*, vol. 18, pp. 6602–6605, 2018.

[18] Y. Guan, Z. Li, W. Zhang, Y. Zhang, and F. Liang. "An Analytical Model of Gate-All-Around Heterojunction Tunneling FET," *IEEE Transactions On Electron Devices*, vol. 65, pp. 776–782, 2018.

[19] P. Saha and S. Kumar Sarkar, "Drain current modeling of proposed Dual Material Elliptical Gate-All-Around Heterojunction TFET for enhanced device performance," *Superlattices and Microstructures*, vol. 130, pp. 194–207, 2019.

[20] C. Usha and P. Vimala, "Analytical Drain Current Modeling and Simulation of Triple Material Gate-All-Around Heterojunction TFETs Considering Depletion Regions," *Semiconductors*, vol. 54, pp. 1634–1640, 2020.

[21] C. Usha and P. Vimala, "An Electrostatic Analytical Modeling of High-K stacked Gate-All-Around Heterojunction Tunnel FETs Considering the Depletion Regions," *AEUE-International Journal of Electronics and Communication*, vol. 152877, 2019.

[22] C. Usha and P. Vimala, "A new analytical approach to threshold voltage modeling of triple material gate-all-around heterojunction tunnel field effect transistor," *Indian Journal of Physics*, vol. 95, pp. 1365–1371, 2020.

7 Analog/RF Performance Analysis of TFET Device

M. Saravanan, K. Ramkumar,
Eswaran Parthasarathy, and J. Ajayan

CONTENTS

DOI: 10.1201/9781003327035-7

7.1   INTRODUCTION

Most digital application studies have found that TFETs are superior to traditional transistors when used for logic functions. Recent studies have analyzed the effectiveness of TFETs in various analog circuits. TFETs are a popular alternative to CMOS for low-power applications because they have a lot of great features, such as the ability to overcome short-channel effects (SCEs), a low sub-threshold swing (SS), a low threshold voltage (Vt), and a high current ratio (I_{ON}/I_{OFF}) [1,2].

The high potential barrier between source to channel junction is responsible for the low current flow. It is necessary to consider capacitance throughout the design process because it impacts the device's performance at higher frequencies. The gate capacitance is critical when evaluating RF performance. The switching frequency of digital circuits and the amplification capability of analog circuits are both governed by the parasitic gate capacitance. Therefore, these capacitances should be minimized, so that signal distortion is kept to a minimum [3]. Although both the gate and drain capacitances are necessary, decreasing the capacitance between them will increase the device's efficiency. Possible analog and radio frequency (RF) parameters are listed below [2–4].

Analog Parameters
    a. The Transconductance (g_m)
    b. The transconductance generation factor (TGF)
    c. ON-current (I_{ON})
    d. OFF-current (I_{OFF})
    e. Surface potential
    f. Electric field
    g. Tunnel path

RF Parameters
    a. Cut-off frequency (f_T)
    b. Gate-to-drain capacitance (C_{GD})
    c. Gate-to-source capacitance (C_{GS})
    d. Gain Band-width Product (GBW)
    e. Frequency of oscillation (f_{max})
    f. Intrinsic gate delay (τ_{int})
    g. Dynamic power dissipation (P_{dyn})
    h. Transit time (τ)

7.2 ANALOG PARAMETERS

7.2.1 THE TRANSCONDUCTANCE (g_m)

An essential measure of a transistor's performance is its Transconductance, which indicates how efficiently it transforms voltage into current. Amplification gain measurement is its most common use. which is mathematically denoted as;

$$g_m = \frac{\partial I_D}{\partial V_{GS}} \tag{7.1}$$

A device with a higher transconductance creates a channel with better transfer efficiency, making it perfect for use in analog applications.

7.2.2 THE TRANSCONDUCTANCE GENERATION FACTOR (TGF)

The TGF indicates the degree to which the I_D successfully achieved the intended g_m magnitude. When constructing analog circuits, especially those tuned for low power consumption, it is assumed that a larger magnitude of TGF will be necessary.

$$TGF = \frac{g_m}{I_D} \tag{7.2}$$

Although gate-engineered devices have a marginally lower TGF, their low power consumption in the subthreshold region may compensate for this. Higher levels of TGF are associated with weaker linearity.

7.2.3 DRAIN CURRENT

The drain current of the gadget determines its ability to pass current. With a higher drain current, the Transconductance of the device is increased. The more g_m, the better amplification and finer for analog purposes.

$$I_D \propto \exp\left[-\frac{4\sqrt{2m^*}\,E_g^{*\frac{3}{2}}}{3|e|\hbar(E_g^* + \Delta\Phi)}\sqrt{\frac{\varepsilon_{Si}}{\varepsilon_{ox}}t_{ox}t_{Si}} \right]\Delta\Phi \tag{7.3}$$

$\Delta\Phi$ - Tunneling window
m^* - Effective mass
q - Charge of an electron
$\hbar$ - Reduced Plank's constant,
E_g^* - Effective bandgap
λ - Tunneling width
t_{si}, t_{ox}, ε_{si}, and ε_{ox} - The thickness and permittivity of silicon and oxide

7.2.4 SURFACE POTENTIAL

Fermi-level pinning makes it impossible to manage the surface potential of a MOSFET created by a high density of interface states. A popular approach to creating a TFET model involves determining the device's surface potential and then calculating the drain current using a suitable BTBT model. Surface potential grows linearly with V_{GS} at low gate voltages. However, the surface potential saturates and becomes gate bias independent when the gate voltage is significantly increased. This is because the inversion charge in the channel prevents the surface potential from being bent in this bias regime, which is analogous to that of a MOSFET driven into strong inversion [1]. It is known that the way the bands bend at the tunnel junction has a significant effect on the BTBT tunneling current. For this, it is even more important for TFETs to have an accurate model of the surface potential [2].

7.2.5 ELECTRIC FIELD

Every Tunnel FET will go through band-to-band tunneling (BTBT). As a non-local dynamic BTB tunneling model keeps track of the electric field at each location in the tunneling zone, it is superior to local BTB tunneling. This improves the accuracy of the simulation findings [5]. Smaller channel thickness results in a steeper electric field profile at the source side and a flatter profile at the drain. This makes sense if the ON current is large and the OFF current is low.

The energy-band diagrams indicate higher band overlap at the source channel interface when the switch is turned on. More overlap occurs between the energy bands in the OFF state due to the greater channel thickness on the source side. As a result, the OFF currents are larger. A larger electric field at the tunneling junction makes tunneling more likely when the gate voltage is held constant [6]. Drain bias increases the electric field strength at the drain-to-channel interface. This results in increased ambipolar conduction.

7.2.6 TUNNEL PATH

When the channel is narrow at the source end, the electric field at the point of intersection between the source and the channel is amplified. This increases the speed of the tunneling current. There is less ambipolarity and I_{OFF} towards the drain because the channel narrows. When the mid-channel is raised, the ON-current is higher, and the OFF-current is lower [2,5]. As V_{GS} is increased above the onset voltage (V_{onset}), the tunneling current increases because the tunneling probability of electrons is exponentially dependent on the tunneling distance. The transistor is in the ON state when V_{GS} is equal to V_{DD}, and I_{DS} at this stage is I_{ON}. The BTBT loses efficiency for tiny V_{DS} if the tunnel path becomes larger as a result of a greater effective oxide thickness (EOT) [6].

7.3   RF PARAMETERS

7.3.1   CUT-OFF FREQUENCY (f_T)

To evaluate the RF performance of TFET devices, it's critical to know the cut-off frequency (f_T). The frequency is fT, and the current gain reaches its peak at unity. Increases in gate voltage are most closely associated with changes in gm and intrinsic capacitances. High gm and low inherent capacitance are dominated by high RF applications to achieve a superior f_T, and its mathematical definition is as follows:

$$f_T = \frac{g_m}{2\pi\,(C_{GS} + C_{GD})} \tag{7.4}$$

Its strong Transconductance and low parasitic capacitance are primarily responsible for the TFET device's remarkable increase in f_T. Numerous RF applications benefit from the high f_T for high-speed operation.

7.3.2   GATE-TO-DRAIN CAPACITANCE (C_{GD})

The C_{GD} refers to the gate-drain region's parasitic effect. When compared to MOSFETs, TFETs have a smaller potential drop at the channel-drain junction, which raises the C_{GD} value under various bias settings. The potential difference between the channel and the drain decreases as the gate bias increases. As a result, C_{GD}, the primary component of capacitance in TFETs, becomes much larger. For analog/RF components like GBP and cut-off frequency, less C_{GD} is better [7].

7.3.3   GATE-TO-SOURCE CAPACITANCE (C_{GS})

C_{GD} refers to the gate-drain region's parasitic effect. When compared to MOSFETs, TFETs have a smaller potential drop at the channel-drain junction, which raises the C_{GD} value under various bias settings. The potential difference between the channel and the drain decreases as the gate bias increases. As a result, C_{GD}, the primary component of capacitance in TFETs, becomes much larger. For analog/RF components like GBP and cut-off frequency, less C_{GD} is better. And for C_{GS}, two parasitic capacitances are crucial for adjusting RF performance [8]. An effective connection between the channel and the source in MOSFETs results in high Cgs. In TFETs, tunneling is less likely whenever the device is turned off due to the high gate voltage at the source-channel junction. However, when the device is ON, the depletion region shrinks, which makes tunneling more likely. This is one of the factors that contribute to TFETs having a lower Cgs value than MOSFETs. Higher Cgs values mean that the gate over the channel is better controlled, which makes the RF FOMs better.

7.3.4 Gain Band-Width Product (GBW)

For high-frequency transient analysis, the gain bandwidth product (GBW) is taken into account. The high GBW device is excellent for low-bias and high-speed applications. The high GBW is greatly desired for circuit applications that need fast speed and little input bias current, like RF amplifiers [7].

$$GBP = \frac{g_m}{10 \times 2\pi C_{GD}} \tag{7.5}$$

7.3.5 Frequency of Oscillation (f_{max})

The frequency at which power gain equals one is known as the f_{max}.

$$f_{max} = \frac{g_m}{2\pi C_{gs}\sqrt{4\left(g_{ds} + g_m\frac{C_{gd}}{C_{gs}}\right)(R_s + R_{ch} + R_g)}} \tag{7.6}$$

$$f_{max} = \frac{f_T}{\sqrt{8\pi\ C_{gd}\ R_{gd}}} \tag{7.7}$$

C_{gs} and C_{gd} – Gate-to-source and Gate-to-drain capacitance
g_{ds} – Output conductance,
R_s, R_{ch}, and R_g – Source, channel, and gate resistance, respectively

7.3.6 Intrinsic Gate Delay (τ_{int})

When measuring a device's speed, the intrinsic gate delay is a good indicator. Parasitic gate capacitance (C_{gg}) and drain current (I_D) play crucial roles. The drain current variation affects the intrinsic gate delay the most. It is stated mathematically as follows;

$$\tau_{int} = \frac{C_{gg} \times V_{DD}}{I_D} \tag{7.8}$$

Dynamic Power Dissipation (Pdyn):

$$P_{dyn} = C_{gg}\ V_{DD}^2 f \tag{7.9}$$

For which f represents the operating frequency. It is essential to keep the device's dynamic power dissipation as low as possible so that the temperature doesn't rise too much and affects the device's performance, whether it is "ON" or "OFF". The equivalent oxide thickness of the gate capacitance per unit area

$$C_{gg} = \frac{\epsilon_{SiO_2}}{EOT} \qquad (7.10)$$

The total gate capacitance is actually determined by adding the gate-to-channel capacitance and the gate-to-source/drain capacitance (fringing capacitance).

7.3.7 TRANSIT TIME (τ)

Transit time, which is the amount of time it takes for carriers to wander from the source to the drain, is used to determine the switching speed of a device.

$$\tau = \frac{1}{2\pi f_T} \qquad (7.11)$$

7.4 INVESTIGATION OF ANALOG/RF PERFORMANCE OF GATE-ALL-AROUND (GAA) TFET

There are two kinds of GAATFETs: nanowire (NW) TFETs with channels of the same width and thickness and nanosheet (NS) TFETs with channels wide in width. To reduce the impact of short channel effects (SCEs), gate-all-around (GAA) architectures have been implemented. Nanowire-sized GAA devices with an underlap structure are expected to exhibit better tunneling behavior due to their short screening length, volume inversion, and enhanced gate tunability over the channel. The BTBT is maximized in a GAA TFET so because gate wraps around all sides of the channel due to the volume inversion over the source-channel junction [6]. The most significant element that contributed to the increased leakage current in the OFF state at low gate biases was discovered to be lateral tunneling at the drain-channel junction (I_{OFF}).

When using only small-bandgap heterojunctions, it's hard to meet the performance needs of low-power applications. Because lowering the BTBT causes SCEs and makes devices work less well, scaling devices becomes more difficult. In this way, it is essential for low-power modules used in Internet of Things (IoT) applications to find an optimized 3D structure that can improve driving current and analog/RF properties without increasing I_{OFF}. For developing transistors, nanowires made of silicon, germanium, and III-V materials are very appealing [7–9].

Figure 7.1 shows Hetero Junction/Dielectric GAA-TFET (HJ-HD-GAA-TFET) 3D-Structure (left) and Schematic cross-section. Germanium is for source material, and the channel and drain material is silicon. The channel length is 20 nm, source and drain size is 30 nm each. Hereafter, it is represented as GAA Structure-1.

Figure 7.2 shows Dual Material GAA TFET (DM-GAA-TFET) 3D-Structure (left) and Schematic cross-section. InAs is for source material, and channel and drain material are GaAs. The channel length is 20nm, source and drain size are 30nm each. Hereafter, it is represented as GAA Structure-2.

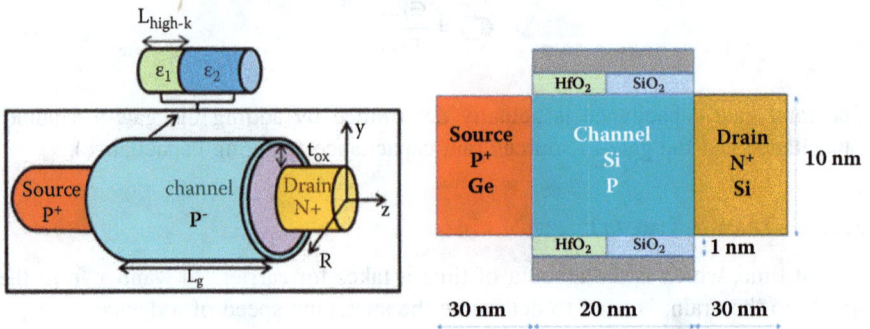

FIGURE 7.1 Heterojunction/dielectric gate all around tunnel field-effect transistors 3D-structure (left) and schematic cross-section.

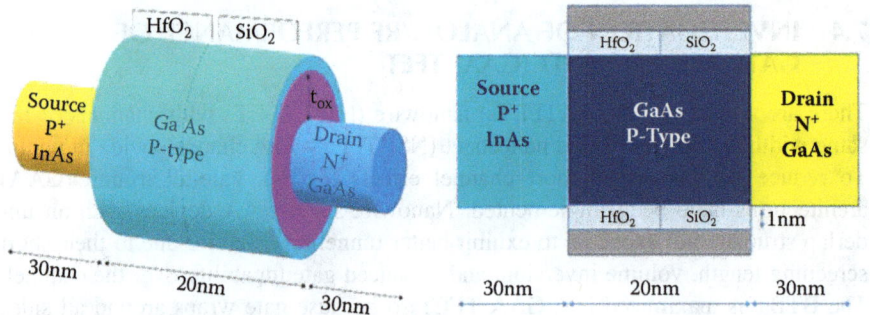

FIGURE 7.2 Dual material GAA-TFET; 3D-structure (left) and schematic cross-section.

7.4.1 Drain Current (I_D)

Thickening the oxide layer (ti) to 1 nm is done for both GAA Structure-1 and GAA Structure-2. The source and drain regions use hafnium oxide (HfO_2) and silicon dioxide (SiO_2), respectively. The following table shows the dopant concentration in each region: Source volume is $1e^{20}$ cm^3 (P$^+$ type), while channel material volume is $1e^{16}$ cm^3 (P-type). Its volume is $5e^{18}$ cm^3 (type N) at its diameter. Using Silvaco ATLAS 2-D simulation, the device was simulated, calibrated, and compared to standard TFET devices.

For GAA Structure-1, drain current characteristics were analyzed for two different dielectric material combinations. $Al_2O_3 + SiO_2$ is one combination, and $HfO_2 + SiO_2$ is another. For GAA Structure-2, dielectric material is $HfO_2 + SiO_2$. The V_{GS} value at which a device's drive current starts to rise is its V_{OFF} value. During this time, the voltage at the gate makes the current growth in a very straight line. When V_{GS} happens in the super-threshold zone, the drive current is much slower than in the threshold region. Since there aren't many charge carriers in this area, making the tunneling window bigger doesn't change the driving current much. Figure 7.3 shows the device I_D and V_{GS} for a simulated device ($V_{DS} = 0.7$ volts).

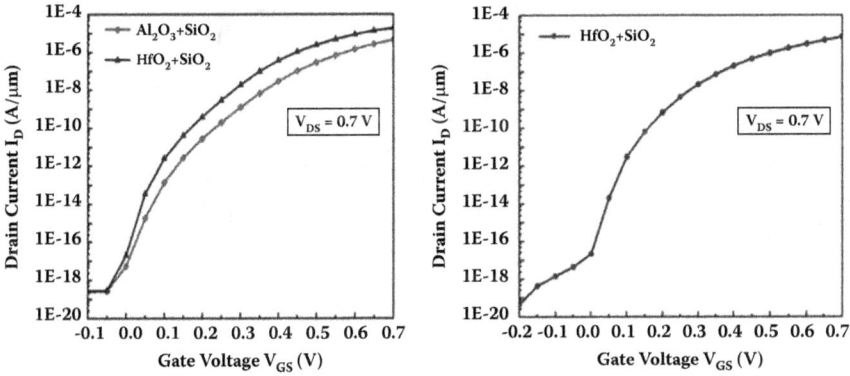

FIGURE 7.3 Drain current (I_D) vs V_{GS} characteristics of GAA structure-1(left) and GAA structure-2.

For GAA Structure-1 drain current is 4.64×10^{-6} A/μm (Al_2O_3 + SiO_2) and 1.92×10^{-5} A/μm (HfO_2 + SiO_2). For GAA Structure-2 drain current value is 7.39×10^{-6} A/μm. Because of the drop in channel resistance and rise in V_{GS}, the lowest channel conduction band became lower than the maximum source valence band, increasing the accumulation of carriers. When the drain voltage is low, exponential growth begins. A higher I_{ON} can be achieved with InAs as the source of GAA Structure-2 since it has a lesser bandgap, low electron effective mass, and high mobility than the germanium of GAA Structure-1 [7–9].

The OFF-current (I_{OFF}) is 8.15×10^{-16} A/μm and 3.28×10^{-18} A/μm for Al_2O_3 + SiO2 and HfO_2 + SiO_2, respectively. The I_{OFF} value for GAA Structure-2 is 2.27×10^{-17} A/μm. The leakage current (I_{OFF}) is a little high in GAA Structure-2 compared to GAA Structure-1.

7.4.2 ELECTRIC FIELD PROFILE

Maximizing the electric field is another thing to think about regarding device performance. This happens when the dielectric material has a high dielectric constant. Figure 7.4 shows the electric field profile of GAA TFETs.

7.4.3 TRANSCONDUCTANCE (GM)

Transconductance (g_m) is an essential part of developing analog circuits such as operational amplifiers (OP-AMPs) require careful consideration of the DC gain, offset, frequency, bandwidth, and noise performance. To get the device's g_m divide the change in I_D by the change in V_{GS} over time. The device's cut-off frequency goes up when it is more efficient at turning input voltage into an output current. Figure 7.5 shows the Transconductance (g_m) characteristics of GAA TFETs.

According to, the g_m component is an essential part of analog circuits. The g_m value for GAA1 is 0.038 mS/μm (HfO_2 + SiO_2) and 0.17 mS/μm (Al_2O_3 + SiO_2), for GAA2 is 0.48 mS/μm. The enhanced performance of InAs/GaSb GAA

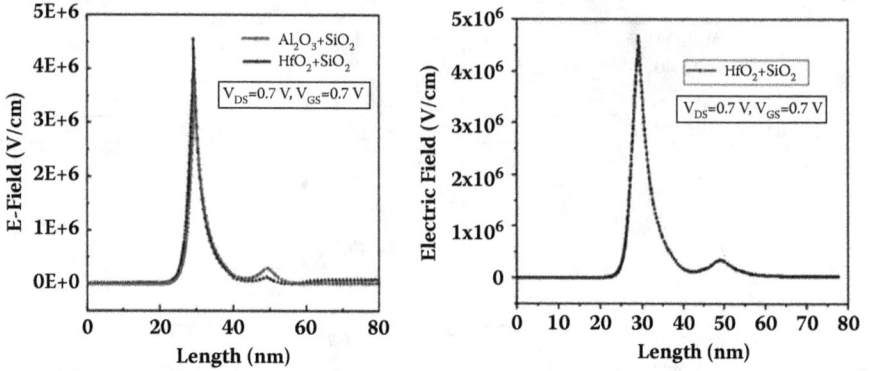

FIGURE 7.4   Electric field profile of GAA structure-1(left) and GAA structure-2.

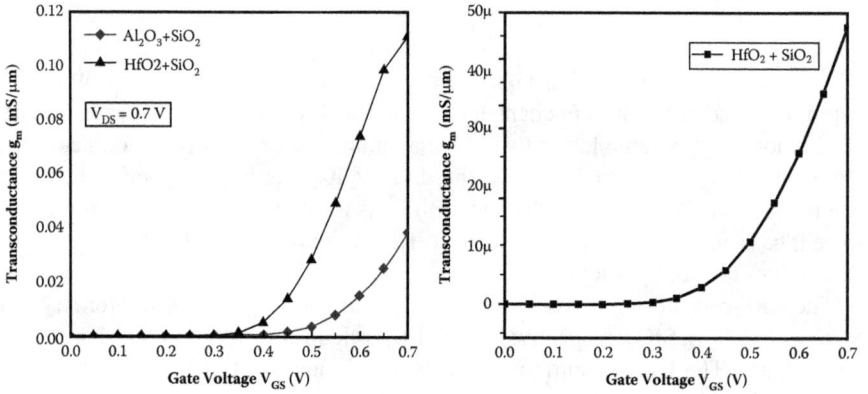

FIGURE 7.5   Transconductance (g_m) vs V_{GS} characteristics of GAA structure-1(left) and GAA structure-2.

Structure-2 is attributed to the material's reduced bandgap, greater conductivity, and greater mobility. After reaching its maximum value, g_m drops owing to mobility degradation when the gate bias is increased.

The C_{GS} and C_{GD}, often known as parasitic capacitances, are additional characteristics used to assess analog/RF performance [9]. This has a significant effect on the operating frequency and controllability of the gate. Figure 7.6 depicts the Gate-Source Capacitance (C_{GS}) characteristics.

7.4.4   GATE SOURCE CAPACITANCE (C_{GS}) AND GATE DRAIN CAPACITANCE (C_{GD})

The C_{GS} value for GAA Structure-1 is 7.08aF (Al_2O_3 + SiO_2) and 7.83aF (HfO_2 + SiO_2). For GAA Structure-2 C_{GS} value is 6.89aF. (1 farad [F] = $1E^{+18}$ *attofarad* [aF]). Most digital application studies shifted away from traditional transistors and

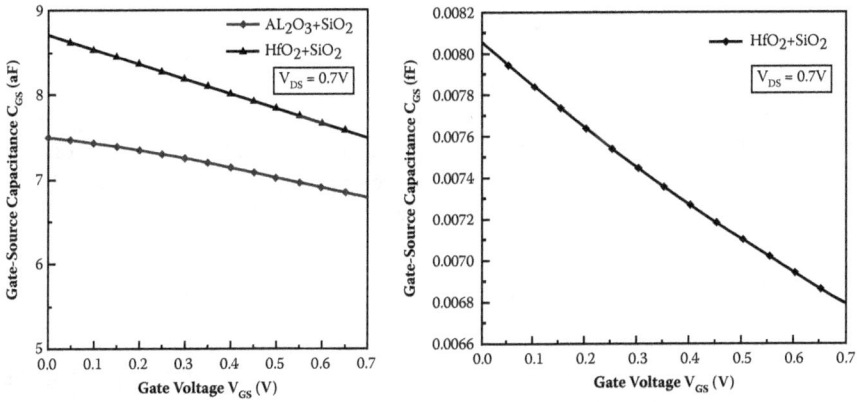

FIGURE 7.6 Gate-source capacitance (CGS) vs VGS characteristics of GAA structure-1(left) and GAA structure-2.

toward TFETs. Many academics have recently looked into the usage of TFETs in analog circuits, and the results are encouraging. Reduced-power analog circuits may benefit TFETs' low subthreshold swing and good saturation features. Capacitance must be taken into account when designing a device that functions at high frequencies. The quantity of capacitance required to produce a decent signal decreases as frequency increases, resulting in distortion. As a result, it is necessary to minimize these capacitances to reduce signal distortion. The capacitance value between the gate and drain, Cgd, must be maintained. However, it can be reduced to improve device efficiency. Figures 7.6 and 7.7 shows how V_{GS} affects C_{GS} and C_{GD}. It was found that possible barriers in drain and source channels went down and up as V_{GS} grew. So, when the V_{GS} level changes, the C_{GD} level, and the C_{GS} level also change [10].

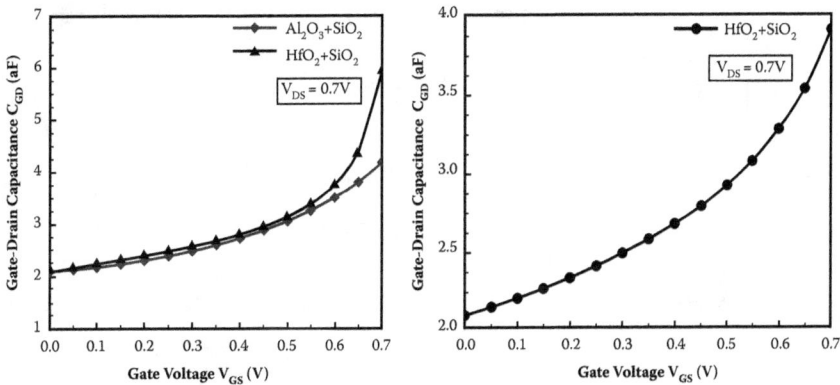

FIGURE 7.7 Gate-drain capacitance (CGD) vs VGS characteristics of GAA structure-1(left) and GAA structure-2.

The C_{GD} increases dramatically when the V_{GS} is increased, as discovered by researchers. With increasing V_{GS}, the barrier at the intersection of the drain decreases, and the barrier at the channel increases. The C_{GD} value for GAA1 is 4.28aF ($Al_2O_3 + SiO_2$) and 6.19aF ($HfO_2 + SiO_2$). For GAA Structure-2 C_{GD} value is 3.84aF. Values of capacitance C_{GD} and C_{GS} are used with g_m to figure out the TFET device's f_T, that's an important RF figure of merit (FOM). Both C_{GS} and C_{GD} are the miller capacitances of the TFET device.

7.4.5 CUT-OFF FREQUENCY (FT)

The cut-off frequency is the most critical factor in determining how quickly an RF circuit operates; the fmax and f_T are the two most essential elements in determining how fast RF circuits can run [10,11]. To put it another way, both C_{GS} and C_{GD} impact the overall performance of the TFET. As shown in Figure 7.8, a relationship between the cut-off frequency and the gate-source voltage.

The f_T value for GAA Structure-1 is 0.56 THz ($Al_2O_3 + SiO_2$) and 1.32 THz ($HfO_2 + SiO_2$). For GAA Structure-2, f_T value is 0.743 THz.

7.4.6 GAIN-BANDWIDTH PRODUCT (GBP)

The Gain-Bandwidth Product (GBP) characteristics as shown in Figure 7.9. The GBP value for GAA Structure-1 is 156 GHz ($Al_2O_3 + SiO_2$) and 362 GHz ($HfO_2 + SiO_2$). For GAA Structure-2 GBP value is 192 GHz.

7.4.7 TRANSIT TIME (τ)

Transit time is the duration of movement from source to drain, known as device switching speed. Transit time is another significant statistic for measuring RF characteristics [10–12]. This field is used to tell us how quickly the gadget is. If the

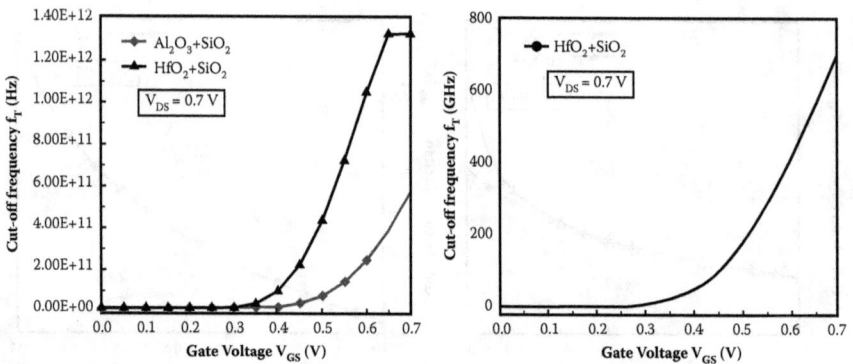

FIGURE 7.8 Cut-off frequency (fT) vs VGS characteristics of GAA structure-1(left) and GAA structure-2.

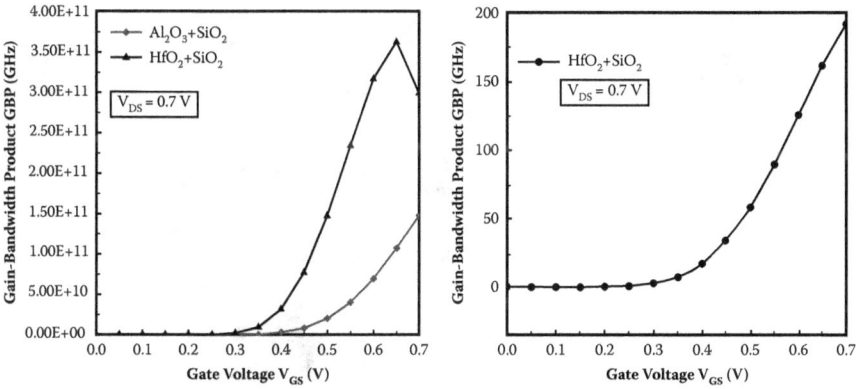

FIGURE 7.9 Gain-bandwidth product (GBP) vs VGS characteristics of GAA structure-1(left) and GAA structure-2.

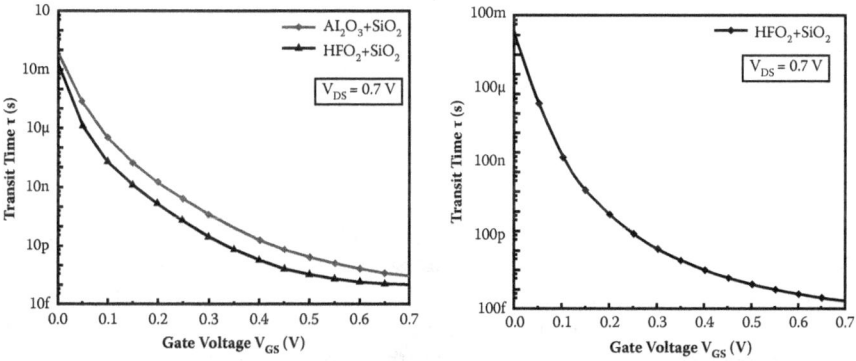

FIGURE 7.10 Transit time(τ) vs VGS characteristics of GAA structure-1(left) and GAA structure-2.

device's transit time is reduced, high-speed applications such as memory design can use it. The VGS response as a proportion of the transit time (τ) is depicted in Figure 7.10. Increased switching speeds enhance the performance of a gadget.

7.5 INVESTIGATION OF DIELECTRIC POCKETS (DP) INFLUENCE IN TFET

This section analyses the role that dielectric pockets (DP) play in improving the analog and radio frequency (RF) performances of vertical tunnel field effect transistors.

Figure 7.11: shows the Ge-Si vertical TFET without DP (left) and with DP. Throughout the chapter, it is denoted as Model-1. The device has a total length of 65 nm and 30 nm thickness, whereas 10 nm is a channel thickness. The source pocket size is roughly 5 nm, with a drain size is 20 nm length, and the source length

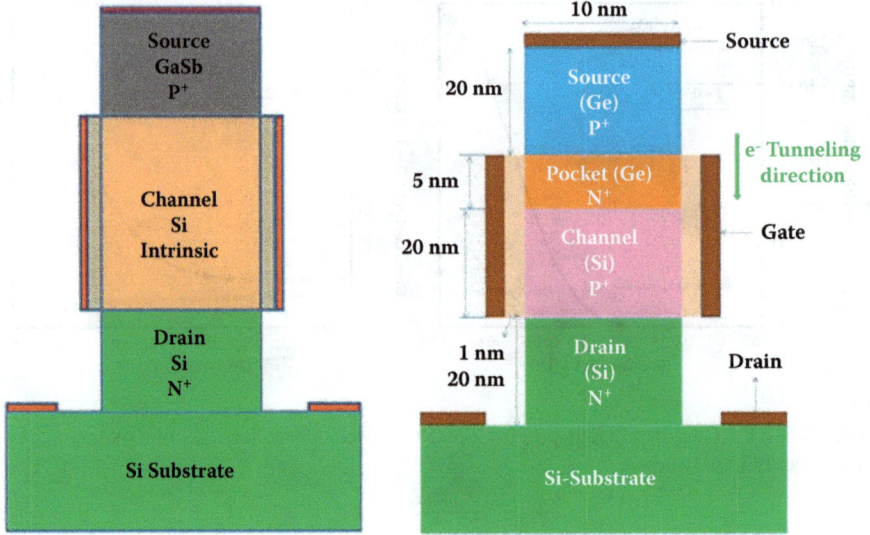

FIGURE 7.11   Ge-Si vertical TFET without DP (left) and with DP.

Source: K. Ramkumar and V. N. Ramakrishnan, 2022.

is 20 nm. The length of the source pocket is carefully selected to achieve an oxide thickness of 1 nm [4].

Figure 7.12 shows SOI-TFET (a) without DP (b) with DP. Throughout the chapter, it is denoted as Model-2. In contrast to conventional SOI-TFETs, a DP of length (LDP) is 30 nm, and a thickness (TDP) of 5 nm is integrated into the drain area at the interface of the drain and channel sections. It has been demonstrated that dielectric pockets in drain areas enlarge the depletion widths at the interfaces between drain and channel regions, hence modifying the energy band topologies at these interface locations. Due to the wider depletion zone, the tunneling width at the interface is expanded, reducing ambipolar conduction [10].

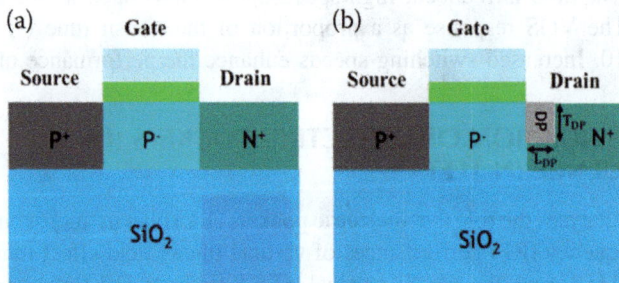

FIGURE 7.12   SOI-TFET (n-channel) (a) without dielectric pocket (b) with dielectric pocket.

Source: C. K. Pandey, D. Dash, and S. Chaudhury, 2020.

7.5.1 DRAIN CURRENT (I_D)

TFETs parameters are measured in the off, sub-threshold, and super-threshold regions. When $V_{GS} = 0$ V, carriers cannot tunnel from the source since there are no energy levels. But leakage current (Unit of Femtoamps) is caused by minimal charge transfer between the drain and source due to thermionic emission. At $V_{GS} = 0$ V, the device enters the sub-threshold region, beyond which the drive current begins to rise. The current increases exponentially when V_{GS} is increased [11–14]. The device enters the super-threshold region at 1×10^{-7} A/µm (V_{th}), where stability is maintained.

Figure 7.13 and Table 7.1 show drain current variation in model-1 due to different dielectric materials. The suggested device's drive current improves when HfO_2 is utilized as a gate dielectric. The maximum I_{ON} and I_{OFF} values are 5.55×10^{-5} A/µm, 2.12×10^{-17} A/µm, respectively, and I_{ON}/I_{OFF} ratio is 2.61×10^{12}.

Figure 7.14 illustrates drain current values by varying LDP (left) and varying TDP. The DP's length and the dielectric pocket's thickness varied from 0 to 40 nm

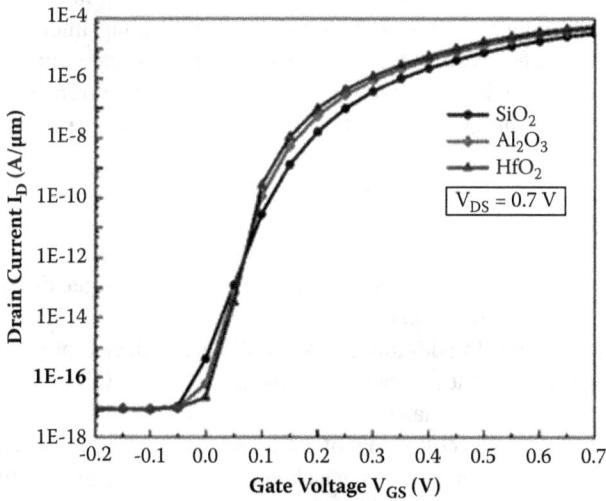

FIGURE 7.13 Drain current (I_D) vs V_{GS} characteristics of model-1.

Source: K. Ramkumar and V. N. Ramakrishnan, 2022.

TABLE 7.1

I_{ON} and I_{OFF} Variation Due to Different Dielectric Materials

OFF-Current I_{OFF} (A/µm)			ON-Current I_{ON} (A/µm)		
SiO_2	Al_2O_3	HfO_2	SiO_2	Al_2O_3	HfO_2
1.11×10^{-17}	6.01×10^{-17}	2.12×10^{-17}	3.37×10^{-5}	4.74×10^{-5}	5.55×10^{-5}

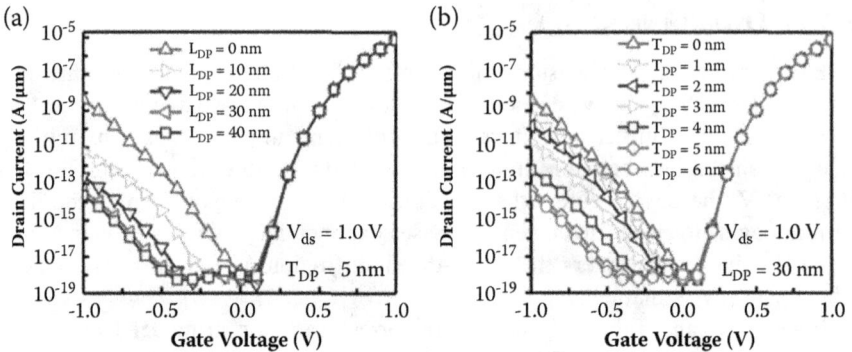

FIGURE 7.14 Drain current (I_D) vs V_{GS} characteristics of model-2; varying length of DP (left) and varying thickness of DP.

Source: C. K. Pandey, D. Dash, and S. Chaudhury, 2020.

and 0 to 6 nm, respectively. DP does not impact the tunneling interface between the source and the channel. Adding DP to the output tunneling interface of an SOI-TFET has no expected performance improvement in ON-state current but got the best results in the ambipolar effect. The final ambipolar current value is 4.15×10^{-14} A/µm, but the initial value is 4.16×10^{-9} A/µm; by altering the length and thickness of the dielectric pocket, this can be achieved.

7.5.2 ELECTRIC FIELD

Figure 7.15 shows the model-1 electric field profile for various gate dielectrics. Due to the depletion of the source pocket, the junction of the source channel records the maximum electric field. In addition, a high-κ dielectric device provides the largest electric field, which facilitates carrier tunneling and raises the ON current [15–17]. As a result, of all the gate oxide materials, HfO_2 has the highest driving current. Positive trap charges enhance the collection of electrons at the silicon substrate. At the same time, the accumulation of holes in a channel is facilitated by negative interfacial charges (ITC). As a result, the I_{ON} either decreases or increases depending on whether the ITC is positive or negative.

Model 2 Electric Field Profile is shown in Figure 7.16. Adding high-κ DP to the drain area of SOI-TFETs has been found to reduce the ambipolar current. To reduce ambipolarity, hafnium oxide (high κ-value dielectric material) is preferable to silicon dioxide (low κ-value dielectric material). Through the drain-to-channel interface of TFETs, the charge carriers must tunnel. Because the depletion width beneath the dielectric pocket rises, the tunneling width at the drain-channel interface also increases, and ambipolarity is reduced [17–19].

Figure 7.16 shows that the space charge width of model-2 is greater than that of the other two devices. When the applied negative gate voltage is higher than the drain voltage, the tunneling width at the output tunneling interface increases, slowing the speed at which charge carriers tunnel and pass through the drain-channel interface.

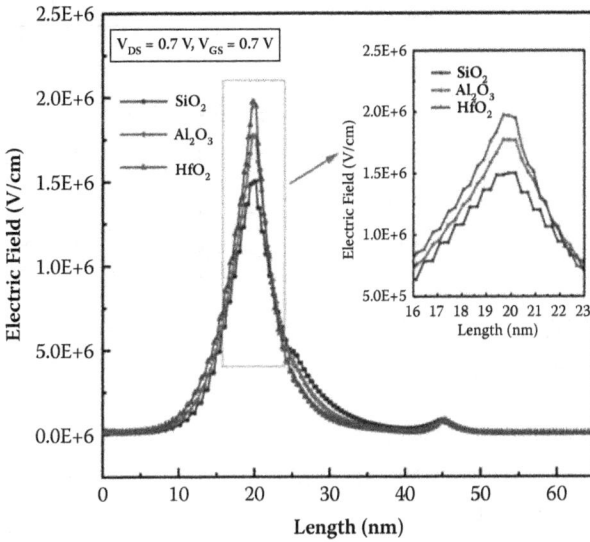

FIGURE 7.15   Electric field profile of model-1.

Source: K. Ramkumar and V. N. Ramakrishnan, 2022.

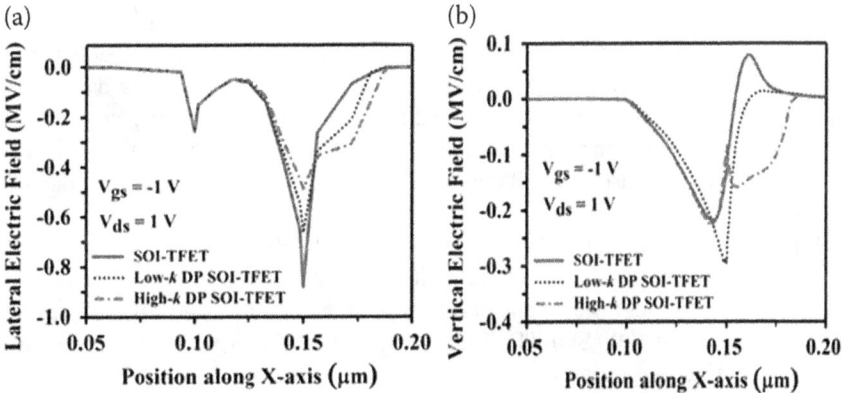

FIGURE 7.16   Electric field profile of model-2; lateral electric field (left) and vertical electric field.

Source: C. K. Pandey, D. Dash, and S. Chaudhury, 2020.

With increasing DP concentration, the tunneling interface between the channel and drain widens, reaching its maximum in the high-k DP SOI-TFET. The high-k DP SOI-TFET exhibits the highest quantity of electron charge carriers floating away from the surface into the drain area based on the maximum vertical field value as well as its polarity [10]. The expanded depletion width in the drain region of high-k DP SOI-TFETs results in the lowest level of ambipolarity possible.

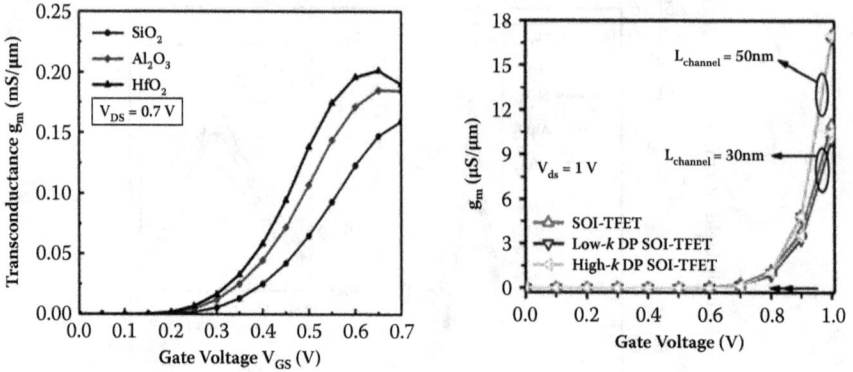

FIGURE 7.17 Transconductance (g_m) vs V_{GS} characteristics of model-1 (left) and model-2.

Source: K. Ramkumar and V. N. Ramakrishnan, 2022 & C. K. Pandey, D. Dash, and S. Chaudhury, 2020.

7.5.3 TRANSCONDUCTANCE (g_m)

Transconductance (g_m) is calculated by dividing the change in I_D by the change in V_{GS}. Figure 7.17 shows gm starts low and subsequently increases as V_{GS} increases because the current is simpler to drive. The g_m decreases at increasing V_{GS} because mobility declines. As g_m increases, the output current may be more efficiently converted from the input voltage, increasing the cut-off frequency in the process. The g_m value is a crucial consideration when designing analog circuits. It has been discovered that gm and the gate oxide's dielectric constant are related [19,20]. When the gate dielectric constant is greater, the tunneling barrier is lower. This increases the likelihood of tunneling, which enhances the I_{ON} and g_m.

Figure 7.17 (b) Model-2 shows how f_T rises with V_{GS} until it hits a maximum at a particular gate voltage, at which point gm rises as V_{GS} rises as more charge carriers tunnel through the input interface. Due to the low gate-to-drain capacitance in the low-κ DP SOI-TFET relative to the other two devices, the most negligible value of f_T may also be seen in this device.

7.5.4 GATE-DRAIN CAPACITANCE (C_{GD}) AND GATE-SOURCE CAPACITANCE (C_{GS})

Figures 7.18 and 7.19 shows how C_{GD} and C_{GS} respond to V_{GS}. The potential barrier just at the drain-channel junction decreases and the potential barrier just at the source-channel junction increases as the V_{GS} is increased. As a result, the C_{GD} increases, and the C_{GS} decreases as the amount of V_{GS} increases.

A Tunnel FET's channel does not have an inversion layer when V_{GS} is less than V_{OFF} (V_{OFF}-OFF state voltage), similar to MOSFETs. As a result, parasitic capacitances are the only capacitances seen in C_{GS} and C_{GD}. As opposed to MOSFETs, when V_{GS} is raised further, the inversion layer first forms at the drain

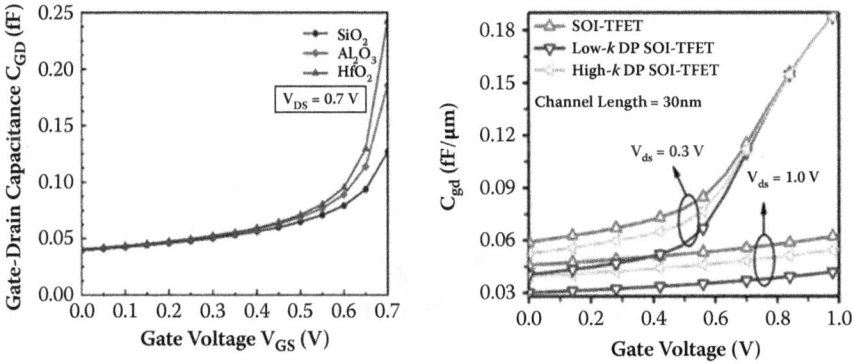

FIGURE 7.18 Gate-drain capacitance (C_{GD}) vs V_{GS} characteristics of model-1 (left) and model-2.

Source: K. Ramkumar and V. N. Ramakrishnan, 2022 & C. K. Pandey, D. Dash, and S. Chaudhury, 2020.

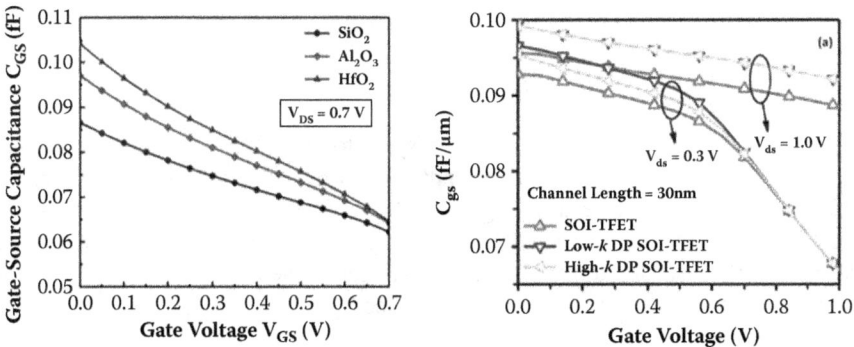

FIGURE 7.19 Gate-source capacitance (C_{GS}) vs V_{GS} characteristics of model-1 (left) and model-2.

Source: K. Ramkumar and V. N. Ramakrishnan, 2022 & C. K. Pandey, D. Dash, and S. Chaudhury, 2020.

terminal before moving toward the source terminal. Because C_{GS} is lesser than C_{GD}, changes in V_{GS} do not affect its value [20]. This is because parasitic capacitance, rather than inversion capacitance, makes up most C_{GS}. As channel length increases, the existence of a dielectric pocket (DP) in an SOI-TFET seems to have no discernible impact on the source-channel interface. The error rate rises when the channel length is shortened to 30 nm, though [10].

Because of the interplay between the fringing field created by DP as well as the source-channel interface, as shown in Figures 7.18 and 7.19(b) in Model 2, Cgs is slightly higher in SOI-TFETs with DPD. C_{GD}, which includes parasitic as well as inversion capacitances, rises when inversion occurs at the drain terminal, in contrast to C_{GS}, which rises when V_{GS} increases. DP has a greater impact on C_{GD} in SOI-TFETs with a channel length of 30 nm or less due to the dielectric gate and DP producing an opposite-polarity-induced fringing field [17–20]. This reduces the overall C_{GD} value of a drain terminal's natural fringe field. Because parasitic capacitances are dominated by inversion capacitance, DP has little effect on C_{GD}.

7.5.5 CUT-OFF FREQUENCY (f_T)

When thinking about wireless applications, different RF properties, such as cut-off frequency (f_T), gain-bandwidth product (GBP), and transit time (τ) are more important. The cut-off frequency is the short circuit's current gain when it is unity (f_T). The correlation between V_{GS} and the shift in f_T is shown in Figure 7.20. The f_T initially increases with increasing V_{GS} due to the increased g_m, reaches a maximum, and then begins to decrease as a result of the combined action of C_{GD} and gm [10]. It is undoubtedly evident that at the lower gate voltage range, the value of f_T is very low, which results from minimal charge carrier injection through the source-channel interface, which lowers the value of g_m.

7.5.6 GAIN-BANDWIDTH PRODUCT (GBP)

Figure 7.21 shows GBP variation with respect to V_{GS}. Due to a lower tunneling barrier at the source channel junction and a higher g_m than other device topologies, HfO_2-based TFETs have the highest GBP [18–20]. Since transconductance increases when the channel length is scaled down while gate-to-drain capacitance decreases, it is discovered that GBW, which is determined by the ratio of these two, is highest for the shorter channel length devices.

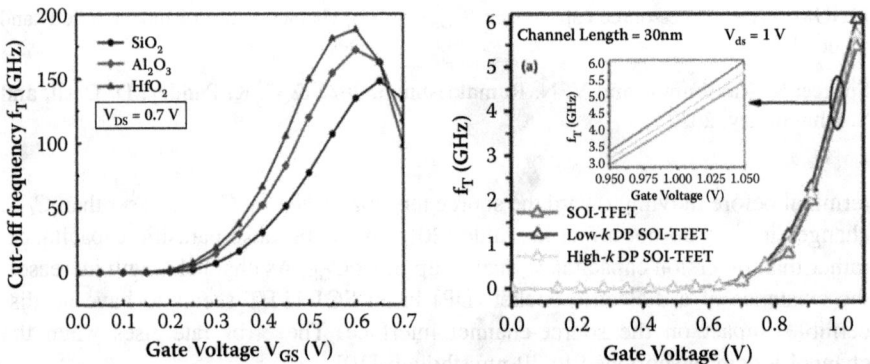

FIGURE 7.20 Cut-off frequency (f_T) vs V_{GS} characteristics of model-1 (left) and model-2.

Source: K. Ramkumar and V. N. Ramakrishnan, 2022 & C. K. Pandey, D. Dash, and S. Chaudhury, 2020.

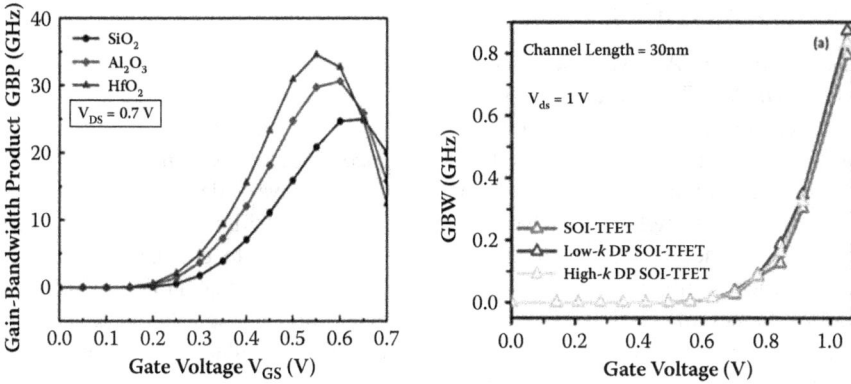

FIGURE 7.21 Gain-bandwidth product (GBP) vs V_{GS} characteristics of model-1 (left) and model-2.

Source: K. Ramkumar and V. N. Ramakrishnan, 2022 & C. K. Pandey, D. Dash, and S. Chaudhury, 2020.

7.5.7 TRANSIT TIME (τ)

The device's switching speed is based on the transit time (τ), and the time it takes for carriers to travel from the source to the drain. Figure 7.22 shows that the transit time decreases as V_{GS} increases, and it stops decreasing when the V_{GS} is high enough. This means that faster switching speeds lead to better device performance.

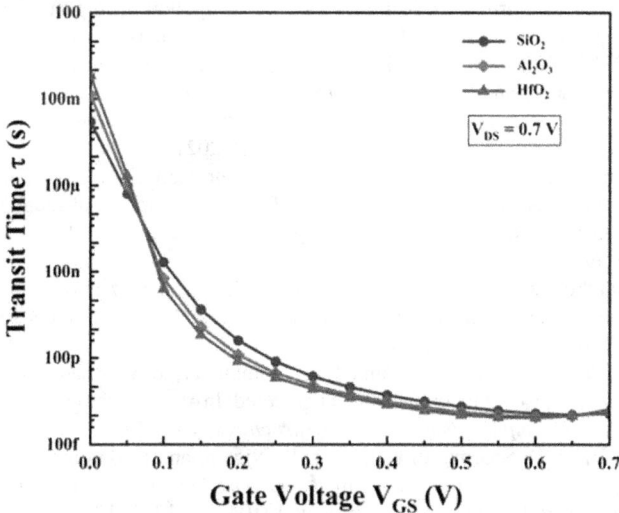

FIGURE 7.22 Transit time(τ) vs V_{GS} characteristics of model-1.

Source: K. Ramkumar and V. N. Ramakrishnan, 2022.

7.6 CONCLUSION

Silvaco-Atlas TCAD was used to determine the RF, analog, and overall perform-ance of GAA-TFETs in section 7.4. A thinner layer of gate oxide frequently leads to a greater ON current for a particular device because of a narrow potential barrier at the source-channel interface. Due to the all-encompassing dual-material gate, en-gineers can employ TFET device topology can improve a device's efficiency in more ways. The high-κ dielectric material could be positioned beneath the gate metal to enhance the tunneling of electrons from source to channel. The analog performance significantly improved as a result of the higher Transconductance. GaAs channel material has a reduced bandgap, higher mobility, and a lower effective electron mass. The results of the GAA-TFET simulation can be related to the improved RF/analog capabilities.

In section 7.5, When HfO_2 is employed as the gate dielectric, the Ge-Si Vertical TFET with dielectric pocket (DP) exhibits improvements within drive current (I_D), Transconductance (g_m), subthreshold swing (SS), as well as other RF properties. This result analysis concludes that compared to conventional and high-κ DP SOI-TFETs, adding DP to SOI-TFETs having low k-values enhances analog/RF performance. This improvement is also compatible with channel length downscaling due to the promise of better RF capabilities from low-k DP SOI-TFET. It was discovered that low-κ DP SOI-TFETs have better analog/RF performance characteristics, such as cut-off frequency, gain-bandwidth product, and gate-to-drain capacitance. When the channel size is lowered to below 30 nm, it has been seen that both the analog and RF systems have higher performance.

REFERENCES

[1] J. Ajayan, D. Nirmal, D. Kurian, et al., "Investigation of impact of gate underlap/overlap on the analog/RF performance of composite channel double gate MOSFETs," *Journal of Vacuum Science & Technology*, vol. 37, p. 062201, 2019.

[2] M. Saravanan and E. Parthasarathy, "A Review of III-V Tunnel Field Effect Transistors for Future Ultra Low Power Digital/Analog Applications," *Microelectronics Journal*, vol. 114, p. 105102, 2021.

[3] V. Darshana, N. B. Balamurugan, and T. S. Arun Samuel, "An Analytical Modeling and Simulation of Surrounding Gate TFET with an Impact of Dual Material Gate and Stacked Oxide for Low Power Applications," *Journal of Nano Research*, vol. 57, p. 68, 2019.

[4] K. Ramkumar and V. N. Ramakrishnan, "Performance Analysis of Germanium-Silicon Vertical Tunnel Field-Effect Transistors (Ge-Si-VTFETs) for Analog/RF Applications," *Silicon*, vol. 1, 2022.

[5] M. Sathish Kumar, T. A. Samuel, K. Ramkumar, I. V. Anand, and S. B. Rahi, "Performance evaluation of gate engineered InAs–Si heterojunction surrounding gate TFET," *Superlattices and Microstructures*, vol. 162, p. 107099, 2022.

[6] B. R. Raad, D. Sharma, P. Kondekar, K. Nigam, and S. Baronia, "DC and analog/RF performance optimisation of source pocket dual work function TFET," *International Journal of Electronics*, vol. 104, no. 12, 2017.

[7] M. Saravanan and E. Parthasarathy, "Investigation of RF/Analog Performance of InAs/InGaAs Channel Based Nanowire TFETS," 2021, 10.1109/ICCISc52257.2021.9484973

[8] G. Naima, S. B. Rahi, and G. Boussahla, "Impact of Dielectric Engineering on Analog/RF and Linearity Performance of Double Gate Tunnel FET," *International Journal of Nanoelectronics & Materials*, vol. 14, p. 3, 2021.

[9] Bhagwan, D. Sharma, K. Nigam, and P. Kondekar, "Group III–V ternary compound semiconductor materials for unipolar conduction in tunnel field-effect transistors," *Journal of Comp. Electronics*, vol. 16, p. 24, 2017.

[10] C. K. Pandey, D. Dash, and S. Chaudhury, "Improvement in Analog/RF Performances of SOI TFET Using Dielectric Pocket," *International Journal of Electronics, International Journal of Electronics*, vol. 107. no. 11, pp. 1844–1860, 2020.

[11] M. Saravanan and E. Parthasarathy, "Investigation of RF/Analog performance of Lg=16nm Planner In0,80Ga0,20As TFET," 2021, 10.1109/ICECCT52121.2021. 9616769

[12] Pravin, D. Nirmal, P. Prajoon, and J. Ajayan, "Implementation of nanoscale circuits using dual metal gate engineered nanowire MOSFET with high-k dielectrics for low power applications," *Phys. E Low-dimensional Systems and Nanostructures*, vol. 83, p. 95, 2016.

[13] M. Vadizade, "Digital Performance Assessment of the Dual-Material Gate GaAs/InAs/Ge Junctionless TFET," *IEEE Transactions on Electron Devices*, vol. 68, p. 1986, 2022.

[14] T. S. Arun Samuel, N. B. Balamurugan, S. Bhuvaneswari, D. Sharmila, and K. Padmapriya, "Analytical modelling and simulation of single-gate SOI TFET for low-power applications," *International Journal of Electronics*, vol. 106, no. 6, pp. 779–788, 2014.

[15] N. Guenifi, S. B. Rahi, and M. Larbi, "Suppression of Ambipolar current and analysis of RF performance in double gate tunneling field effect transistors for low-power applications," *Int J nanoparticles nanotech*, vol. 6, p. 033, 2020.

[16] K. Ramkumar and V. N. Ramakrishnan, "Investigation of Hetero Buried Oxide and Gate Dielectric PNPN Tunnel Field Effect Transistors," *Silicon*, pp. 1–8, 2020.

[17] K. Ramkumar, S. R. Shailendra, and V. N. Ramakrishnan, "Performance Analysis of Carbon Nanotube and Graphene Tunnel Field-Effect Transistors," *In Semiconductor Devices and Technologies for Future Ultra Low Power Electronics*, 87–113. CRC Press, 2021.

[18] I. V. Anand, T. S. Samuel, V. N. Ramakrishnan, and K. Ram Kumar, "Influence of trap carriers in SiO2/HfO2 stacked dielectric cylindrical gate tunnel FET," *Silicon*, pp. 1–12, 2021.

[19] C. Usha, P. Vimala, K. Ramkumar, and V. N. Ramakrishnan. "Electrostatic characteristics of a high-k stacked gate-all-around heterojunction tunnel field-effect transistor using the superposition principle," *Journal of Computational Electronics*, pp. 1–10, 2022.

[20] M. Saravanan, E. Parthasarathy, J. Ajayan, and Nirmal. *Impact of Semiconductor Materials and Architecture Design on TFET Device Performance*. CRC Press, 2022. https://www.taylorfrancis.com/chapters/edit/10.1201/9781003240778-5/

8 DC Analysis and Analog/ HF Performances of GAA-TFET with Dielectric Pocket

Chandan Kumar Pandey, Diganta Das,
Umakant Nanda, Debashish Dash,
Saurabh Chaudhury, Young Suh Song, and
Shiromani Balmukund Rahi

CONTENTS

8.1 INTRODUCTION

In this chapter, the DC and RF/analog performances of Gate-all-around Tunnel Field Effect Transistors (GAA-TFETs) are thoroughly investigated where a Dielectric Pocket (DP) is inserted at the channel-drain (C-D) interface into the drain region. Available literatures demonstrate that the GAA-TFET structures are more immune to short-channel effects (SCEs) during device downscaling as they have superior electrostatic control of the channel potential (electric field) by gate terminal than the planner gate structures [1–4]. However, GAA-TFETs continue to face significant challenges due to ambipolar conduction, which causes significant standby power dissipation mainly in circuit-based applications. The majority of the

DOI: 10.1201/9781003327035-8

149

techniques proposed by researchers are found to significantly reduce ambipolarity, but at the cost of increasing fabrication complexity and/or deteriorating HF/analog performances [5–16].

In this chapter, a novel GAA-TFET with a dielectric pocket (DP GAA-TFET) is investigated which has been proposed by the author in [17], which exhibits a substantial reduction in ambipolar conduction (I_{amb}) and subthreshold leakage current (I_{OFF}) without affecting the ON-state current (I_{ON}) and/or subthreshold swing (SS) while enhancing the HF performance of GAA-TFETs. In order to explore the influence of DP on overall device performance, the DC and RF/analog performances of the proposed DP GAA-TFET is compared with conventional GAA-TFET using 3-D TCAD simulation. This comparison was carried out for both high-k and low-k dielectric materials that were employed in DP. Moreover, we have analyzed the influence of variation in thickness and length of the DP on several electrical parameters of DP GAA-TFET to achieve the optimum values for the same. Since the subthreshold leakage current and ambipolar conduction rely on band-to-band overlapping of drain and channel together, the DP GAA-TFET with high-k is found superior to suppress (I_{amb}) more effectively compared to low-k DP GAA-TFET. It also observed through the simulation-based results that the proposed structure does not affect SS, I_{ON}, and/or HF parameters while narrowing the drain region. In fact, numerous RF parameters such as drain-to-gate capacitance (C_{dg}), source/gate capacitance (C_{gs}), and cut-off frequency (f_T) are found to be improved after insertion of DP to GAA-TFET at the C-D interface, essentially for low-k DP GAA-TFET.

As the probability of BTBT mainly relies on the length of the screening in channel λ_{ch}, which is regulated by gate bias, BTBT generation of electron/holes is observed larger at tunneling interface nearer to dielectric interface between channel/ gate. BTBT generation of charges in TFETs can be mathematically represented using Kane's models as given in [18] and expressed as:

$$G(E) = A\frac{|\varepsilon|^2}{\sqrt{E_g}}\exp\left(-\frac{BE_g^{3/2}}{|\varepsilon|}\right) \tag{8.1}$$

where A and B are called to be as Kane's BTBT model parameters depending on properties of material and ε represents the electric field at S/C interface. To enhance BTBT generation rate in TFETs, the electric field in tunneling area at S/C interface should be made large. Based on the Landauer approach [19], drain current in TFETs can be expressed as:

$$I_D = \int_0^{\Delta\Phi} dED(E)v(E)T(E)[f_s(E) - f_{ch}(E)] \tag{8.2}$$

where D(E), v(E), and T(E) are density of states, group velocity and tunneling probability, respectively. $f_s(E)$ and $f_{ch}(E)$ are Fermi-Dirac distribution function in source and channel regions, respectively.

By integrating BTBT rate of charge carriers generating at S/C interface, tunneling current can be easily modeled as given in eq. (8.3) over the volume of device [18] and can be expressed as:

$$I_D = q \int G(E)dV \qquad (8.3)$$

8.2   STRUCTURE OF THE DEVICE AND SIMULATION SET-UP

The 3D simulated structure of DP-GAA-TFET is shown in Figure 8.1(a) and also a 2-D cross-sectional representation of GAA-TFET with and without DP (n-type) are represented in Figures 8.1(b) and (c), respectively, to understand better device structure. In comparison to the conventional GAA-TFET shown in Figure 8.1(b), structural asymmetry between drain and source terminals is introduced in the proposed structure by replacing a portion of drain with optimized values of T_{pocket} and L_{pocket}. A DP of the same L_{pocket} and T_{pocket} has also been inserted over that partially scaled drain region. Here three main design parameters of the dielectric pocket, i.e., k-value, T_{pocket}, and L_{pocket} are optimized to enhance the overall device performance. Additionally, a DP with same thickness and length is inserted over the portion of removed drain. One can find it as the only difference between GAA-TFET with and without DP in structural way. To improve the overall device performance, three design parameters of the dielectric pocket (k-value, T_{pocket}, and L_{pocket}) need to be optimized. During the designing procedure, a 3 nm thick SiO$_2$ layer is employed as gate dielectric and the gate-metal work function is set to 4.3 eV. In DP GAA-TFET, the diameter of drain, source and channel is fixed for 20 nm and channel length is used to be 50 nm. Drain and source regions are uniformly doped with 5×10^{18} and 1×10^{20} cm^{-3}, respectively. Similarly, a uniform and light doping of acceptors with concentration of 1×10^{16} cm^{-3} is used in channel region of the device. The dielectric constant for high and low- k DP is set at 25 (HfO$_2$) and 3.9 (SiO$_2$) in this work.

Synopsys Sentaurus TCAD simulator [20] is expansively used to study and analyze the proposed DP GAA-TFET structure. For inclusion of continuous change in electric field inside tunneling region, a non-local BTBT model is used in this work [21], which helps in calculating the exact drain current. In order to obtain the accurate analysis, few other models like bandgap narrowing and Shockley-Red-Hall recombination models, concentration- and electric field-dependent mobility, and auger recombination are also activated during our simulation. While conducting the simulation, no defect and leakage current through gate are taken into consideration. For the authenticity of the simulation work, BTBT model parameters [20] have been calibrated by reproducing the result reported in experimental work [22].

8.3   FABRICATION POSSIBILITY

Dielectric pockets (DP) in MOSFETs were first reported in [23] to eliminate 2nd order effects. The following year, Donaghy et al. [24] studied a vertical structure of MOSFET having DP with a channel of 50 nm. The vertical MOSFETs containing

(a)

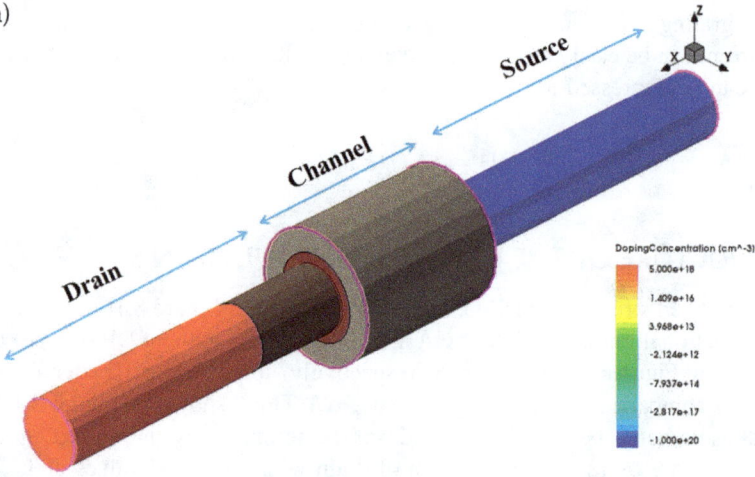

(b)

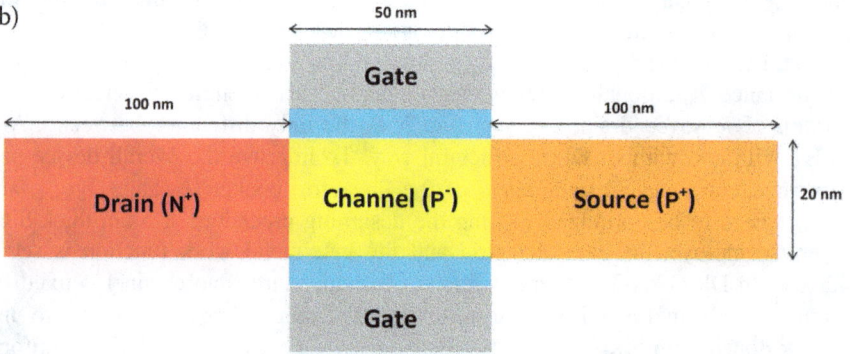

(c)

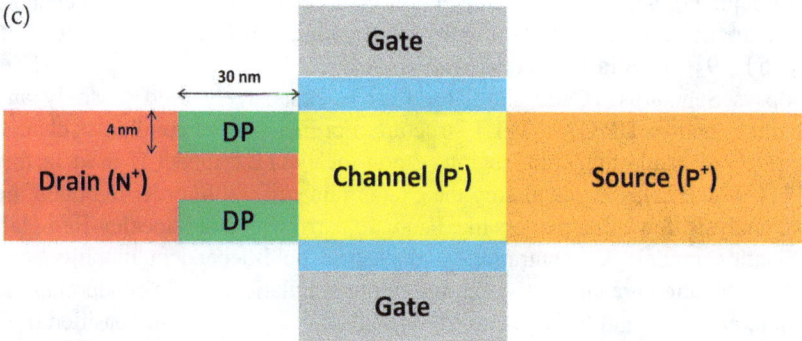

FIGURE 8.1 (a) Simulated structure, and cross-sectional view of GAA-TFET (b) with and (c) without DP.

DP between the source/drain and channel regions to eliminate SCEs as DIBL and V_T lowering was fabricated by Jayanarayanan *et al.* [25]. Gili *et al.* [26] suggested a method of fabrication close to the fabrication steps used for CMOS, using a two-step column etching technique to fabricate vertical n-channel MOSFETs with DP.

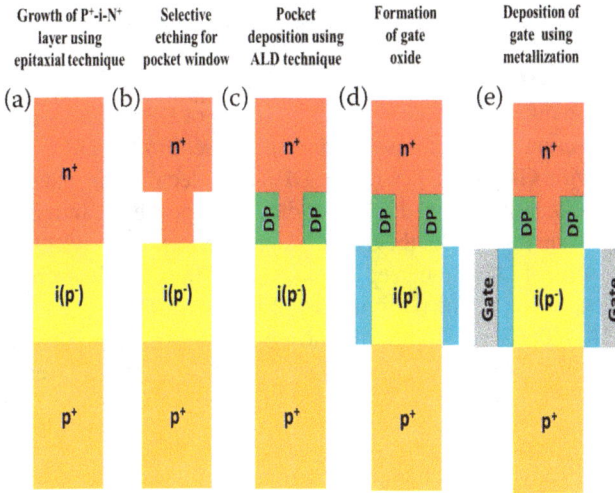

FIGURE 8.2 Predictable steps of fabrication for GAA-TFET with DP.

The possibility of fabricating devices using DP was also reviewed by Wang *et al.* [27] in which a dielectric plug-in was introduced into the MOSFETs. Then, Kumar *et al.* [28] came up with a full-gate Schottky barrier MOSFET fabricated along with DP to eliminate the ambipolarity. Though fabricating this device is complicated as compared with TFET without DP, it can be performed using modern nanolithography techniques based on the abovementioned methodologies. Recently, SOI-TFETs having DP was fabricated by Luo *et al.* [29] in which DP was embedded in the channel at source-channel interface to enhance the on-state current and SS of TFET. Here, the probable short and simplified methods of DSDP-GAA-TFET fabrication are shown in Figure 8.2. Figure 8.2(a) depicts that a P^+-i-N^+ portion of DP-GAA-TFET can be grown in vertical direction using the technique of epitaxial growing, as shown in Figure 8.2(a). To embed a DP on the drain side, proper masking along with selective etching method is used to form a U-shaped pocket window, which is next followed by ALD (atomic layer deposition) technique to deposit the dielectric pocket. Finally, an oxidation process is used to deposit a gate-oxide around the channel region, followed by process of metallization to deposit the gate-metal of DP-GAA-TFET, as depicted in Figures 8.2(d-e).

8.4 SIMULATION RESULTS AND DISCUSSION

This section presents a brief analysis of DP-GAA-TFET on several electrical parameters (subthreshold leakage and ambipolar currents, cut-off frequency, and various parasitic capacitances) and a detailed comparison with conventional GAA-TFET. This section also demonstrates the influence of varying different parameters like dielectric constant, dielectric pocket length, and thickness. Finally, a brief analysis of numerous electrical parameters is performed even for a smaller channel length of 30 nm also to validate the reliability of the proposed structure while scaling.

8.4.1 Optimization of Dielectric Constant (k) of DP

8.4.1.1 DC Performances

In this section, the influence of dielectric constant (k) of DP is mainly investigated for ambipolar conduction and subthreshold leakage current in the DP GAA-TFET, and are compared with GAA-TFET without DP. The electrostatic nature of the C-D interface is significantly changed when a dielectric pocket is introduced over the partially scaled drain region. So, the overall device performance must be examined by varying the k-value of DP while keeping L_{pocket} and T_{pocket} at their optimal value for 30 and 4 nm, respectively. The performance of DP GAA-TFET is compared with the structure of GAA-TFET for both high and low-k value of DP. The minimum tunneling width (W_{min}) increased due to employing high-k dielectric material (i.e., HfO_2) which induces more depletion at the drain region under DP around C-D junction. An increase in W_{min} reduces the ambipolar conduction by a large order compared to low-k structure and conventional one. Figure 8.3 compares the I_D-V_{GS} of high-k DP-GAA-TFET with the other two structures. It is found that DP with higher dielectric constant can reduce ambipolar conduction more effectively. Figure 8.3 clearly shows that the high-k DP GAA-TFET facilitates larger ambipolar window in comparison with the conventional and low-k DP GAA-TFET. Table 8.1 displays a comparison of various performance parameters of the proposed device for both high and low-k DP at different channel lengths as obtained by TCAD synopsis simulator.

Further, distribution of electron density, lateral electric field, and the energy band profile is plotted for all the three above-mentioned structures with a biasing voltage under ambipolar state (i. e. $V_{ds} = 1V$ and $V_{gs} = -0.5$ V). Figure 8.4(a) shows the energy-band profile where it can be observed that the tunneling width (W_{min}) at C/D interface is observed to be large in GAA-TFET with high-k DP in comparison to other structures, which enables more reduction in the ambipolar current (I_{amb}). Additionally, the fringing field adjacent to the C-D interface is found higher when DP of high-k is used in GAA-TFET. As a result, depletion width is enhanced in the drain region. This happens mainly due to the falling of carriers from the surface to the bulk

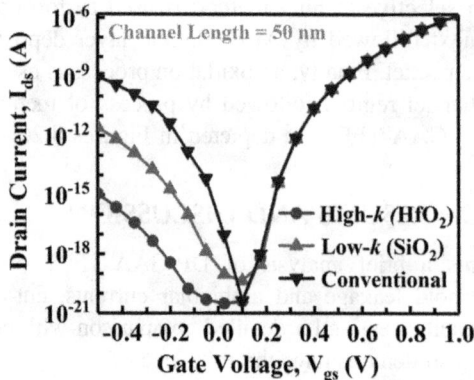

FIGURE 8.3 Comparison of I_d-V_{gs} between GAA-TFET without and with DP of high and low-k.

TABLE 8.1

Comparison of Performance Parameters Between GAA-TFET Without and with DP of High and Low-k

Important Parameters	GAA-TFET with High-k DP		GAA-TFET with Low-k DP		Conventional	
	30 nm	50 nm	30 nm	50 nm	30 nm	50 nm
$L_{channel}$						
I_{ON} (µA)	8.01	8.38	8.01	8.38	8.01	8.39
I_{amb} (A)	4.62×10^{-16}	8.13×10^{-15}	8.75×10^{-10}	2.71×10^{-12}	8.81×10^{-10}	8.78×10^{-10}
I_{OFF} (A)	8.9×10^{-19}	3.5×10^{-21}	$8.5 \times 10-16$	2.37×10^{-19}	8.1×10^{-15}	8.41×10^{-17}
SS_{avg} (mV/decade)	58.8	52.3	57.4	53.2	58.6	54.1
V_{th} (V)	0.797	0.784	0.8	0.785	0.796	0.784

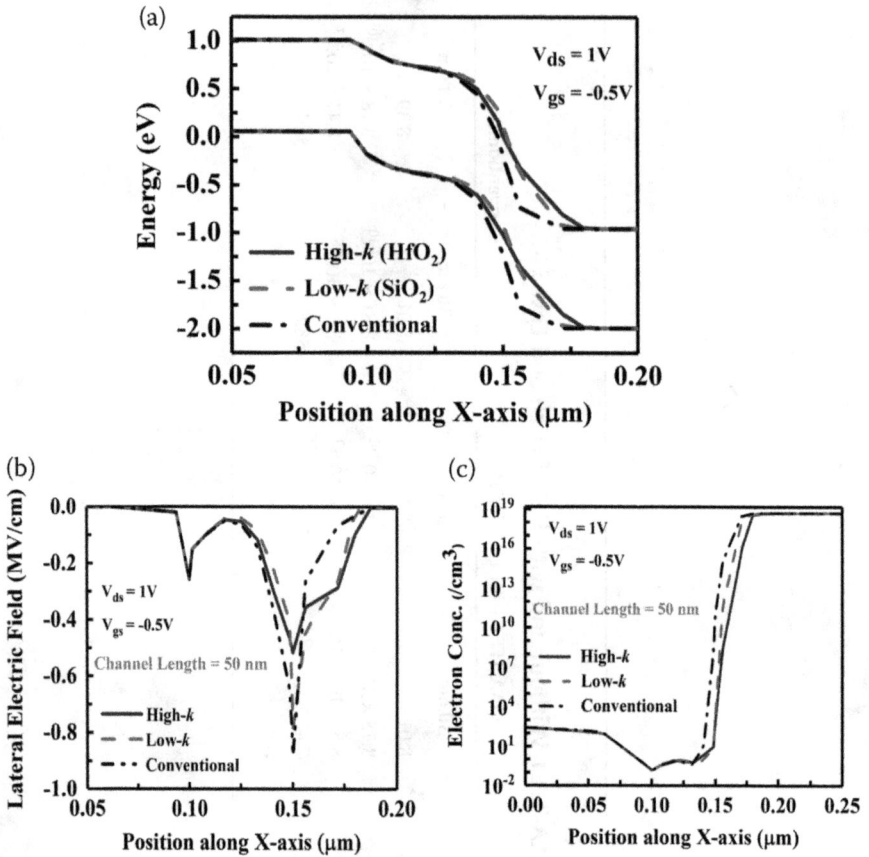

FIGURE 8.4 Comparison of (a) bandgap energy, (b) electric field in lateral direction, and (c) concentration of electron between GAA-TFET without and with DP of high and low-k.

region, where the minimum tunneling width is more and lateral electric field is less compared to the surface [30]. As a consequence, a substantial reduction in ambipolar current can be observed in GA-TFET after inclusion of DP. Due to employing of a high-k dielectric material (i.e., HfO_2) as DP, the carriers downfallen to the bulk from surface are further increased. Figure 8.4(b) depicts a comparison of lateral electric field of DP GAA-TFET with that of conventional structure, both for high- and low-k dielectric constant, and cut line is set at the device's center. This can be noticed that insertion of DP causes the redistribution of electric field as it is developed across the depleted region of drain beneath DP. Therefore, peak value of electric field near C/D interface is reduced in high-k DP GAA-TFET, which significantly reduces the tunneling probability at C/D interface during ambipolar operation. For further detailing, the distribution of electrons in high-k DP-GAA-TFET is compared with low-k DP and conventional GAA-TFET. Figure 8.4(c) shows that concentration of electrons at C/D interface is found less in GAA-TFET when DP of high-k is used, which ultimately reduces ambipolar conduction in GAA-TFET.

(a)

(b)

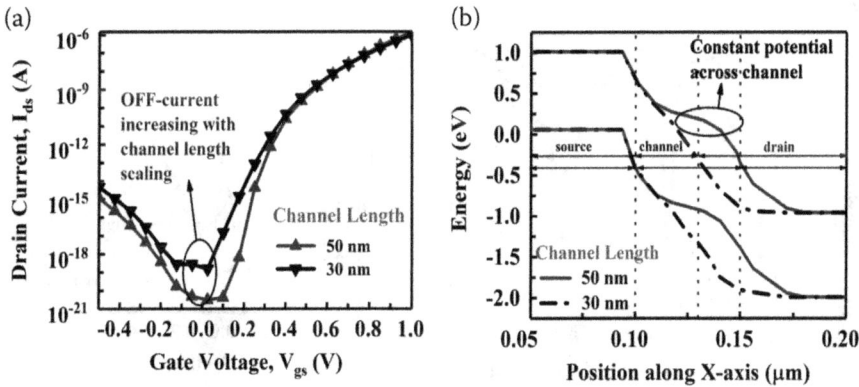

FIGURE 8.5 (a) I_D-V_{gs} and (b) bandgap energy of GAA-TFET with high-k DP at OFF-state while scaling.

While downscaling length of the channel, the impact of DP's inclusion on current-switching ratio (I_{ON}/I_{OFF}) is further explored. The ON-state current mainly relies on the rate of charge carriers tunneling through source/channel (S/C) interface. Instead of being dependent on the channel length, this tunneling rate actually depends on W_{min} [19], which ensures that I_{ON} is insignificantly affected by downscaling of $L_{channel}$, and the same is depicted in Figure 8.5(a). The I_{OFF} found increased significantly in contrast to I_{ON}, when the channel length is scaled down [31]. Consequently, I_{ON}/I_{OFF} varies with change in only I_{OFF}. For an instance, when $L_{channel}$ of high-k DP GAA-TFET is downscaled to 30nm from 50nm, if I_{OFF} is increased from 3.5×10^{-21} to 2×10^{-19} A where the I_{ON} decreased marginally from 8.38×10^{-6} to 8.01×10^{-6} A, shown in Figure 8.5(a) and also tabulated in Table 8.2. During OFF-state, the band diagram for both 50 and 30 nm of $L_{channel}$ is compared in Figure 8.5(b) for the same dimensions of DP, source, and drain regions. A constant potential across the channel region is observed as band diagram of the channel region is found to be nearly flat if $L_{channel}$ is considered to be long like 50 m. This causes weak lateral electric field between drain and source terminals. To achieve better band bending at the channel region, $L_{channel}$ is downscaled to 30 nm, which significantly induces electric field across the entire region of

TABLE 8.2

The Values of and Referred to Figure 8.10(a)

L_{pocket} (nm)	I_{amb} (A)	I_{OFF} (A)
0	8.78×10^{-10}	8.41×10^{-17}
10	2.32×10^{-13}	2.17×10^{-20}
20	3.7×10^{-14}	8.56×10^{-20}
30	8.13×10^{-15}	3.5×10^{-21}
40	8.62×10^{-16}	2.93×10^{-21}

(a)

(b)

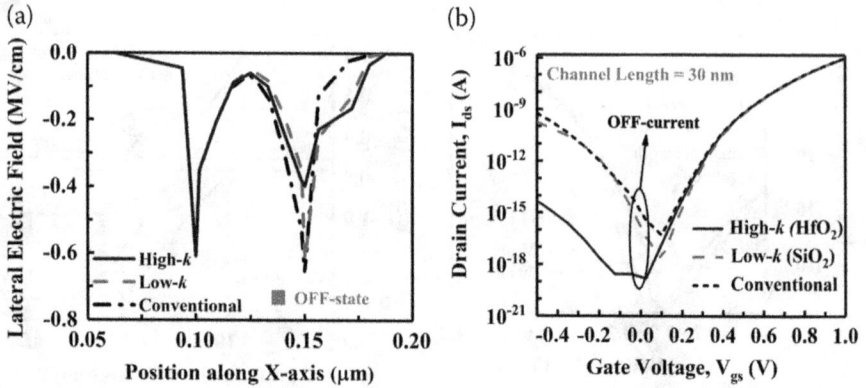

FIGURE 8.6 (a) Electric field and (b) I_D-V_{GS} comparison GAA-TFET without and with DP of high and low-k with 30 nm of channel length.

channel in lateral direction. Next, electric field induced inside the channel triggers a large number of charge carriers drifting towards drain from source, thereby causing more standby leakage current. Figure 8.6(a) depicts that field intensity near C/D interface is notably reduced if DP of high-k is used in GAA-TFET. So, I_{OFF} is observed to be reduced in GAA-TFET with DP of high-K, which consequently increases I_{ON}/I_{OFF} when $L_{channel}$ is downscaled to 30 nm, as illustrated in Figure 8.6(b).

To study the influence of DP on output characteristics, GAA-TFET with and without DP is compared by varying gate voltage from 0.5 to 0.7 V. From Figure 8.7, it can be clearly observed that there is no change in the output characteristics after insertion of DP to GAA-TFET. Thus, this can be concluded that DP GAA-TFET offers smaller value of I_{amb} and I_{OFF} currents compared to conventional GAA-TFET without deteriorating I_{ON}, SS, and/or the output characteristics.

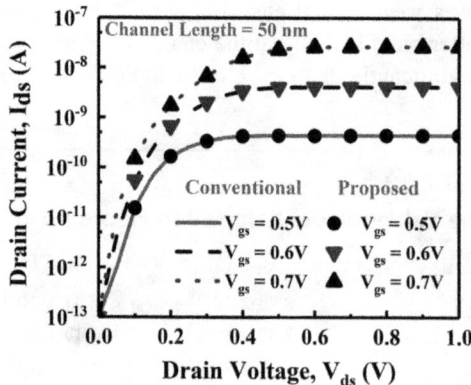

FIGURE 8.7 Output characteristics comparison between GAA-TFET without and with DP of high-K for different gate biases.

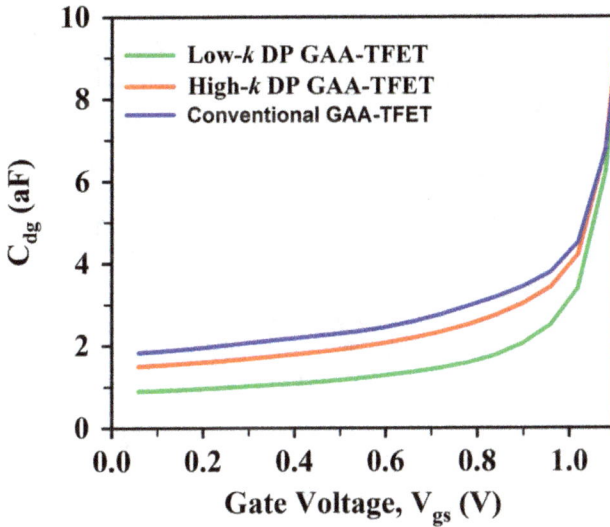

FIGURE 8.8 Comparison of C_{gd} between GAA-TFET without and with DP of high and low-k as a function of V_{gs}.

8.4.1.2 Effect of k-value on High-Frequency Parameters

In this section, the comparison of HF performances of conventional GAA-TFET with DP GAA-TFET for low and high-k dielectric pocket is presented. The figure of merits considered for analyses is mainly gate-to-drain and gate-to-source capacitance, and cut-off frequency. Figure 8.8 shows a comparison of the gate/drain capacitance (C_{gd}) between GAA-TFET without and with low and high-k DP for varying V_{gs}. It can be observed that C_{gd} attains the lowest value for DP of low-K if compared with rest two structures while the conventional GAA-TFET offers the highest value for the same. Since low-k DP offers minimal value of parasitic gate capacitance so it is evident that HF performances will be superior in low-k DP GAA-TFET as compared with other two devices. The influence of k-value of DP on cut-off frequency as a function of V_{gs} is presented in Figure 8.9, and compared with the conventional GAA-TFET. The cut-off frequency (f_T) primarily relies on parameters, such as C_{gd}, C_{gs}, and g_m, and it is represented as $f_T = g_m/2\pi(C_{gs} + C_{gd})$. As, DP has no significant influence on C_{gs}, and g_m the value of f_T is mainly governed by C_{gd} in DP GAA-TFETs. It is clearly seen from Figure 8.9 that low-k GAA-TFET offers the highest value of f_T due to minimal value of C_{gd} among all three devices being considered here. In the previous sections, it was observed that high-k DP leads to more suppression in ambipolarity and OFF-state current in GAA-TFET as compared to low-K DP. So, there is a trade-off between superior dc and HF performances while taking low or high-k DP GAA-TFET. So, it is the application requirement that will decide either to choose low or high-k DP GAA-TFET.

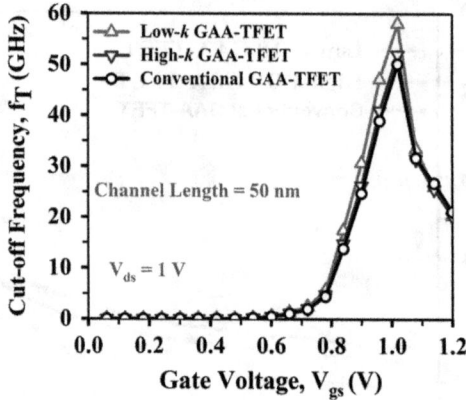

FIGURE 8.9 Comparison of f_T between GAA-TFET without and with DP of high and low-k as a function of V_{gs}.

8.4.2 IMPACT OF VARYING LENGTH OF DP

The length of DP (L_{pocket}) is varied up to 40nm from 10nm to find its optimum value for high-K DP-GAA-TFET. Throughout these simulations, DP thickness has been set for its optimum dimension of 4 nm. K-value of DP is also taken to be $25(HfO_2)$, which has been found its optimum value. Rest of configurations including $L_{channel}$, concentration of dopant atoms, and so on have been left unaffected. I_D-V_{GS} shown in Figure 8.10(a) depicts the dependence of I_{amb} on L_{pocket}. It is clearly visible that when I_{amb} reduces to 10^{-15}A from 2.3×10^{-13} A if L_{pocket} is enlarged to 30 nm from 10nm. It has been shown that I_{amb} is being continuously reduced up to 30 nm of L_{pocket} and only 10%–12% reduction is noticed if L_{pocket} is increased more than 30 nm. Figure 8.10(a) and Table 8.2 show that I_{OFF} attains the lowest value of 3.5×10^{-21}A at 30 nm while increasing L_{pocket}.

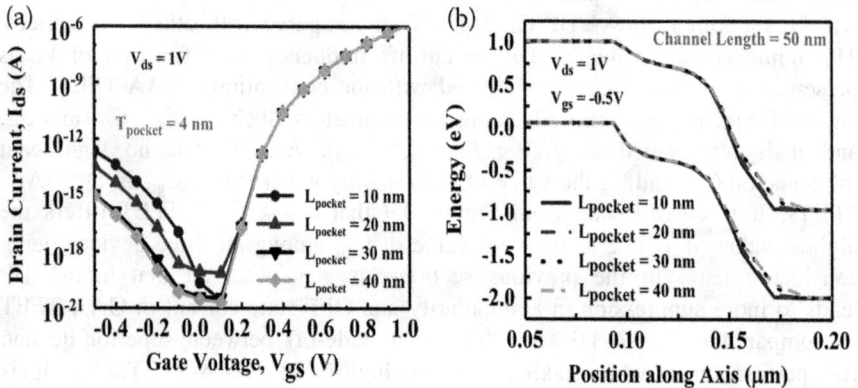

FIGURE 8.10 (a) I_D-V_{GS}, and (b) band diagram of GAA-TFET with high-k DP at different L_{pocket}.

If L_{pocket} is enlarged even more, insignificant variation in drain current is observed for the proposed structure of GAA-TFET. That is why 30 nm of L_{pocket} has been chosen to be its optimum value.

The dependency of I_{amb} on L_{pocket} can be realized from the plot of Figure 8.10(b) by scrutinizing the energy band diagram of high-k DP GAA-TFET biased at am-bipolar state for varying L_{pocket}. It is clearly evident that if L_{pocket} is varied to 40 nm from 10 nm, width of tunneling at C/D interface is enhanced, which causes huge suppression in I_{amb}. The broadening of W_{tunnel} happens because space charge width inside drain beneath DP near C/D interface for enhancement in L_{DP}. As, 30 nm of L_{pocket} achieves the maximum space charge width at C/D interface, I_{amb} is found to be insignificantly decreased if L_{pocket} is enlarged more than 30 nm.

8.4.3 EFFECT OF VARYING THICKNESS OF DP

The current section illustrates the influence of changing thickness of DP (T_{pocket}) on HF/analog and dc performances of GAA-TFET with high-k DP and compared the results with GAA-TFET without DP (i.e., $T_{pocket} = 0$).

8.4.3.1 Effect of T_{pocket} on DC Performances

In this section, the dc performances of high-k DP GAA-TFET is investigated by varying T_{pocket} up to 5 nm. While simulating the device, k-value and L_{pocket} are kept unchanged for 25 and 30 nm, respectively. Figure 8.11(a) displays the transfer characteristics of high-k DP GAA-TFET from which it can be observed that varying T_{pocket} has no impact on SS and I_{on}. The magnitude of I_{amb} is suppressed to 2.13×10^{-16} from 8.78×10^{-10} A if T_{pocket} is varied up to 5 nm. Suppression in I_{amb} can be seen to be persistent till 4 nm, as listed in Table 8.3. As, further enhancement in T_{pocket} does not significantly reduces I_{amb} (approximately 9%) as shown in Figure 8.11(a), 4 nm of T_{pocket} is therefore regarded as its optimum value. Figure 8.11(b) shows the band diagram to show the influence of T_{pocket} on ambi-polarity. At low T_{pocket}, $E_{V,channel}$ comes in the line to $E_{C,drain}$, which causes an increment in BTBT of electrons/holes between drain and channel at C-D interface.

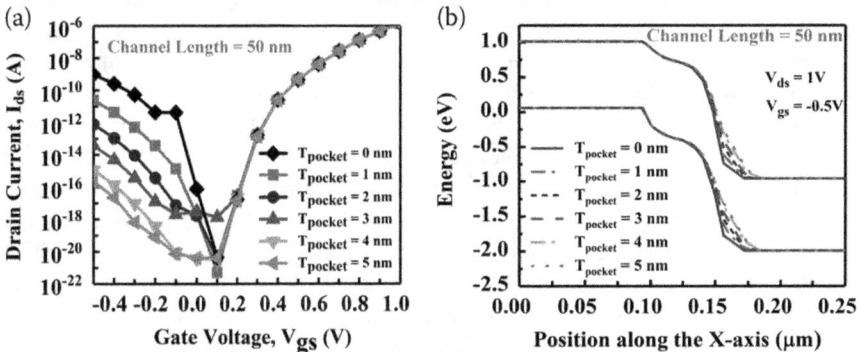

FIGURE 8.11 (a) I_D-V_{GS}, and (b) band diagram of GAA-TFET with high-k DP at different L_{pocket}.

TABLE 8.3

Extracted Values of and from Figure 8.11(a)

L_{pocket} (nm)	I_{amb} (A)	I_{OFF} (A)
0	8.78×10^{-10}	8.41×10^{-17}
1	2.48×10^{-11}	8.88×10^{-18}
2	7.91×10^{-13}	8.73×10^{-18}
3	3.88×10^{-14}	2.99×10^{-18}
4	3.13×10^{-15}	3.78×10^{-21}
5	2.13×10^{-16}	4.03×10^{-21}

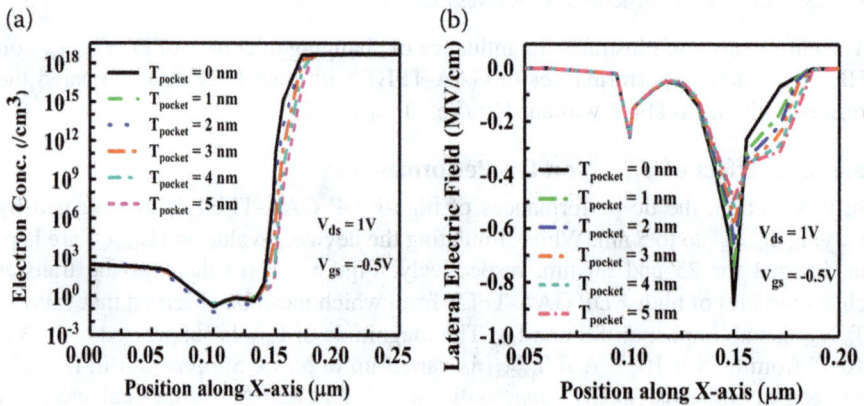

FIGURE 8.12 (a) Concentration of electrons, and (b) $E_{Lateral}$ in high-k DP GAA-TFET at different L_{pocket}.

It is clearly visible in Figure 8.11(b) that with increasing T_{pocket}, $W_{min,tunnel}$ at C/D interface gets enlarged, thereby reducing the band-overlapping between channel and drain. This is why BTBT probability is significantly reduced. 4nm of T_{pocket} has been considered as its optimum one as largest value of $W_{min,tunnel}$ is attained for the same. For further analysis, concentration of electron and $E_{Lateral}$ at C/D interface have been shown in Figures 8.12(a) and (b), respectively. It has been observed from the plots that concentration of electron and $E_{Lateral}$ attain a lower value at its optimum value of 4 nm when T_{pocket} is varied. Moreover, the investigation of the influence of varying T_{pocket} on I_{OFF} in high-k DP-GAA-TFET is performed while biasing at $V_{gs} = 0V$, i.e., OFF-state. As we know that I_{OFF} becomes more prominent in device when channel length is made short like 30 nm, and it has been considered and analyzed in this work. This can be seen from Figure 8.13(a) that I_{OFF} is reduced to 8.9×10^{-19} from 8.12×10^{-15} A if T_{pocket} is enlarged to 4nm from 0nm. This also can be noticed from the plot of Figure 8.13(b), that shows a variation in $E_{lateral}$ for increasing T_{pocket}. The magnitude of $E_{Lateral}$ is particularly reduced with the increment in T_{pocket} at C-D interface. This is primarily because of widening of

(a)

(b)

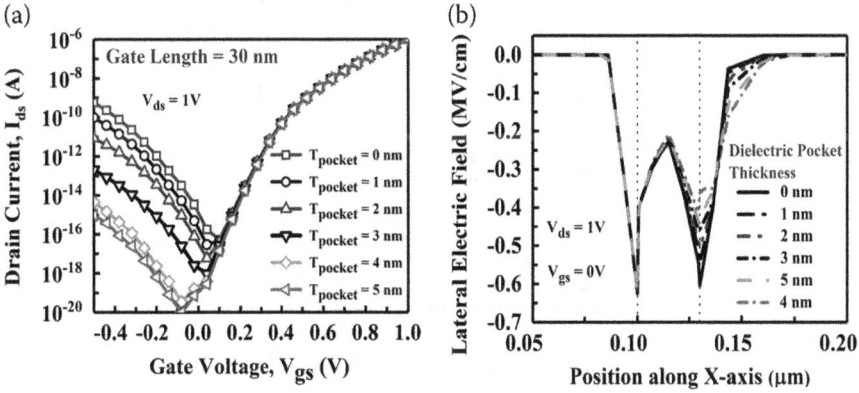

FIGURE 8.13 (a) I_D-V_{GS}, and (b) $E_{lateral}$ (during OFF-state) in GAA-TFET with high-k DP at 30nm of $L_{channel}$ for changing T_{pocket}.

space-charge width inside drain beneath DP. The same has been shown in Figure 8.13(b) where distribution of field is found to extended in drain region. Consequently, a decreased $E_{Lateral}$ diminishes charge carriers drifting towards drain from source through channel, thus reducing I_{OFF}.

8.4.3.2 Impact of T_{pocket} on Analog/High-frequency Performance

This section illustrates, the influence of T_{pocket} on analog/HF performances of high-k DP GAA-TFET, and the comparison with GAA-TFET is performed by using the results obtained by simulation.

No substantial change in transconductance is obtained in high-k DP-GAA-TFET if compared to GAA-TFET without DP and the same is found and depicted in Figure 8.14. It reveals that g_m does not get affected much with varying T_{pocket}. Since the application of DP on drain side has no direct relation

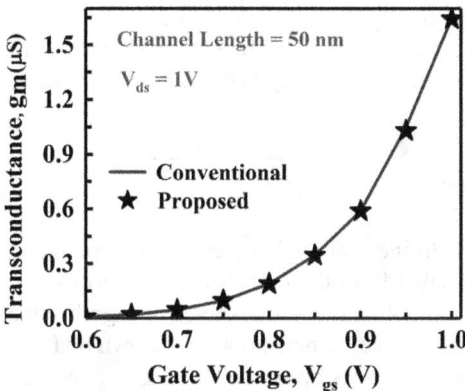

FIGURE 8.14 Comparison of g_m between GAA-TFET with high-k and without DP versus V_{gs}.

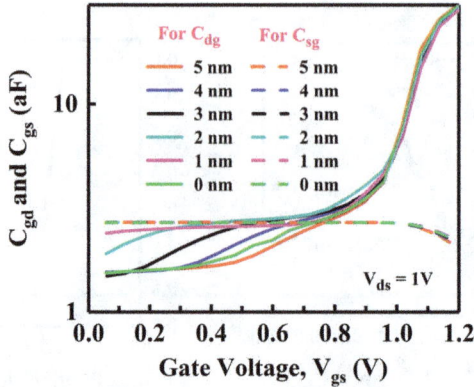

FIGURE 8.15 Optimization of T_{pocket} with the help of C_{gs} and C_{gd} at varying V_{gs}.

with S/C interface at which BTBT mainly happens and regulates the ON-state characteristics of the device, the characteristics of g_m are found to be not changing with varying T_{pocket}.

Moreover, the variation in C_{gs} and C_{gd} is also analyzed when T_{pocket} is varied. It is observed from Figure 8.15 that C_{gd} does not change with varying T_{pocket} after it attains a maximum value at higher V_{gs} when it is raised up to ≈ 8.0 V. The way C_{gd} has behaved may be understood with the following expiation: C_{gd} and C_{gs} are mainly constituted with parasitic capacitances when gate is biased at lower voltage and this mainly happens because inversion layer is not formed towards the drain side. In contrast to MOSFET, the layer is firstly created nearby drain terminal when V_{gs} is raised for a constant value of V_{ds}. The region of inversion starts coming closer to source terminal and as a result of this, C_{gd} starts getting dominated by the inversion created between drain and gate when V_{gs} is kept increasing further [18]. Additionally, inversion region created on drain side is degraded by the depletion charge-sheet caused by DP, which keeps increasing with increasing T_{pocket}. That is why C_{gd} shows a lower value at larger T_{pocket}. Meanwhile, C_{gs} is shown almost unaffected with varying T_{pocket} when V_{gs} is increased up to ≈ 8.1V, and marginally decreases with increasing T_{pocket} for $V_{gs} > 8.1V$.

Next, the impact of varying T_{pocket} is investigated on cut-off frequency (f_T) that is calculated for current gain to be unity [18], and represented as:

$$f_T = \frac{g_m}{2\pi(C_{gs} + C_{gd})} \qquad (8.4)$$

The variation in f_T with increasing T_{pocket} can be interpreted by result presented in Figure 8.16. It is revealed from the plot that $f_{T,max}$ becomes larger with increasing T_{pocket} to 5 nm from 0 nm. It is found that there is insignificant improvement/change in $f_{T,max}$ when T_{pocket} is increased beyond 4 nm (shown by Figure 8.16) and 4 nm of T_{pocket} should be chosen to be its optimum value as I_{amb} was also shown unaffected for $T_{pocket} > 4\ nm$. It can be realized that $f_{T,max}$ has been improved from 38 to 43 GHz when T_{pocket} was increased to 4nm from 0nm. Since, g_m and C_{gs} were found

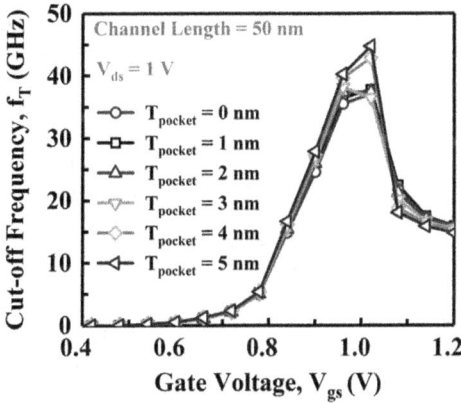

FIGURE 8.16 Optimization of T_{pocket} with the help of f_T at varying V_{gs}.

almost unaffected with varying T_{pocket}, f_T only relies on C_{gd}. Based on the above-mentioned analysis, it can be concluded that the device discussed in this chapter can withstand against the structure of various TFET and MOSFET devices proposed in [32–52] in terms of performance parameters like dc and analog/HF.

8.5 CONCLUSION

In this chapter, the influence of dielectric pocket's inclusion to GAA-TFET is investigated on its various performance parameters. Based on the results obtained through simulation, it is found that ambipolarity and OFF-state leakage conduction get notably reduced in GAA-TFET when DP of any dielectric material either high and low-k is included in GAA-TFET. Anyway, the ambipolarity is found to be minimum in GAA-TFET with high-k DP when it is compared with rest of two structures. Moreover, it is also observed that this reduction in OFF-state leakage and ambipolarity is obtained without getting ON-state characteristics like sub-threshold swing and I_{ON} deteriorated. To achieve improved performance parameters, important dimensions of dielectric pocket-like thickness and length have been optimized and found to be 4 and 30 nm, respectively. The impact of DP's k-value on various analog/HF parameters in DP-GAA-TFET is also analyzed in this work and found that GAA-TFET with low-k DP may offer superior HF performances than those of GAA-TFET without DP and with high-k DP. It is noticed that HF parameters including drain/gate capacitance along with cut-off frequency are improved in DP GAA-TFET while source/gate capacitance and transconductance are shown to be nearly unchanged with the inclusion of DP as expected.

REFERENCES

[1] Upasana, R. Narang, M. Saxena, and M. Gupta, "Investigation of dielectric pocket induced variations in tunnel field effect transistor," *Superlattices and Microstructures*, vol. 92, pp. 380–390, 2016.

[2] Upasana, R. Narang, and M. Saxena, "Influence of dielectric pocket on electrical characteristics of tunnel field effect transistor: A study to optimize the device efficiency," *2015 IEEE International Conference on Electron Devices and Solid-State Circuits (EDSSC)*, Singapore, 2015, pp. 762–768.

[3] R. Gandhi, Z. Chen, N. Singh, K. Banerjee, and S. Lee, "CMOSCompatible Vertical-Silicon-Nanowire Gate-All-Around p-TypeTunneling FETs With ≤ 50-mV/ decade Subthreshold Swing," *IEEE Electron Device Lett.*, vol. 32, no. 11, pp. 1504–1506, Nov. 2018.

[4] S. K. Sinha and S. Chaudhury, "Impact of Oxide Thickness on GateCapacitance-A Comprehensive Analysis on MOSFET, Nanowire FET and CNTFET Devices," *IEEE Trans. Nanotechnol.*, vol. 12, no. 6, pp. 958–964, Nov. 2013.

[5] W. Y. Choi and W. Lee, "Hetero-gate-dielectric tunneling field effect transistors," *IEEE Trans. Electron Devices*, vol. 57, no. 9, pp. 2317–2319, 2010.

[6] P. -F. Wang *et al.*, "Complementary tunneling transistor for low power application," *Solid-State Elec.*, vol. 48, pp. 2281–2286, May 2004.

[7] Y. A. S. Verhulst, W. G. Vandenberghe, K. Maex, and G. Groeseneken, "Tunnel field-effect transistor without gate-drain overlap," *Appl. Phys. Lett.*, vol. 91, no. 5, p. 053102, Jul. 2007.

[8] J. Wan, C. Le Royer, A. Zaslavsky, and S. Cristoloveanu, "SOI TFETs: Suppression of ambipolar leakage and low-frequency noise behavior," *2010 Proceedings of the European Solid State Device Research Conference*, Sevilla, 2010, pp. 341–344.

[9] J. Wan, C. L. Royer, A. Zaslavsky, and S. Cristoloveanu, "Tunneling FETs on SOI: Suppression of ambipolar leakage low-frequency noise behavior and modeling," *Solid-State Electron*, vols. 65–66, pp. 226–233, Dec. 2018.

[10] S. Ghosh, K. Koley, S. K. Saha, and C. K. Sarkar, "High-Performance Asymmetric Underlap Ge-pTFET With Pocket Implantation," *IEEE Trans. Electron Devices*, vol. 63, no. 10, pp. 3869–3875, 2016.

[11] J. Madan and R. Chaujar, "Gate Drain Underlapped-PNIN-GAA-TFET for Comprehensively Upgraded Analog/RF Performance," *Superlattices Microstruct.*, vol. 102, pp. 17–26, 2017.

[12] D. B. Abdi and M. J. Kumar, "Controlling ambipolar current in tunneling FETs using overlapping gate-on-drain," *IEEE J. Electron Devices Soc.*, vol. 2, no. 6, pp. 187–190, 2014.

[13] Hraziia, A. Vladimirescu, A. Amara, and C. Anghel, "An analysis on the ambipolar current in Si double-gate tunnel FETs," *Solid. State. Electron.*, vol. 70, pp. 67–72, 2012.

[14] S. Sahay and M. J. Kumar, "Controlling the Drain Side Tunneling Width to Reduce Ambipolar Current in Tunnel FETs Using Heterodielectric BOX," *IEEE Trans. Electron Devices*, vol. 62, no. 11, pp. 3882–3886, 2018.

[15] A. Beohar and S. K. Vishvakarma, "Performance enhancement of asymmetrical underlap 3D-cylindrical GAA-TFET with low spacer width," *Micro & Nano Letters*, vol. 11, no. 8, pp. 443–445, 8 2016.

[16] A. Shaker, M. El Sabbagh, and M. M. El-Banna, "Influence of Drain Doping Engineering on the Ambipolar Conduction and High-Frequency Performance of TFETs," *IEEE Trans. Electron Devices*, vol. 64, no. 9, pp. 3541–3547, 2017.

[17] C. K. Pandey, D. Dash, and S. Chaudhury, "Impact of Dielectric Pocket on Analog and High-Frequency Performances of Cylindrical Gate-All-Around Tunnel FETs," *ECS J Solid Stae Sc.*, vol. 7, no. 5, pp. 59–66, May 2018.

[18] Y. Yang, X. Tong, L.-T. Yang, P. -F. Guo, L. Fan, and Y. -C. Yeo, "Tunneling field-effect transistor: Capacitance components and modeling," *IEEE Electron Device Lett.*, vol. 31, no. 7, pp. 752–754, Jul. 2010.

[19] J. Wu, and Y. Taur, "Reduction of TFET OFF-Current and Subthreshold Swing by Lightly Doped Drain," *IEEE trans. Electron Devices*, vol. 63, no. 8, pp. 3342–3345, Aug. 2016.

[20] Sentaurus Device User Guide. Version M-2016.12, Synopsys, Inc., Dec 2016.

[21] A. Biswas, S. S. Dan, C. Le Royer, W. Grabinski, and A. M. Ionescu, "TCAD simulation of SOI TFETs and calibration of non-local band-toband tunneling model," *Microelectron. Eng.*, vol. 98, pp. 334–337, Oct. 2012.

[22] Z. X. Chen, H. Y. Yu, N. Singh, N. S. Shen, R. D. Sayanthan, G. Q. Lo, and D.-L. Kwong, "Demonstration of tunneling FETs based on highly scalable vertical silicon nanowires," *IEEE Electron Device Lett.*, vol. 30, no. 7, pp. 754–756, Jul. 2008.

[23] M. Jurczak, T. Skotnicki, R. Gwoziecki, M. Paoli, B. Tormen, P. Ribot, D. Dutartre, S. Monfray, and J. Galvier, "Dielectric pockets-a new concept of the junctions for deca-nanometer CMOS devices," *IEEE trans. Electron Devices*, vol. 48, no. 8, pp. 1770–1774, Jul. 2008.

[24] D. C. Donaghy, S. Hall, D. Kunz, K. de Groot, and P. Asburn, "Investigating 50 nm channel length vertical MOSFET containing a dielectric pocket, in a circuit environment," in *proc. ESSDERC*, 2002, pp. 499–503.

[25] S. K. Jayanarayanan, S. Dey, J. P. Donnelly, and S. K. Banerjee, "A novel 50 nm vertical MOSFET with dielectric pocket," *Solid State Electron.*, vol. 50, no. 5, pp. 897–900, 2006.

[26] E. Gili, T. Uchino, M. M. A. Hakim, C. H. De Groot, O. Buiu, S. Hall, and P. Ashburn, "Shallow junctions on pillar sidewalls for sub-100-nm vertical MOSFETs," *IEEE Electron Device Lett.*, vol. 27, no. 8, pp. 692–695, Aug. 2006.

[27] Dielectric plug in MOSFETs to suppress short-channel effects, by Wang H, and Wang Z. (2006, Dec 26). US Patent 7154146 B2.

[28] M. Kumar, Y. Pratap, S. Haldar, M. Gupta, and R. S. Gupta, "Cylindrical gate all around Schottky barrier MOSFET with insulated shallow extensions at source/drain for removal of ambipolarity: a novel approach," *J. Semicond.*, vol. 38, no. 12, pp. 124002-1–124002-6.

[29] Z. Luo, H. Wang, N. An, and Z. Zhu, "A Tunnel Dielectric-Based Tunnel FET," *IEEE Electron Device Letters*, vol. 36, no. 9, pp. 966–968, Sept. 2018.

[30] C. Anghel, P. Chilagani, A. Amara, and A. Vladimirescu, "Tunnel field effect transistor with increased ON current, low-k spacer and high-k dielectric," *Appl. Phys. Lett.*, vol. 96, no. 12, p. 122 104, Mar. 2010.

[31] K. Boucart, and A. M. Ionescu, "Double-gate tunnel FET with high-k gate dielectric," *IEEE Trans. Electron Devices*, vol. 54, no. 7, pp. 1725–1733, Jul. 2007.

[32] Y. S. Song, J. H. Kim, G. Kim, H. -M. Kim, S. Kim, and B. -G. Park, "Improvement in Self-Heating Characteristic by Incorporating Hetero-Gate-Dielectric in Gate-All-Around MOSFETs," in IEEE Journal of the Electron Devices Society, vol. 9, pp. 36–41, 2021, 10.1109/JEDS.2020.3038398.

[33] Y. S. Song, S. B. Rahi, S. Tayal, A. Upadhyay, and J. H. Kim, "Design Techniques for High Reliability FET by Incorporating New Materials and Electrical/thermal Co-optimization," in *Emerging Materials* pp. 133–154. Singapore: Springer, 2022.

[34] M. Sathish Kumar, T. A. Samuel, K. Ramkumar, I. V. Anand, and S. B. Rahi, "Performance evaluation of gate engineered InAs–Si heterojunction surrounding gate TFET," *Superlattices and Microstructures*, vol. 162, p. 107098, 2022.

[35] S. Tayal, A. Kumar Upadhyay, D. Kumar, and S.B. Rahi (Eds.). *Emerging Low-Power Semiconductor Devices: Applications for Future Technology Nodes* (1st ed.). CRC Press, 2022, 10.1201/9781003240778.

[36] A. Singh and C.K. Pandey, "Improved DC Performances of Gate-all-around Si-Nanotube Tunnel FETs Using Gate-Source Overlap," *Silicon*, vol. 14, pp. 1463–1470, 2022. 10.1007/s12633-021-00957-0.

[37] A. Singh, C.K. Pandey, S. Chaudhury *et al.*, "Tuning of Threshold Voltage in Silicon Nano-Tube FET Using Halo doping and its Impact on Analog/RF Performances," *Silicon*, vol. 13, pp. 3871–3877, 2021. 10.1007/s12633-020-00698-6.

[38] S. B. Rahi, P. Asthana, and S. Gupta, "Heterogatejunctionless tunnel field-effect transistor: future of low-power devices," *J Comput Electron*, vol. 16, no. 1, pp. 30–38, 2017.

[39] V. B. Srinivasulu and V. Narendar, "Characterization and optimization of junctionless gate-all-around vertically stacked nanowire FETs for sub-5 nm technology node," *Microelectronics Journal*, vol. 116, p. 105214, 2028.

[40] V. B. Sreenivasulu and V. Narendar, "Performance improvement of spacer engineered n-type SOI FinFET at 3-nm gate length," *AEU - International Journal of Electronics and Communications*, vol. 137, p. 153803, 2021.

[41] C. K. Pandey and S. Chaudhury, "Dual-Metal Graded-Channel Double-Gate Tunnel FETs for Reduction of Ambipolar Conduction," *2018 IEEE Electron Devices Kolkata Conference (EDKCON)*, pp. 572–576, 2018, doi: 10.1109/EDKCON.2018.8770448.

[42] K.R.N. Karthik and C.K. Pandey, "A Review of Tunnel Field-Effect Transistors for Improved ON-State Behaviour," *Silicon*, 2022. 10.1007/s12633-022-02028-4.

[43] A. Singh, C. K. Pandey, and U. Nanda, "Performance analysis of silicon nanotube dielectric pocket Tunnel FET for reduced ambipolar conduction," *Microelectronics Journal*, vol. 126, 2022, 10.1016/j.mejo.2022.105512.

[44] C.K. Pandey, A. Singh, and S. Chaudhury, "A simulation-based analysis of effect of interface trap charges on dc and analog/HF performances of dielectric pocket SOI-tunnel FET," *Microelectron Reliab*, vol. 122, pp. 114166, 2021.

[45] S.K. Das, U. Nanda, S.M. Biswal *et al.* "Performance Analysis of Gate-Stack Dual-Material DG MOSFET Using Work-Function Modulation Technique for Lower Technology Nodes," *Silicon*, vol. 14, pp. 2965–2973, 2022, 10.1007/s12633-021-01095-3.

[46] S. Tayal and A. Nandi, "Optimization of gate-stack in junctionless si-nanotube FET for analog/rf applications," *Mater Sci Semicond Process*, vol. 80, pp. 63–67, 2018.

[47] S. Panda, B. Jena, and S. Dash, "Ambipolarity Suppression of a Double Gate Tunnel FET using High-*k* Drain Dielectric Pocket," *ECS Journal of Solid State Science and Technology*, vol. 11, no. 1, 2022.

[48] S. Moparthi, P. K. Tiwari, V. R. Samoju, and G. K. Saramekala, "Investigation of Temperature and Source/Drain Overlap Impact on Negative Capacitance Silicon Nanotube FET (NC Si NTFET) with Sub-60mV/decade Switching," in *IEEE Transactions on Nanotechnology*, vol. 19, pp. 800–806, 2020, 10.1109/TNANO.2020.3033802.

[49] S. Dubey, A. Santra, G. Saramekala, M. Kumar, and P. K. Tiwari, "An Analytical Threshold Voltage Model for Triple-Material Cylindrical Gate-All-Around (TM-CGAA) MOSFETs," in *IEEE Transactions on Nanotechnology*, vol. 12, no. 5, pp. 766–774, Sept. 2013, 10.1109/TNANO.2013.2273805.

[50] P. Kumar and B. Bhowmick, "2-D analytical modelling for electrostatic potential and threshold voltage of a dual work function gate Schottky barrier MOSFET," *J Comput Electron*, vol. 16, no. 3, pp. 658–665, Sep 2017.

[51] J. Talukdar, G. Rawat, K. Singh *et al.*, "Low Frequency Noise Analysis of Single Gate Extended Source Tunnel FET," *Silicon*, vol. 13, pp. 3971–3980, 2021. 10.1007/s12633-020-00712-x.

[52] A. Lahgere, C. Sahu, and J. Singh, "PVT-Aware Design of Dopingless Dynamically Configurable Tunnel FET," in *IEEE Transactions on Electron Devices*, vol. 62, no. 8, pp. 2404–2409, Aug. 2015, 10.1109/TED.2015.2446615.

9 Investigation on Ambipolar Current Suppression in Tunnel FETs

Mohammad Ehteshamuddin, S. Manikandan, and Adhithan Pon

CONTENTS

DOI: 10.1201/9781003327035-9

9.1   INTRODUCTION

The extraordinary improvement in power, switching speed, density, and cost of IC production has resulted from aggressive CMOS scaling. However, the inability to further reduce power consumption has put forward the need for steep-slope devices [1]. In this context, TFET has emerged as a potential candidate that can bring down the supply voltage V_{DD} of ICs below 0.5V. Unlike MOSFETs, TFET utilizes Band-to-Band tunneling (BTBT) as a current injection mechanism and explores the possibility of reducing V_{DD} by increasing the turn-on steepness of the device characteristics. TFETs, in essence, are the reverse-biased p-i-n structure that follows the tunneling mechanism of the highly-doped p^+-n^+ Zener diode. They became indispensable as they do not have any theoretical limitation on the subthreshold slope (SS), unlike conventional MOSFETs. However, their major impediments, such as low drive current, high ambipolar current, and high miller capacitance, must be overcome before competing against existing CMOS. Low I_{ON} negatively impacts the switching speed. The increased miller capacitance further aggravates this disadvantage in the case of TFETs. It is then expected that the focus for further TFET optimization will be to achieve the highest ON-current possible. Source-side optimization for better carrier injection includes low bandgap material, strained-source, source-side delta doping, and alignment of source with gate, to count a few. Steeper SS, low $I_{OFF,}$ and suppressed ambipolarity are other directions in which the research community has made a modest effort. The subsequent research followed in the case of TFETs is divided into two categories: 1) use of performance boosters and structural engineering to improve ON-state performance. 2) Utilization of different techniques to suppress the OFF-state current and ambipolar conduction.

9.2   WORKING MECHANISM

TFETs differ from MOSFETs in terms of their current transport mechanism. The charge carriers must tunnel through the potential barrier instead of surmounting it. This inter-band tunneling in TFETs, in contrast to thermally injected carriers in MOSFET, reduces the current drivability of the device. However, due to the possibility of abrupt on-off switching via controlling the band bending in the channel region, TFET can offer steeper SS. One challenge in TFET is to obtain ON-current on par with MOSFET at sub-0.5V supply voltage. For that, a higher tunneling rate at the source-channel junction is required, which can be achieved by optimizing the transmission probability of the source tunneling barrier as close to unity as possible for a smaller gate voltage bias. Applying positive gate bias helps create a path for electrons to tunnel from the source valence band to the channel conduction band by bending the bands downward in the channel region. This opens the window for BTBT. Figure 9.1(a) and (b) illustrate the schematic of the ON-state and ambipolar state band diagrams of the TFET.

In principle, the tunneling mechanism is possible at the source-channel and channel-drain junction, making TFET an ambipolar device. Ambipolarity indicates conduction in both directions for V_{DS} polarized only in one direction (positive for n-TFET and negative for p-TFET). For an n-TFET, changing the gate bias from the positive voltage ($V_{GS} > 0$) to negative voltage changes the tunnel junction from

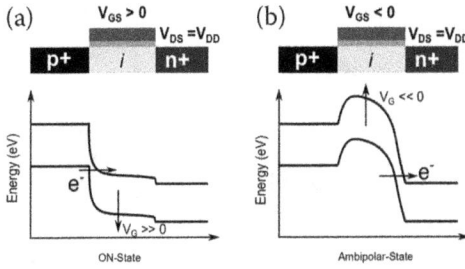

FIGURE 9.1 The ON-state and ambipolar state behavior of TFET schematically using band diagrams.

source-channel to channel drain at which carrier tunneling occurs. This can be observed in Figure 9.1(b). Multiple ways of asymmetry have been demonstrated in the literature to suppress ambipolarity. The widening of the tunnel barrier at the drain side can be done in various ways. Different doping profile strategies at the source and drain side, using heterostructures, etc., are standard methods to suppress ambipolar conduction. It also helps in minimizing the OFF-state current. It is worth mentioning that ambipolarity is an inherent characteristic of a TFET device that limits its usage in the complementary logic circuit. To be suitable as an ideal switch in a digital circuit, TFET should conduct only in one direction. Therefore, if ambipolarity can be suppressed for a more significant part of negative gate bias (for an n-TFET), TFET can become a reliable device for circuit operation.

Inverter-based logic using TFETs will work well for a particular technology node if the leakage and ambipolar current associated with it will be less than the IRDS-defined limit [2]. This will help in achieving the minimum static power leakage. In general, changes in a TFET structure to minimize the ambipolar conduction also affect the ON-state performance and the miller capacitance. Therefore, before choosing a suitable architecture for TFET, all trade-offs should be studied in detail.

9.3 AMBIPOLAR I-V CHARACTERISTICS OF TFET

Before quantifying the ambipolar behavior in physical parameters, it is better to characterize them by the graphical representation for better clarity and understanding, as shown in Figure 9.2. For the case of n-TFET, the parameters are chosen based on the ideal reference behavior and how other characteristics deviate from it. This can be observed from Figure 9.2(a), as rectangular geometry is the ideal case scenario. The other two graphical representations are trapezoid and triangle. The reference case device remains in the OFF state for negative gate bias like MOSFET, and I_{AMB} is eliminated for negative V_{GS} ($\theta_2 = 180$). For the second trapezoid case, the I-V characteristics are inferior to Figure 9.2(a), wherein I_{AMB} remains lower than the current reference level depicted by the dashed line for the voltage range A, as seen in Figure 9.2(b). Here, the desirable trend would be to have higher θ_2 compared to θ_1 ($\theta_2 > \theta_1$). This will ensure that the I_{AMB} remains lower than the reference level as much as possible. The third and worst-case scenario is the transformation of a trapezoid into a triangle. Here

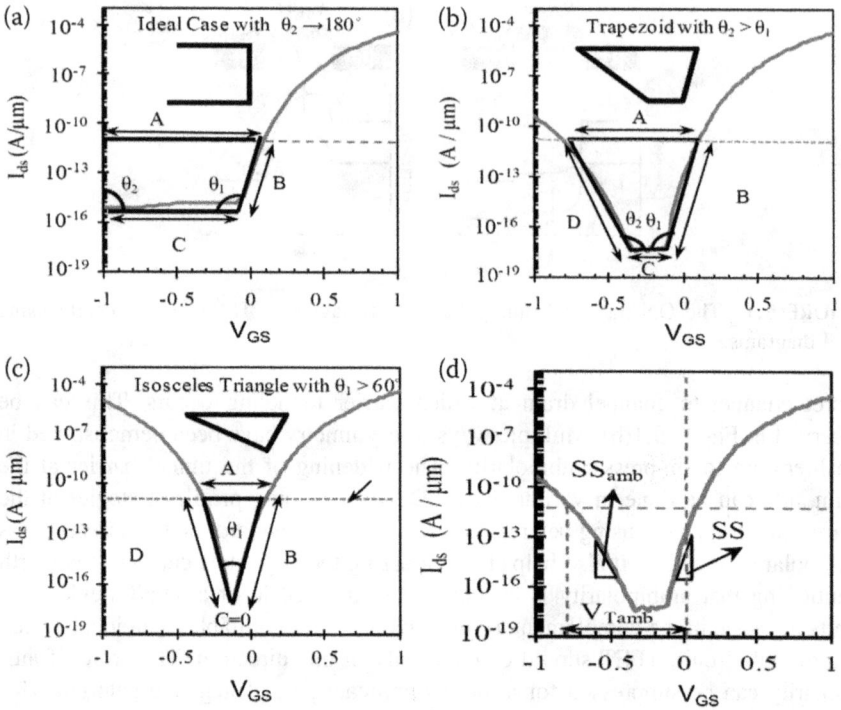

FIGURE 9.2 Three cases of ambipolar characteristics ((a), (b), (c)) and their analogy with geometrical shapes for quantification. (d) shows the SS and SS_{amb} in TFET characteristics.

the I-V characteristics become symmetric along the drain current axis, and θ_2 does not exist. The side C becomes zero, and I_{AMB} variation is sensitive to negative gate bias. Here, the θ_1 should be as large as possible so that the I_{AMB} remains low than the reference level for the larger part of the negative gate bias. Based on the discussion above, we can quantify the ambipolar characteristics of TFET using two parameters: V_{Tamb} and SS for the ambipolar region (SS_{amb}). V_{Tamb} signifies the range of negative V_{GS} for which the I_{AMB} remains below the defined level. At the same time, SS_{amb} tells us how quickly the device goes from OFF-state into significant ambipolar conduction. It is a subthreshold slope but in the negative gate bias direction. Unlike the conventional SS, which should be small for the steep turn-on of the device, SS_{amb} should be as high as possible so that I_{amb} remains below the reference level. Note that the SS is inversely proportional to the actual slope of the log (I_D) versus the V_{GS} curve.

9.4 COMMON I_{AMB} REDUCTION TECHNIQUES

9.4.1 GATE-DRAIN UNDERLAP

Some of the earlier techniques proposed in the literature for the suppression of the ambipolar current are the use of gate-drain underlap design [3] and a decrease in the doping concentration [4]. The use of underlap structure and the reduced drain concentration minimizes the electric field at the channel-drain junction. This widens the drain-side tunnel barrier. Both these methods decrease the ON-current of the device by increasing the series resistance of the channel and drain region, respectively. In addition, the first technique requires more than the minimum channel length necessary. In contrast, the second technique increases contact resistance and needs more masks when a complementary TFET technology is used.

9.4.2 REDUCED DRAIN DOPING CONCENTRATION

Drain doping must be carefully optimized to minimize the ambipolar current of a TFET while keeping its value realistic from a device fabrication perspective. It is very well understood in the literature that reducing drain doping concentration decreases the electric field on the drain side. Thus, it widens the tunnel barrier width and minimizes the off-state and ambipolar current. This method also allows us to reduce the off-state leakage and uncover a part of the I-V curve that has a steep slope. Drain doping reduction would require some silicide or other technique to form an excellent Ohmic drain-side contact [5].

9.4.3 USE OF LOW-K SPACER

Hetero-gate dielectric TFET was proposed [6] for higher I_{ON} and lower I_{OFF} and ambipolarity. Using the combination of low-k dielectric on the drain side and high-k dielectric on the source side, controlled ambipolar current and high BTBT in the ON-state can be ensured. This can be seen in Figure 9.3, which shows the use of a low-k dielectric on the drain effectively suppresses ambipolarity.

9.4.4 INFLUENCE OF THE CONTACT LAYOUT

The impact of ambipolar current in the silicon-based double gate (DG) TFETs [7] has been explored by placing the contact in a different configuration. The layout of drain

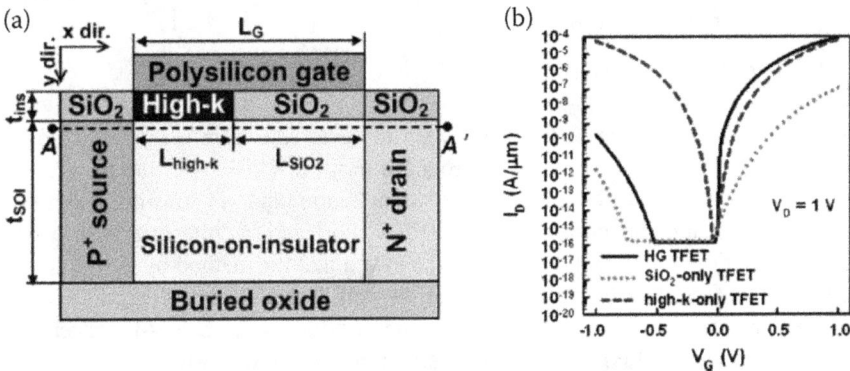

FIGURE 9.3 (a) Hetero-dielectric tunnel FET architecture (b) and their corresponding transfer characteristics.

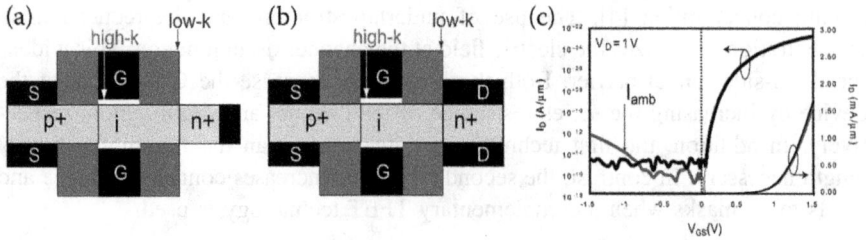

FIGURE 9.4 Silicon double gate TFET: (a) contact placed on the side, (b) contact on top/bottom. (c) Corresponding I-V characteristics.

contact can either be on the sides or the top/bottom configuration, as depicted in Figures 9.4(a) and (b). It was shown in the study that placing the contact on the top/bottom configuration eliminates the ambipolar current. The two-dimensional potential distribution shows that the potential lines are highly localized at the channel/drain interface, resulting in a large electric field for the side contact. On the other hand, in the case of the top/bottom configuration, potential lines are distributed uniformly in the spacer and the channel-drain interface. This leads to an almost constant electric field and reduces the parasitic ambipolar current, as shown in Figure 9.4(c).

9.5   AMBIPOLARITY IN HOMOJUNCTION TFET

Homojunction TFETs are devices that use single device material for all the junctions present in the structure. The band structures are not broken at the junctions and remain continuous throughout the device layer. All silicon, germanium, In As-based TFETs, or any other single material TFET can be classified as Homojunction TFETs. The earlier design of TFET started with homojunction TFET because of the enormous literature present related to the MOSFET. The majority of homojunction TFETs made structural modifications and doping changes to meet the criteria of suppressed ambipolarity. Some of them are discussed below.

9.5.1   STRUCTURAL ENGINEERING

A. Gate-drain underlap

The concept of using the underlap between the gate and drain region of TFET to suppress the ambipolar current was proposed early in 2007 by Anne S. Verhulst et al. [3]. They used the double gate architecture and demonstrated that the reduction of ambipolar current for a 100 nm short gate architecture with a high source/drain doping is possible. They also showed that the method of short-gate can be employed for the all-germanium TFET as well. Later, this method was adopted by many researchers worldwide. One example is the use of a low-k spacer with high-k dielectric in TFET (LSHG TFET structure), as shown in work [8]. Since the drain implant is done after the drain side spacer formation, they have shown that the physical gate-drain distance will automatically work as an underlap between them and decrease the undesired ambipolar current.

B. Intrinsic region adjacent to drain

TFET with asymmetric source and drain side have been fabricated in work demonstrated by [9]. They have added an intrinsic region (L_{IN}) as an extension of the channel region on the drain side. This region is useful in suppressing both the OFF-state and the ambipolar current while maintaining the same I_{ON} in the device, as shown in Figure 9.5. Through simulations, they have shown that the electric field on the drain side varies significantly with the change in the L_{IN}. The electric field is inversely proportional to the length of L_{IN} [Figure 9.5(c)]. If observed closely, this work is very similar to the gate-drain underlap structure since we inadvertently add an intrinsic channel region on the drain side by reducing the gate length.

C. Gate-over-drain overlap

So far, we have observed either the creation of the gate-drain underlap structure or the introduction of the intrinsic region on the drain side to minimize the impact of the gate-field on the drain side, as mentioned in the last section. However, an overlapping gate-over-drain as a potential method to solve the ambipolar issue has been proposed in work [10]. The authors have shown that using this method; the ambipolar current can be controlled up to the drain doping of 1×10^{19} cm^{-3}. By overlapping the highly doped drain region using the gate metal, the bands near the channel-drain junction are pulled up for the negative gate bias over the entire

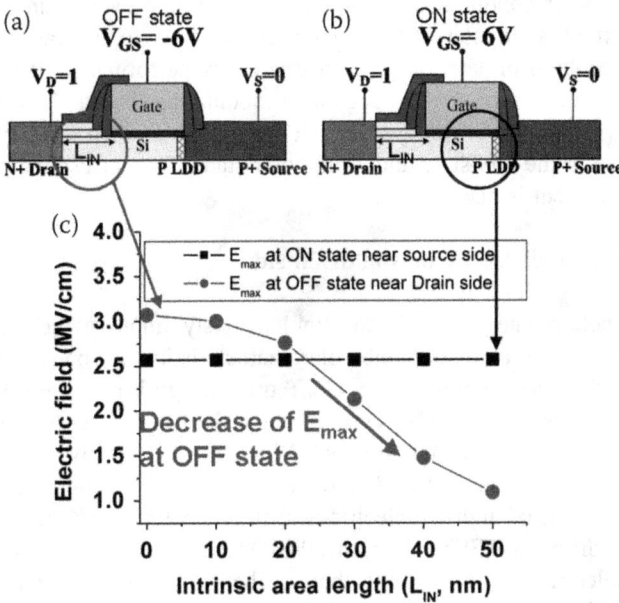

FIGURE 9.5 Bias polarity in the (a) ON-state and (b) Ambipolar state with the extended intrinsic region (L_{IN}). (c) Shows the electric field dependence vs. L_{IN} length on the drain side.

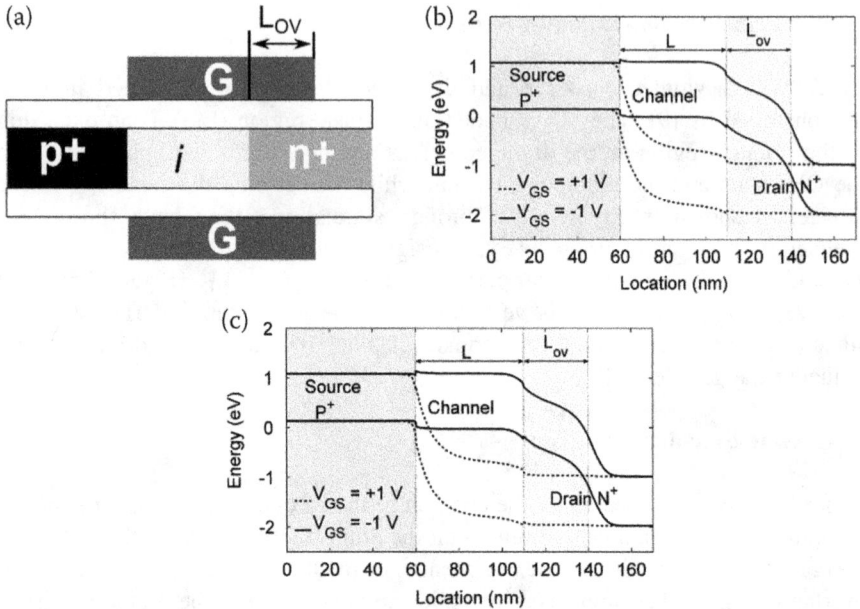

FIGURE 9.6 (a) Gate-over-drain DG TFET structure. (b) Band diagrams in the ON-state and ambipolar state. (c) Their corresponding I-V characteristics.

overlap region (L_{OV}). This controls the tunnel barrier width and can effectively suppress ambipolar conduction. The overlap of the gate over drain seems useful in the context of ambipolar reduction but is expected to increase the parasitic capacitance resulting in deteriorated high-frequency performance. It has also been shown that the impact of overlap region in modulating the bands to control ambipolar behavior. Figure 9.6 shows the gate-over-drain overlap DG TFET with the band diagrams in the ON-state and Ambipolar state. The corresponding reduction in ambipolar current is also shown.

D. Low-k gate dielectric on the drain side

The use of hetero-gate dielectric to simultaneously improve the ON-state performance without degrading the ambipolar characteristics was one of the few early ideas proposed in the literature. The TFET uses a high-k gate dielectric near the source side to improve the BTBT current, while the low-k gate dielectric near the drain side tends to minimize the ambipolar current. The negative impact of low-k dielectric on the ON-state current due to an increase in the channel region is taken care of with the use of high-k dielectric near the tunneling junction. This work [6] studied three different TFETs with an all-SiO2, all-HfO2, and a combination of both gate dielectrics. It was shown that the hetero-dielectric TFET has a drive current very close to that of the all-HfO2 dielectric TFET, while its ambipolar current is at the level of all-SiO2 TFET. In this way, a hetero-gate dielectric TFET is a device with the advantages of both reference structures.

E. Dual-metal-gate work function

The concept of dual metal work function (DMG) has been used extensively in the case of MOSFETs. Later, its implications on nanoscale TFET were also studied [11]. This method uses a gate of two different work function materials adjacent to each other. One is called the tunnel gate (ϕ_{tunnel}), and the other is called the auxiliary gate (ϕ_{aux}). As the name suggests, the tunnel gate is the one that is near the source, while the gate away from the source is called the auxiliary gate. For the case of n-TFET, the tunnel gate typically has a smaller value than the auxiliary gate. The smaller the ϕ_{tunnel}, the higher the drive current of the device. However, the value of the ϕ_{aux} should be chosen with care. The relationship between the I_{AMB} and ϕ_{aux} is not that straightforward. It has been shown in this study that with an increase in $\phi_{aux} > 4.0$ eV, the band overlap near the source side decreases, leading to a significant reduction in the OFF-state current. However, as $\phi_{aux} > 4.4$ eV, the band overlap starts appearing near the channel-drain junction, allowing the carriers to tunnel in the off- as well as ambipolar state. So, an optimized value of ϕ_{aux} is required in relation to ϕ_{tunnel} for effective suppression of the I_{AMB} in the DMG-based TFET. The above method can be more useful if combined underlap gate-drain structure or the low-k spacer on the drain side structure.

F. Tunnel dielectric-based TFET

Usually, it has been observed that TFET ON-state parameters are often compromised to fix ambipolarity. The effort to address one problem negatively impacts the other problem. In this work [12], a p-type tunnel-based dielectric TFET has been proposed that addresses this challenge. Instead of tunneling carriers through continuous band bending like normal TFETs, the carrier tunnel through a thin dielectric from the source side into the channel region. The device is referred to as tunnel-dielectric TFET (TD-TFET). The presence of tunnel dielectric provides an additional potential barrier to the carriers in the off and ambipolar state. This significantly reduces the off-state as well as the ambipolar leakage in TFET. It has been demonstrated that the drive currents can be modulated by varying the tunnel dielectric thickness. However, no undesired ambipolar currents appear in any case [Figure 9.7].

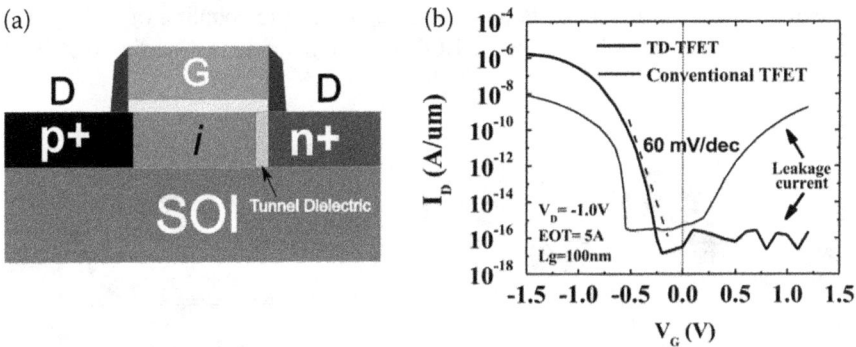

FIGURE 9.7   (a) Tunnel-dielectric-based TFET. (b) Shows suppressed I_{AMB}.

G. Orthogonal gate dielectric

TFETs have seen many structural changes to address the issue of ambipolarity. This work uses an orthogonal gate structure with a drain extension on top of the T-shaped channel to limit ambipolarity [13]. Two orthogonal gates are used with asymmetric gate dielectric thickness to control the electric field at the channel drain junction. The gate dielectric thickness on the source side ($t_{ox, h}$) is small, and it controls the I_{ON} of the device, while the dielectric gate thickness on the drain side ($t_{ox, v}$) is large and controls the off-state and ambipolar current. It has been demonstrated that this methodology provides flexibility in tuning the ON-state and ambipolar-state behavior independent of each other, which is not the case with the solutions proposed earlier. The orthogonal control of the I_{ON} and I_{AMB} became possible due to the different screening lengths at the source-channel ($\lambda_{source-channel}$) and channel-drain junction ($\lambda_{channel-drain}$), wherein $\lambda = \sqrt{\varepsilon_{Si} t_{Si} t_{ox}/2\varepsilon_{ox}}$. Smaller λ will result in higher BTBT current on the source while higher λ will suppress the BTBT current contributing to ambipolar behavior on the drain side. This TFET structure can be optimized by modulating the vertical thickness of the gate dielectric to mitigate the ambipolar current without affecting the ON-current of the device. In summary, depending upon the application of the device, we can independently tune the dielectric thickness to modulate the ambipolar current. Figure 9.8 shows the TFET structure and corresponding transfer characteristics.

H. Use of hetero-dielectric BOX

In this work [14], an SOI-based *pnpn* TFET is studied to control ambipolarity using the concept of hetero-dielectric buried oxide (HD BOX TFET). An HD BOX combined with a highly-doped ground plane increases the tunnel barrier width on the channel-drain interface in an SOI TFET. Low-k SiO_2 is used as a BOX dielectric beneath the source and the channel region, while a high-k HfO_2 dielectric is used under the drain. The idea of using high-k is to create a depletion width near the bottom of the drain region, effectively increasing the tunneling width. An optimum BOX dielectric width of 25 nm is shown to suppress the ambipolar current for most of the applied negative bias; however, it affects the ON-current. The depletion region formed in the bottom drain region due to coupling of high ground plane doping and the high-k dielectric BOX results in a potential barrier to electron

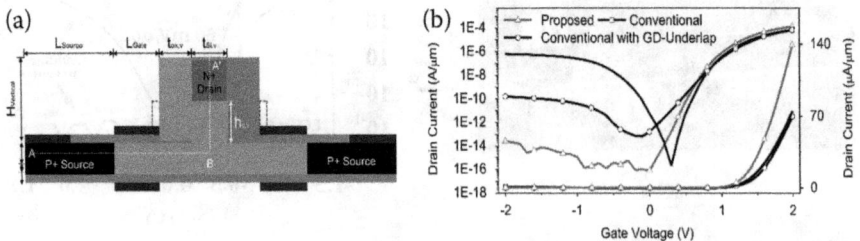

FIGURE 9.8   (a) Tunnel-dielectric-based TFET. (b) Shows suppressed I_{AMB}.

flow from the bottom of the drain region. This negatively impacts the ON-current for BOX thickness of 25 nm or below. However, if the BOX thickness is chosen to be 30 nm and above, the impact on the ON-state current is negligible, and the ambipolar current is reasonably under control.

I. Use of metal-silicide drain

Most of the methods to alleviate ambipolar conduction is achieved at the cost of some other process complexity. In this work [15], ambipolarity is suppressed without sacrificing the ON-state performance of conventional n-TFET, with the p-i-n structure being replaced by the p-i-m structure, where the drain side is a metal silicide of a suitable work function. The device can retain the good ON-state current with improved I_{ON}/I_{OFF} ratio and ambipolarity. Further, the traditional drain implant method is eliminated. The initial work function was chosen to be 4.25 eV, slightly higher than the work function of the Si-channel region (4.215 eV).

Further, it is then varied from the 4.1 eV up to 4.5 eV to see its impact on the device performance. It has been observed that the tunnel barrier width in the case of p-i-m TFET is large compared to the conventional p-i-n TFET. This translates into a larger BTBT difference between these two devices in the OFF- and ambipolar state. It has been further stated that work function below 4.1 eV or higher than 4.5 eV degrades the off- and ambipolar characteristics. This means there should be an optimal design methodology to choose the work function. For the smaller work function below the channel work function, the metal silicide forms the Ohmic contact, whereas 4.25 eV shows the Schottky contact, which is the reason for the higher ambipolar current. Similarly, for a higher work function of 4.5 eV, even though it is a Schottky contact, the band bending is such that there is a lot more injection of hole carriers from the drain metal silicide. For this case study, 4.25 eV is the optimal work function of the metal silicide used as a drain. The corresponding structure and the ambipolar characteristics have been shown in Figure 9.9.

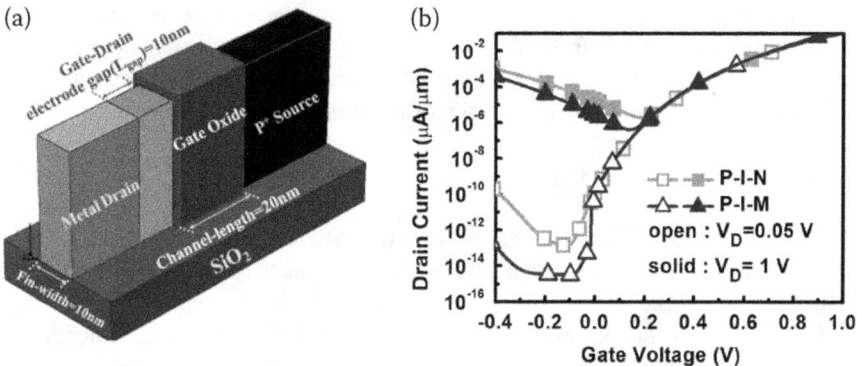

FIGURE 9.9   (a) Fin-TFET based on metal silicide drain. (b) Shows suppressed I_{AMB} using silicide.

9.5.2 DOPING ENGINEERING

A. Lateral Gaussian Doping

It has been shown in work [16] that lateral Gaussian doping in the drain region can mitigate the ambipolar conduction for the double-gate TFET. Drain doping profiles with different characteristics length (CL) have been compared with uniform drain doping. It has been demonstrated on the ATLAS device simulator that the best possible reduction of OFF-state and ambipolar current is observed for CL = 0.05. The result obtained is the non-abrupt channel-drain junction formed due to Gaussian doping, unlike uniform doping. This non-abrupt junction creates a low electric field resulting in low tunneling of carriers in the ambipolar state. For peak Gaussian doping density up to 1×19 cm^{-3} in the drain region, the ambipolarity is conservative; however, as the peak doping is further increased, a significant amount of conduction is observed. However, it is still low by several orders of magnitude compared to the uniformly doped drain. This method highlights the fact that if one can obtain a non-abrupt junction on the drain side using asymmetric doping, it is possible to reduce ambipolar conduction. Though Gaussian doping is an effective method, physical realization is a difficult task.

B. Stacked Drain Doping

In this method [17], a highly doped and lightly doped drain region is stacked together in which the latter is placed below the former. The presence of an intrinsic or low-doped drain region is intended to suppress the ambipolarity without compromising on the high-frequency switching of the device. Compared to the conventional SOI-TFET, the proposed structure in this work has two drain-one is lightly doped, and the other is heavily doped. The thickness t_h represents the height of the highly doped region. It has been observed that the lower the heavily doped top portion of the device, the lower the ambipolar current. This can be attributed to the reduced area of a high electric field in the channel drain junction as we increase the lightly doped portion in the drain region. Hence the number of carriers that can tunnel through the highly doped region reduces, and we observe a reduction in the ambipolar current. Figure 9.10 shows the cross-sectional schematic of the drain-engineered TFET and its corresponding I-V characteristics exhibiting suppressed ambipolarity for different t_h thicknesses of the highly doped drain region.

C. Vertical Gaussian Doping

The work proposed [18] derives the concept of vertical gaussian doped silicon film thickness from the junction-less transistor. It uses it in the SOI-TFET architecture for the very first time. Here, as shown in Figure 9.11(a), an n-type SOI TFET with vertically gaussian doped silicon film is proposed (VG-SOI-TFET), wherein the peak doping density (N_P) is chosen to be at the top of the film. Later, the source can be realized using the high work function metal to induce the p$^+$ source region. This gaussian doped silicon film allows the non-abrupt channel-drain junction to form.

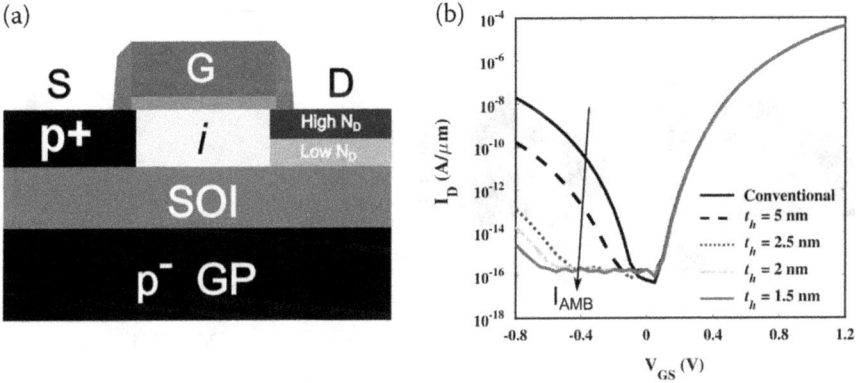

FIGURE 9.10 (a) Cross-section of the multi-doped drain SOI TFET. (b) Corresponding I-V characteristics show suppressed ambipolarity.

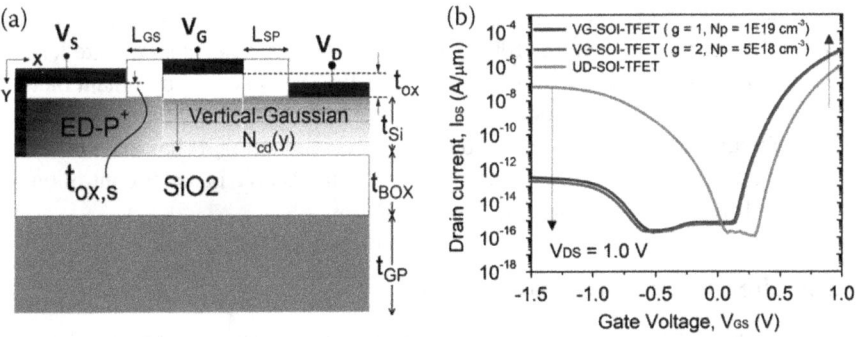

FIGURE 9.11 (a) Vertical Gaussian doped SOI TFET. (b) Corresponding I-V characteristics show suppressed ambipolarity.

The study shows that the ambipolar current is a sensitive function of the peak doping density (N_P) alongside the doping gradient of the Gaussian function (g). A rigorous study has concluded that the best suppression in the ambipolar current observed for the N_P of 1×19 cm^{-3} and g of 1 nm/decade, as illustrated in Figure 9.11(b).

D. Drain Pocket Engineering

Various doping methods engineering on the drain side have been adopted to suppress the electric field at the channel-drain junction. Another method that emerged is using a drain pocket (DP) near the channel-drain junction [19]. Combined with the existing methods, this method can potentially improve the overall performance of the TFET. The presence of the DP decreases the band overlap and increases the tunnel barrier width. This, in turn, eliminates ambipolar conduction to a great extent. The proposed method effectively suppresses the I_{AMB} even when drain doping

(a)

S   G   D

p+   *i*   **n+**

G   Drain
pocket

(b)

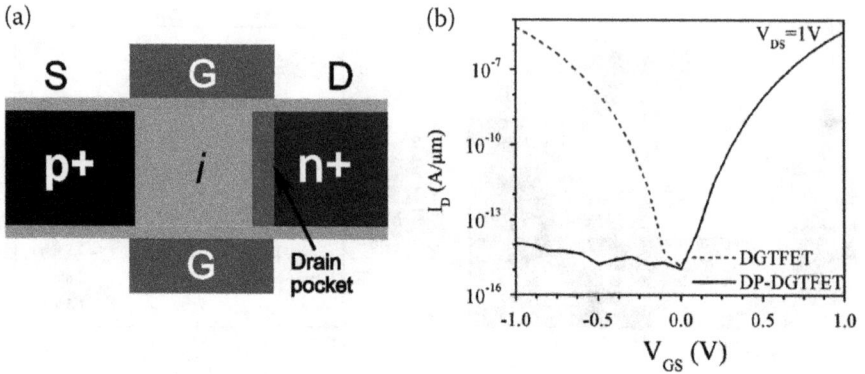

FIGURE 9.12 (a) Double gate TFET with drain pocket. (b) Corresponding I-V characteristics show suppressed ambipolarity.

is high and equal to the source doping. DP as the method, effectively controls the I_{AMB} for longer channel length (100 nm) devices. It is because, at short gate lengths of 20 nm and below, direct source-to-drain tunneling becomes the dominant method of ambipolar conduction. In such a case, this method must be combined with some other techniques to minimize direct source-to-drain tunneling. It has been observed that at an optimum DP doping of 1.5×10^{19} cm^{-3}, the BTBT is under control on the drain side [Figure 9.12(a) and (b)].

9.5.3 JUNCTION-LESS/DOPING-LESS DESIGN

TFETs based on junction-less and doping-less designs are a class of electrostatically-doped semiconductor (ED) devices. Lately, they have been explored in the literature to include the advantage of field-effect transistors without junctions [20]. The choice to not worry about the formation of metallurgical at advanced node FETs has been made possible using ED. Junction-less (JL) TFET differs from the doping-less TFET or charge-plasma (CP) TFET because, in JL TFET, the semiconductor film has an initial doped to a value. It could either be of n-type or p-type. However, DL TFET has no initial doping. All dopings are induced using the source/gate/drain metal or silicide of suitable work function. The absence of a metallurgical junction indicates the non-abrupt junction presence on the drain side. This will translate into a controlled ambipolar effect in these devices. Few of which are discussed below:

A. Electrostatically Doped Drain-Engineered TFET

In this work, unlike a conventional double gate TFET with a p-i-n structure created using doping, the proposed device has constant n-type doping in the channel-drain region [21]. The device is termed drain-engineered DG-TFET (DE-DG-TFET). The initial structure resembles the p^+-n-n structure, later modified to p^+-n-n$^+$. This way, an inbuilt drain pocket is created on the drain side of length L_{GD} (gate-drain gap).

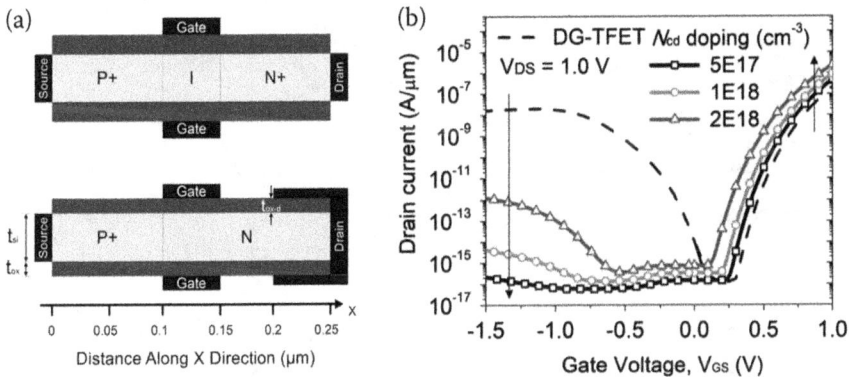

FIGURE 9.13 (a) Drain-engineered electrostatically-doped double gate TFET. (b) Corresponding I-V characteristics show suppressed ambipolarity.

The initial doping of film decides the doping of the pocket. The low work function of the drain side metal makes the n-type doping beneath it to n^+. The transfer characteristics for various N_{CD} (channel-drain doping) has been shown. It has been observed that the N_{CD} value of 1×18 cm^{-3} gives a reasonably good I_{ON} in addition to suppressed ambipolar conduction, as shown in Figures 9.13(a) and (b).

B. PNPN Tunnel FET

A detailed investigation to obtain a controllable drain-side tunnel barrier width in a *pnpn* TFET is done in this work [22]. The starting structure is a p^+-n doped silicon film. Using the concept of ED, both the p+ source and n+ drain can be realized without needing chemical-doped junctions. On the drain side, the gap between the gate and drain metal will act as a design parameter that can be tuned to optimize the tunnel barrier width at the channel-drain junction. The source-side work function of 5.93 eV is taken to induce the hole concentration, while the drain-side work function of 3.93 eV is taken to induce the electron concentration. A *pnpn* structure is then created, with a high drive current and simultaneously a reduced OFF-state and ambipolar current. It is well known that the tunnel barrier width on the drain side is controlled by depletion layer width, which should be large to control the ambipolar behavior. This is impossible for a chemically-doped TFET without reducing the drain doping or using the gate-drain underlap. However, for ED-doped TFETs, L_{gap} can be tuned to increase the depletion layer width. It is observed that for $L_{gap} = 30$ nm or above, the ambipolar current is wholly suppressed for all values of drain-to-source bias (V_{DS}). Figure 9.14 shows the schematic cross-section of the charge plasma-based PNPN TFET and corresponding I-V characteristics.

9.6 AMBIPOLARITY IN HETEROJUNCTION TFET

So far, we have discussed various methods to control the ambipolarity in homo-junction TFETs that use a single material device layer. The methods to control the

(a)

(b)

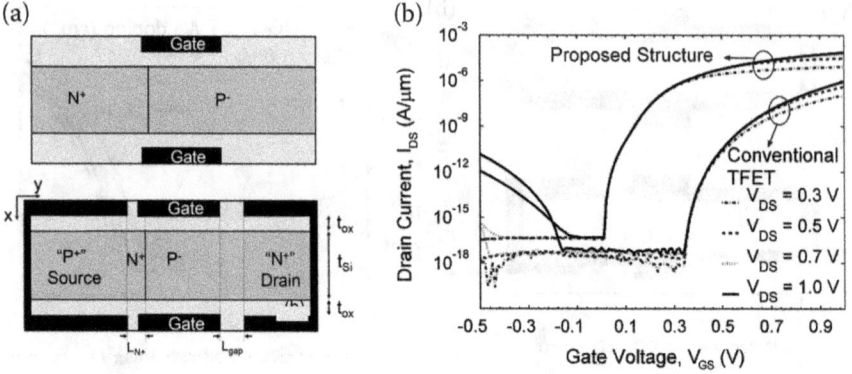

FIGURE 9.14 (a) Charge plasma-based PNPN tunnel FET. (b) Corresponding I-V characteristics show suppressed ambipolarity.

ambipolarity basically constitutes the doping engineering on the drain side alongside the use of low-k spacer and low-k gate dielectric on the drain side. This section will examine the existing methods to suppress ambipolarity in heterojunction TFETs. One standard approach is using large bandgap material in a TFET on the drain side to increase the bandgap and thus the tunneling width effectively. Along these lines, a few of the methods are discussed below.

9.6.1 Si/GaAs TFET

Most ambipolar current-reducing techniques are plagued by the I_{ON} degradability, process variation complexity, and compromised frequency response of the device due to increased capacitances. High bandgap material for the drain region can be used to increase the drain side tunnel barrier width. In this context, this work [23] introduces a Silicon/Gallium Arsenide material system-based heterojunction TFET (HTFET), as schematically shown in Figure 9.15(a). The device has a Si source region and GaAs channel-drain region. The results of the proposed device were compared with the two

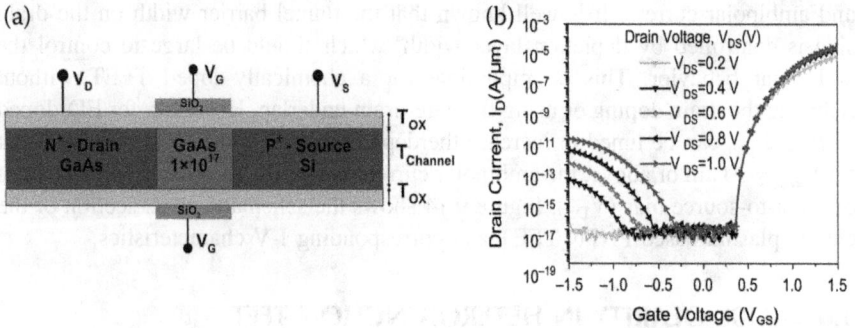

(a)

(b)

FIGURE 9.15 (a) Si/GaAs DG heterojunction TFET. (b) Corresponding I-V characteristics show suppressed ambipolarity for various V_{DS}.

reference devices: all-GaAs TFET and all-Si TFET. From the band energy diagrams, Si/GaAs TFET has larger tunnel barrier width due to the increased bandgap of GaAs. The tunneling probability has a negative exponential dependence on the bandgap; thus, we observe a significant reduction in the ambipolar current. Figure 9.15(b) plots the transfer characteristics for different V_{DS}. It is observed that ambipolarity is completely suppressed until VGS = −0.4V and then starts depending upon the V_{DS} voltage. The device has the added advantage of smaller gate capacitance compared to the all-Si TFET, which then reflects in terms of better AC performance of the device.

9.6.2 Si/GaAs Doping-less TFET

Silicon as a source material and Gallium Arsenide (GaAs) as a channel-drain material has been proposed in this work, in addition to the device being doping-less [24]. This means that the charge plasma concept will be used on the Si/GaAs layer to realize the p^+-i-n^+ doping. The authors reported 8 orders of reduction in ambipolar current at V_{GS} = −1.5V compared to an all-Si doping-less device. The source-side work function is 5.93 eV, and the drain-side work function is 3.93 eV. Gate-drain gap (L_{GD}) of 15 nm has been taken as it will play a crucial role in deciding the ambipolar current. As seen in Figure 9.16, the reduced band overlaps can be attributed to multiple reasons. First, the device is doping-less, so there is a non-abrupt channel-

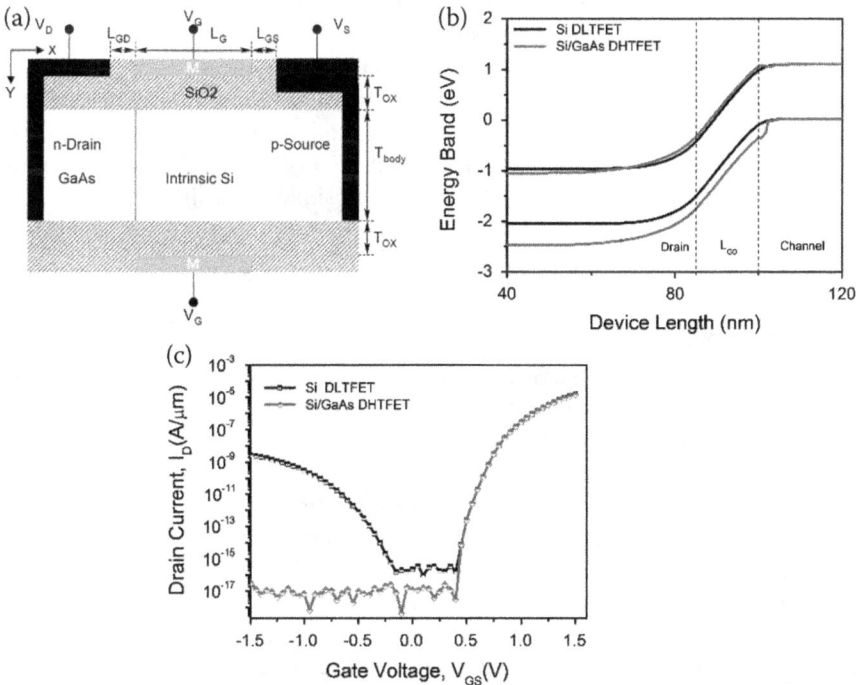

FIGURE 9.16 (a) A doping-less Si/GaAs heterojunction TFET. (b) Corresponding band diagrams in the ambipolar state. (c) I-V characteristics show suppressed ambipolarity.

drain junction. Second, L_{GD} is reasonably high. Finally, the GaAs have a high bandgap compared to silicon. So cumulatively, the drain-side tunnel barrier width is large enough to suppress the ambipolarity for the entire range of V_{GS} up to -1.5 V.

9.6.3   PSEUDO SPLIT GATE IN$_{0.53}$GA$_{0.47}$AS/INP HETEROJUNCTION TUNNEL FET

This work [25] studies a heterojunction structure of In0.53Ga0.47As/InP-based TFET. The device uses two gates: one is the conventional control gate, and the other one is the gate on top of the drain, termed a pseudo-split gate (PSG). Work function and the equivalent gate oxide (EOT) are the same for both gates. PSG is located at an optimized distance from the main gate. The simulation study shows that the ambipolar current is suppressed by 10 orders of magnitude compared to the non-overlapped structure (Lovr = 0 nm).

In contrast, in comparison to the overlapped gate-on-drain (OGD HetTFET) structure (Lovr = 30 nm), the improvement is 7 orders of magnitude. The device also exhibits lower gate capacitances compared to the overlapped gate-on-drain structure. PSG-HetTFET offers better ambipolar suppression than overlapped structures even at higher drain doping and smaller EOT. It has been observed that ambipolarity is well controlled for LPSG of 30 nm and above; however, increasing L_{SG} beyond 30 nm also increases the gate capacitance (C_{gg}). Also, to have a fair comparison between the three different structures in this study, the length of the overlapping gate in the conventional structure is kept at 30 nm. It has been observed that OGD HetTFET has comparatively higher C_{gg} compared to the other two. The reason for suppressed ambipolar current is the same in this structure as that of overlapped gate-on-drain structure; however, this method gives us additional parameters such as L_{PSG}, the work function of PSG (Φ_{PSG}), to further control the ambipolar current as well as the gate capacitance simultaneously Figure 9.17.

9.7   AMBIPOLAR SUPPRESSION IN ADVANCED TFET

9.7.1   Z-SHAPED TFET

This work utilizes the concept of vertical BTBT (line TFET) to scale the device and simultaneously provide higher I_{ON}, I_{ON}/I_{OFF} ratio, and suppressed ambipolar behavior. A Z-shaped TFET is proposed in which the source, channel, and gate regions are elevated vertically [26]. The device has a tunnel direction perpendicular to the channel direction, which results in the formation of a relatively large tunnel area. At the same time also allows for the suppression of any lateral tunneling using raised BOX dielectric, making this device a unique and interesting concept. Because the channel has an elevated design, the physical length of the channel region is effectively increased, resulting in the suppression of the ambipolar characteristics. Also, the buried layer beneath the source helps mitigate the ambipolar behavior since it prevents parasitic lateral tunneling of carriers from source to drain. Figure 9.18 shows the schematic cross-section of the conventional and Z-shaped PNPN TFET and their ambipolar characteristics Figure 9.19.

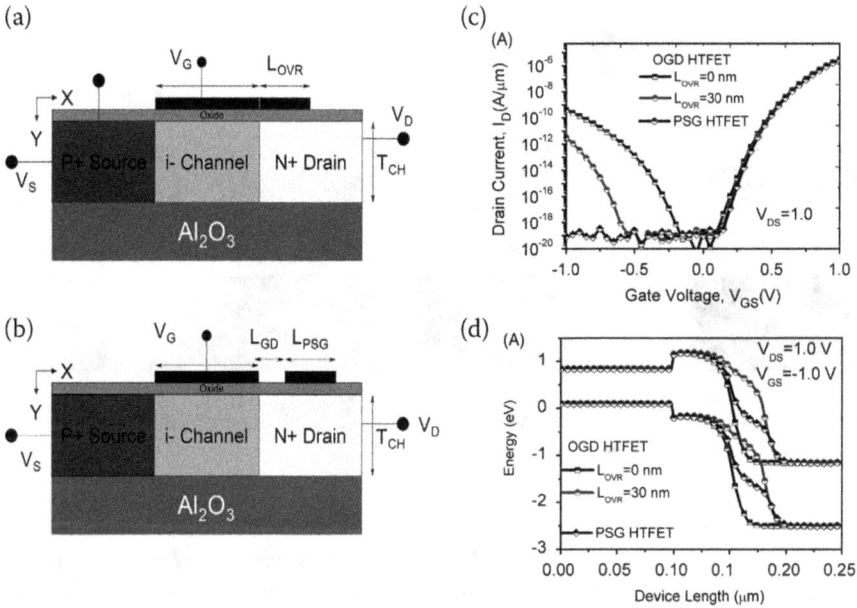

FIGURE 9.17 (a), (b) Overlapped gate-on-drain and pseudo-split-gate $In_{0.53}Ga_{0.47}As/InP$ Heterojunction Tunnel FET. (c), (d) Compares the drain current characteristics and the band diagrams in the ambipolar state.

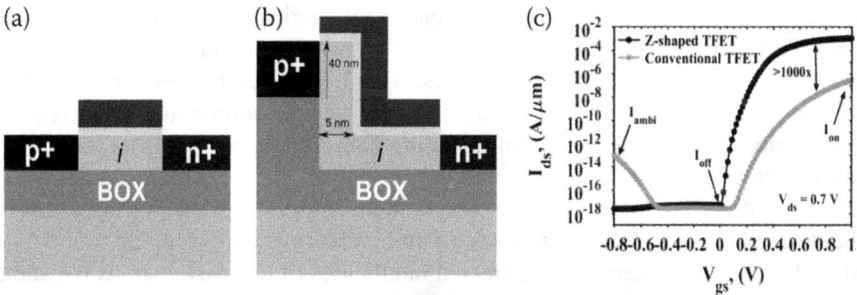

FIGURE 9.18 (a) Conventional p-i-n TFET, (b) Z-shaped PNPN TFET, and (c) corresponding I-V characteristics showing ambipolar for both the devices.

9.7.2 DUAL-MOS TFET

The TFET proposed in this work is a line tunneling device having dual MOS-capacitor extensions (D-MOS TFET). It is compared against the line tunneling planar single MOS-capacitor-based TFET [27]. Both TFETs can inhibit ambipolar conduction using the concept of gate-drain underlap, as shown in the adjoining figure. However, in the case of raised D-MOS TFET, the lateral tunneling is relatively low compared to the planar conventional line TFET because of the smaller gate field at the source edges. Thus, the source carriers which can tunnel into the

(a)   **D**   (b)

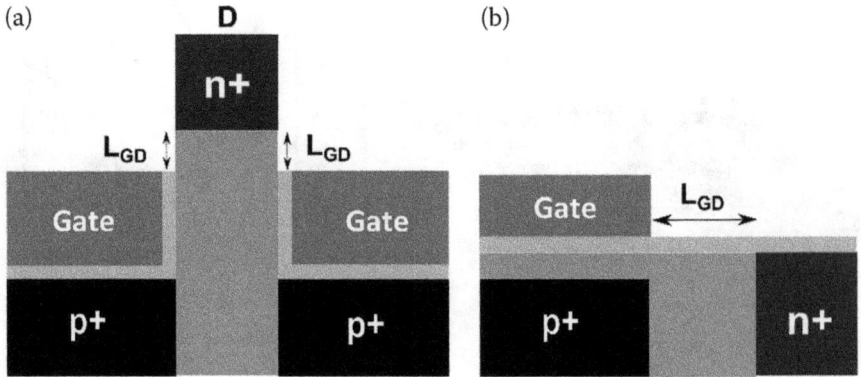

FIGURE 9.19 (a) Dual MOSCAP TFET with elevated channel-drain structure and (b) conventional single gate line TFET.

channel as a parasitic lateral leakage are small in the D-MOS TFET. Due to this, for a similar value of L_{GD}, the raised D-MOS structure has a lower probability of carriers tunneling from the channel into the drain region during the negative gate bias operation. Also, the D-MOS TFET is better in terms of I_{ON}, SS, and I_{ON}/I_{OFF} ratio compared to planar line TFET.

9.7.3 RECESSED-GATE TFET

The work proposed here comprises a trench gate vertically elevated source-drain line TFET [28]. The device footprint is limited owing to its raised structure. Also, since the device has two p+ Ge sources, a high drive current is expected. The device also exhibits the oxide separation between the source and drain region, which cuts off the parasitic lateral tunneling path for the carriers from the source toward the drain. Only vertical tunneling is allowed, due to which sharper SS is observed. Also, due to its intrinsic design, there is gate-on-drain overlap exists in the device because of which suppressed ambipolarity is expected. Up to the V_{GS} of -0.4V, the ambipolarity is suppressed in all cases of N_D; however, as N_D starts increasing above the value of 5×10^{18} cm^{-3}, the band modulation on the drain side due to overlapped gate starts to lose control, as evident from ambipolar current in the Figure 9.20.

9.7.4 ELECTRON-HOLE BILAYER TFET

TFETs, based on the idea of line tunneling between the induced 2-D electron and hole gas layers, were proposed by Lattanzio et al. in 2012 [29]. The device is termed an electron-hole bilayer TFET (EHBTFET). Since line tunneling TFETs are better than conventional TFETs in terms of SS and I_{ON}. EHBTFET capitalizes on that and increases the net tunneling area by inducing the electron and hole concentration in the intrinsic channel region. It is achieved by selectively choosing the appropriate top and bottom gate work function with asymmetric bias

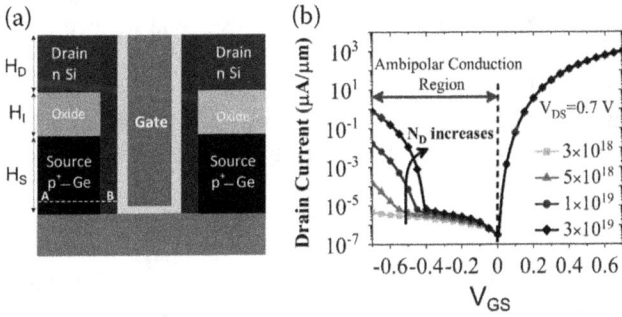

FIGURE 9.20 (a) Schematic of the proposed recessed TFET and (b) corresponding I-V characteristics showing ambipolar behavior for various drain doping.

on both gates. Once done, the tunnel junction is formed parallel to the gate inside the channel. The entire length of the channel now acts as a tunneling area. High ON-currents with steep SS are obtained in Ge-based EHBTFET. However, multiple leakage paths are present in EHBTFETs due to geometric misalignment that degrades the SS of the device. It was later reported that the leakage is a sensitive function of underlap length and can be mitigated using the hetero-gate technique. An additional leakage path exists between the channel and drain region for negative gate bias and causes ambipolar leakage. Figure 9.21(a)–(c) shows the two proposed structures of hetero-gate EHBTFET to mitigate the ambipolar

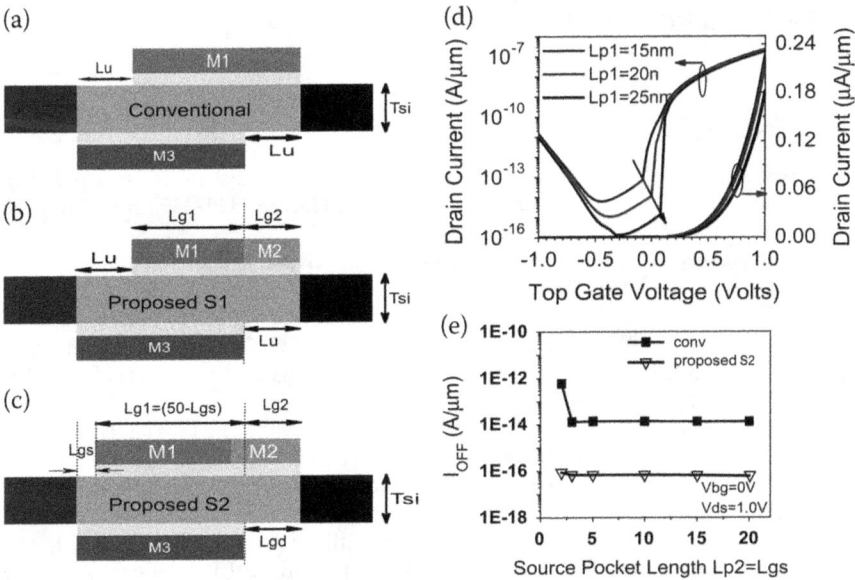

FIGURE 9.21 Schematic of the (a) conventional, (b) proposed S1, and (c) proposed S2 EHBTFET. (d) Corresponding I-V characteristics show ambipolar characteristics for all three devices. (d) Shows improved I_{OFF} using proposed structure S2.

leakage paths. The proposed structure 1 (S1) uses one dielectric pocket on the drain side and helps suppress the OFF-state and ambipolar leakage. The proposed structure 2 (S2) further helps improve the I_{ON}, I_{ON}/I_{OFF}, SS, and I_{AMB} using the source side underlap dielectric pocket. The results from the simulation for both structures are shown in Figures 9.21(d) and (e).

9.8 SUMMARY

In summary, we have dealt with the recent advances in TFET ambipolar behavior from different perspectives. The fundamental of TFET ambipolar characteristics and why it is a potential candidate to replace the conventional MOSFET is discussed briefly. Theoretically, it has the advantage of sub-kT/q SS, low leakage, and high I_{ON}/I_{OFF} ratio but low drive current and ambipolar behavior hinder its adoption in digital circuit design. To make TFET a viable alternative, different device engineering strategies of ambipolar suppression have been employed in the literature. This chapter reviews techniques for minimizing ambipolarity, broadly classified into 1) structural engineering and 2) doping engineering. The impact of various strategies on other parameters of importance is also studied. The authors have tried to gather all the possible state-of-the-art techniques of ambipolar suppression in TFET and present them in one place. The beginners will find it helpful to kick-start their research endeavors. This review will also help researchers working on advanced TFET architectures who want to control ambipolar conduction in their work. Either they can improvise a single technique or use their amalgamation as needed. In essence, this review highlights the recent structural and semiconductor material engineering to suppress the ambipolar behavior in tunnel transistors.

REFERENCES

[1] A. M. Ionescu and H. Riel, "Tunnel field-effect transistors as energy-efficient electronic switches," *Nature*, vol. 479, no. 7373, pp. 329–337, 2011, 10.1038/nature10679.

[2] R. Narang, M. Saxena, R. S. Gupta, and M. Gupta, "Assessment of ambipolar behavior of a tunnel FET and influence of structural modifications," *J. Semicond. Technol. Sci.*, vol. 12, no. 4, pp. 482–491, 2012, 10.5573/JSTS.2012.12.4.482.

[3] A. S. Verhulst, W. G. Vandenberghe, K. Maex, and G. Groeseneken, "Tunnel field-effect transistor without gate-drain overlap," *Appl. Phys. Lett.*, vol. 91, no. 5, p. 53102, 2007, 10.1063/1.2757593.

[4] K. Boucart and A. M. Ionescu, "Double-Gate Tunnel FET With High-κ Gate Dielectric," *IEEE Transactions on Electron Devices*, vol. 54, no. 7, pp. 1725–1733, 2007. [Online]. Available: http://ieeexplore.ieee.org/lpdocs/epic03/wrapper.htm?arnumber=4252356

[5] K. Boucart, "Simulation of Double-Gate Silicon Tunnel FETs with a High-k Gate Dielectric," *Techniques*, vol. 4729, p. 136, 2010, [Online]. Available: http://infoscience.epfl.ch/record/148596

[6] W. Y. Choi and W. Lee, "Hetero-Gate-Dielectric Tunneling Field-Effect Transistors," *IEEE Trans. Electron Devices*, vol. 57, no. 9, pp. 2317–2319, Sep. 2010, 10.1109/TED.2010.2052167.

[7] Hraziia, A. Vladimirescu, A. Amara, and C. Anghel, "An analysis on the ambipolar current in Si double-gate tunnel FETs," *Solid. State. Electron.*, vol. 70, pp. 67–72, 2012, 10.1016/j.sse.2011.11.009.

[8] C. Anghel, Hraziia, A. Gupta, A. Amara, and A. Vladimirescu, "30-nm Tunnel FET With Improved Performance and Reduced Ambipolar Current," *IEEE Trans. Electron Devices*, vol. 58, no. 6, pp. 1649–1654, Jun. 2011, 10.1109/TED.2011.212 8320.

[9] J. Wan, C. Le Royer, A. Zaslavsky, and S. Cristoloveanu, "Tunneling FETs on SOI: Suppression of ambipolar leakage, low-frequency noise behavior, and modeling," *Solid. State. Electron.*, vol. 65–66, no. 1, pp. 226–233, 2011, 10.1016/j.sse.2011 .06.012.

[10] D. B. Abdi and M. J. Kumar, "Controlling Ambipolar Current in Tunneling {FETs} Using Overlapping Gate-on-Drain," *IEEE J. Electron Devices Soc.*, vol. 2, no. 6, pp. 187–190, Nov. 2014, 10.1109/JEDS.2014.2327626.

[11] S. Saurabh and M. J. Kumar, "Novel attributes of a dual material gate nanoscale tunnel field-effect transistor," *IEEE Trans. Electron Devices*, vol. 58, no. 2, pp. 404–410, 2011.

[12] Z. Luo, H. Wang, N. An, and Z. Zhu, "A Tunnel Dielectric-Based Tunnel FET," *IEEE Electron Device Lett.*, vol. 36, no. 9, pp. 966–968, Sep. 2015, 10.1109/LED.2 015.2458932.

[13] M. R. Uddin Shaikh and S. A. Loan, "Drain-Engineered TFET With Fully Suppressed Ambipolarity for High-Frequency Application," *IEEE Trans. Electron Devices*, vol. 66, no. 4, pp. 1628–1634, Apr. 2019, 10.1109/TED.2019. 2896674.

[14] S. Sahay and M. J. Kumar, "Controlling the Drain Side Tunneling Width to Reduce Ambipolar Current in Tunnel FETs Using Heterodielectric BOX," *IEEE Trans. Electron Devices*, vol. 62, no. 11, pp. 3882–3886, 2015, 10.1109/TED.2015. 2478955.

[15] S.-C. Teng, Y.-S. Su, and Y.-H. Wu, "Design and Simulation of Improved Swing and Ambipolar Effect for Tunnel FET by Band Engineering Using Metal Silicide at Drain Side," *IEEE Trans. Nanotechnol.*, vol. 18, pp. 274–278, 2019, 10.1109/ TNANO.2019.2902251.

[16] V. Vijayvargiya and S. K. Vishvakarma, "Effect of Drain Doping Profile on Double-Gate Tunnel Field-Effect Transistor and its Influence on Device RF Performance," *IEEE Trans. Nanotechnol.*, vol. 13, no. 5, pp. 974–981, Sep. 2014, 10.1109/TNANO.2014.2336812.

[17] A. Shaker, M. El Sabbagh, and M. M. El-Banna, "Influence of Drain Doping Engineering on the Ambipolar Conduction and High-Frequency Performance of TFETs," *IEEE Trans. Electron Devices*, vol. 64, no. 9, pp. 3541–3547, Sep. 2017, 10.1109/TED.2017.2724560.

[18] M. Ehteshamuddin, S. A. Loan, and M. Rafat, "A vertical-gaussian doped soi-tfet with enhanced dc and analog/rf performance," *Semicond. Sci. Technol.*, vol. 33, no. 7, p. 75016, Jun. 2018, 10.1088/1361-6641/aac97d.

[19] S. Garg and S. Saurabh, "Suppression of ambipolar current in tunnel FETs using drain-pocket: Proposal and analysis," *Superlattices Microstruct.*, vol. 113, pp. 261–270, Jan. 2018, 10.1016/j.spmi.2017.11.002.

[20] G. Gupta, B. Rajasekharan, and R. J. E. E. Hueting, "Electrostatic Doping in Semiconductor Devices," *IEEE Trans. Electron Devices*, vol. 64, no. 8, pp. 3044–3055, Aug. 2017, 10.1109/TED.2017.2712761.

[21] M. R. U. Shaikh, S. A. Loan, and A. Alshahrani, "Electrostatically doped drain engineered DG-TFET: Proposal and analysis," *Int. J. Numer. Model. Electron. Networks, Devices Fields*, vol. 33, no. 6, p. e2769, 2020.

[22] D. B. Abdi and M. Jagadesh Kumar, "PNPN tunnel FET with controllable drain side tunnel barrier width: Proposal and analysis," *Superlattices Microstruct.*, vol. 86, pp. 121–125, Oct. 2015, 10.1016/j.spmi.2015.07.045.

[23] M. Haris, S. A. Loan, and others, "Si/GaAs hetero junction tunnel FET: design and investigation," *J. Nanoelectron. Optoelectron.*, vol. 14, no. 10, pp. 1434–1444, 2019.

[24] M. Haris, S. A. Loan, and others, "An Ambipolar immune Si/GaAs hetero-junction doping-less TFET," in *2017 International conference on Microelectronic Devices, Circuits and Systems (ICMDCS)*, 2017, pp. 1–4.

[25] M. Haris, S. A. Loan, and Mainuddin, "Pseudo Split Gate $In_{0.53}Ga_{0.47}As$/InP Hetero-Junction Tunnel FET: Design and Analysis," *Int. J. Numer. Model. Electron. Networks, Devices Fields*, vol. 34, no. 2, Mar. 2021, 10.1002/jnm.2826.

[26] R. Molaei Imenabadi, M. Saremi, and W. G. Vandenberghe, "A Novel PNPN-Like Z-Shaped Tunnel Field- Effect Transistor With Improved Ambipolar Behavior and RF Performance," *IEEE Trans. Electron Devices*, vol. 64, no. 11, pp. 4752–4758, Nov. 2017, 10.1109/TED.2017.2755507.

[27] M. Ehteshamuddin *et al.*, "Investigating a Dual MOSCAP Variant of Line-TFET with Improved Vertical Tunneling Incorporating FIQC Effect," *IEEE Trans. Electron Devices*, vol. 66, no. 11, pp. 4638–4645, Nov. 2019, 10.1109/TED.2019. 2942423.

[28] I. Chahardah Cherik and S. Mohammadi, "Enhanced on-state current and suppressed ambipolarity in germanium-source dual vertical-channel TFET," *Semicond. Sci. Technol.*, vol. 36, no. 4, 2021, 10.1088/1361-6641/abd63e.

[29] L. Lattanzio, L. De Michielis, and A. M. Ionescu, "The electron–hole bilayer tunnel FET," *Solid. State. Electron.*, vol. 74, pp. 85–90, Aug. 2012, 10.1016/j.sse.2012. 04.016.

10 Analysis of Channel Doping Variation on Transfer Characteristics to High-Frequency Performance of F-TFET

Prabhat Singh and Dharmendra Singh Yadav

CONTENTS

10.1 INTRODUCTION

In distinctive for bulk silicon, it becomes exceedingly challenging to devise abrupt metallurgical p-n contacts and to mitigate the SCEs (Short-Channel-Effects) as the physical gate length of MOSFETs downscales to the realm of nm in size. The Double Gate MOSFET technology has become the most scalable of all MOSFET designs due to its outstanding management of SCEs, which is caused by the close interaction between the gates and the channel [1–4]. Because of this, MOSFET has had complete control over semiconductor devices in recent decades. To replace MOSFETs, TFETs have gained a lot of attention due to their ability to achieve subthreshold swings (SS) below 60 mV/decade at ambient temperature and enable further supply voltage reduction without compromising OFF-current [5,6].

For device downscaling, DG-TFET with intrinsic channels is thought to be the ideal option since they provide benefits like the lack of the dopant fluctuation effect, which can cause variations in the threshold voltage and drive current increased carrier mobility because there aren't any depletion charges to add to the effective electric field and reduce the mobility [7–9]. Moreover, because body doping is not a significant means for adjusting the threshold voltage, intrinsic channel TFET must fixate on the gate work function to accomplish specific threshold voltages on an integrated circuit [10].

DOI: 10.1201/9781003327035-10

In this chapter, we present a simulated finger-like source inserted within the channel with single gate TFET (F-TFET) and analyze the outcomes for different values of N_C (ranging from 10^{14} cm^{-3} to 10^{18} cm^{-3}). The F-TFET comprises an enhanced source-channel interface which reflects in terms of higher I_{ON}. Another side, the drain-channel interface is limited and direct tunneling between source-drain has become difficult, which will help minimize the ambipolar conduction. The energy band diagram, E_{fields}, and analog-RF parameters are used to analyze the device performance with drain current and ambipolar conduction. This device is viable for use as a low-power device due to improved DC/analog and high-frequency FOMs (figure of merits).

10.2   SIMULATED DEVICE STRUCTURE AND PARAMETERS

The designed F-TFET is presented in Figure 10.1 (structure has finger-like ultra this source region within the channel region). $Si_{0.5}Ge_{0.5}$ alloy substance is used in the highly Boron-dopant (P-type, 10^{20} cm^{-3}) premised source region. Si is used for the drain (Arsenic-dopant, 5×10^{18} cm^{-3}) and channel (doping variation from 10^{14} to 10^{18} cm^{-3}) materials with a 1nm HfO$_2$ as a gate oxide. The t_S and L_S (thickness and length of source) are set to 3 nm and 35 nm, making the gate more controllable for both vertical and lateral tunneling in between the source and gate region, lateral tunneling length (L_t) is set to 4 nm. Some another device proportions are t_d = 5 nm, L_g = 20 nm, t_g = ($2t_{cu} + t_s$) = 30 nm, t_{cb} = ($t_d + t_{ox}$) = 6 nm, gate work function= 4.50 eV, total length of device L_{total} = ($L_{cb} + L_d$) = (60 nm + 35 nm) = 95 nm and total height of device H_{total} = ($t_{cb} + t_g$) = (6 nm + 30 nm) = 36 nm. The designed F-TFET device is represented in Figure 10.1 and includes multiple colors and physical characteristics. The SILVACO ATLAS tool was used for all simulation experiments. The non-local BTBT framework supports the quantum tunneling characteristic by identifying quantum tunneling areas (qt regions) at the SCI$_{int}$ and DCI$_{int}$.

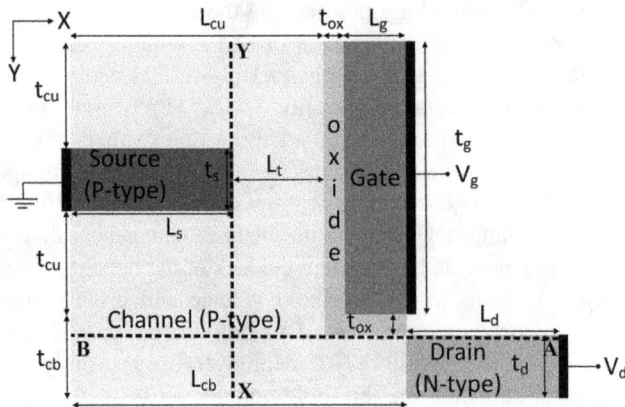

FIGURE 10.1   2D cross-sectional representation of F-TFET with two cut-lines AB and XY.

10.3 DC CHARACTERISTICS

Variations in fundamental DC FOMs such as the EBDs (Energy Band Diagrams), I_{ds}-V_{gs} (Transfer Characteristics), V_{th} (threshold voltage), potential, recombination rate, E_{fields}, and Sub-threshold Swing (SS) are the acceptable approach to examine in the initial stage of the device's vulnerability study under diverse channel doping levels [11,12]. As a result, the effect of increasing N_C on the DC efficacy of F-TFET is extensively deliberated in this work. When N_C increases from 10^{14} cm^{-3}, a high number of empty states are generated within the channel region because increasing dopants create empty states in the band gap [13]. These empty states have small ionization energy; dopants may create a band when further doping increases. Depending on the location of generated band, the band gap will either increase or decrease. The intrinsic (n_i) and impurity (N_a) concentrations are proportionately related to each other, as given by Equ. 10.1. Because of high N_C, a high carrier concentration is present within the channel region [14–16].

$$n_i = N_a \, exp\left(\frac{E_g}{2kT}\right) \tag{10.1}$$

The doping profile of the source region is very high (10^{20} cm^{-3}) as compared to the channel region. From Figure 10.2(a), when $N_C = 10^{14}$ and 10^{15} cm^{-3}, the band alignment significantly improved at SCI$_{int}$, after that the present potential barrier between bands starts increasing because the higher doping of the channel leads to mobility degradation. A similar effect can be seen at the DCI$_{int}$ in Figure 10.2(b). The energy band alignment at both interfaces starts diminishing when N_C is higher than 10^{16} cm^{-3}.

The mobility and the doping density are inversely related to each other; for increasing the doping profile, the mobility of charge carriers gets diminished. Because of the reduced mobility, the movements of the charge carriers become restricted [17,18]. These restricted movements turn into vibration with further increasing doping level; it becomes zero for ideal condition. As a result, the

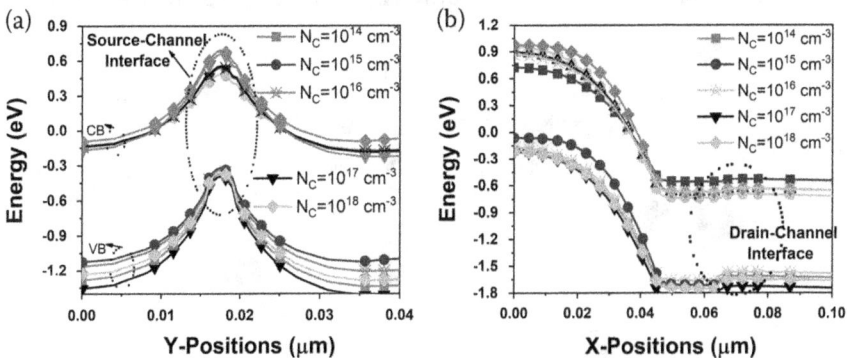

FIGURE 10.2 ON-state EBD at **(a)** Source-Channel-Interface (SCI$_{int}$) along XY cut-line and **(b)** Drain-Channel-Interface (DCI$_{int}$) along AB cut-line for various N_C.

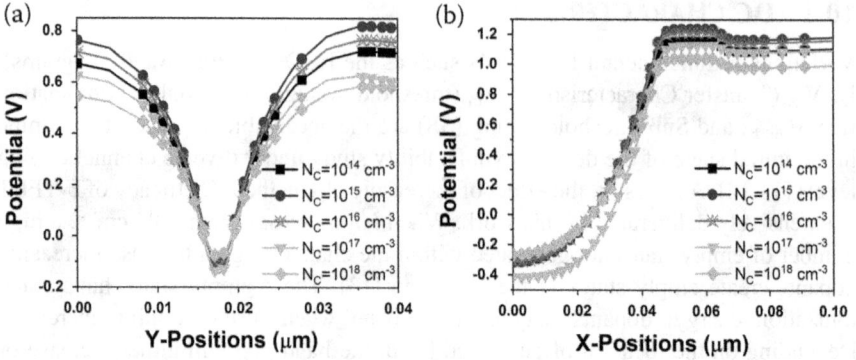

FIGURE 10.3   The ramifications of N_C on potential at (a) SCI_{int} and (b) DCI_{int}.

considered energy between two points lessens, and the potential at SCI_{int} and DCI_{int} started to decline with rising N_C in ON-state conditions, as shown for SCI_{int} and DCI_{int} in Figure 10.3(a) and 10.3(b), respectively.

The E_{fields} at both interfaces are also significantly affected by the change in N_C because E_{fields} mainly depends on the carrier concentration and presented force between them, both change with N_C for the F-TFET. The deviation in E_{fields} at SCI_{int} and DCI_{int} is portrayed in Figure 10.4(a) and 10.4(b). The E_{fields} start improving when N_C varies from 10^{14} to 10^{15} cm^{-3}; it decreases because of the deduced potential between the charge carriers. When $N_C = 10^{15}$ cm^{-3}, $E_{fields} = 3.01 \times 10^6$ V/cm (at SCI_{int}) and $E_{fields} = 6.12 \times 10^5$ V/cm (at DCI_{int}). The high E_{fields} at SCI_{int} help to increase I_{ON} and are beneficial for improving device performance. Another side, E_{fields} at DCI_{int} try to increase the device's leakage/ambipolar current, which is not desirable for optimum device efficacy [19,20]. The E_{fields} and potential significantly affect the generation and recombination rate of charge carriers at both interfaces. In the ideal case, generation and recombination at thermal equilibrium conditions are equal to each other. When external supply is applied to the device, the equilibrium condition gets disturbed and the recombination rate may be different from the generation rate [21,22].

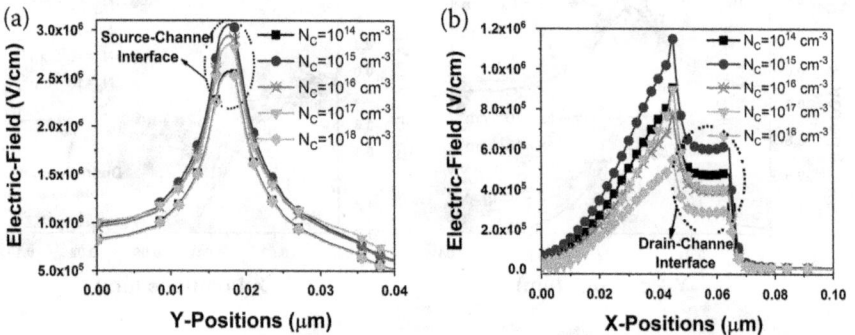

FIGURE 10.4   Variations in Electric Field (E_{fields}) for ON-state at (a) SCI_{int} and (b) DCI_{int}.

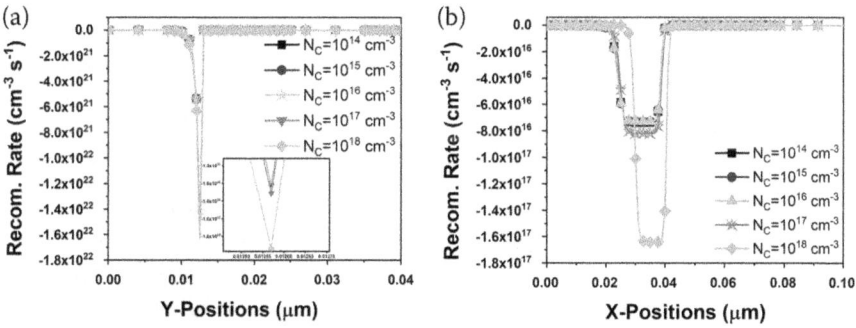

FIGURE 10.5 Variations in Recombination rate for ON-state at **(a)** SCI_{int} and **(b)** DCI_{int}.

When N_C increases, a greater number of charge carriers generated and ready to recombine with carriers present at either source or drain region. Because of this, the recombination rate at SCI_{int} and DCI_{int} enhanced as N_C increased from 10^{14} to 10^{18} cm^{-3}, as depicted in Figure 10.5(a) and 10.5(b).

The drain current is not mainly dependent on the channel doping profile; it depends on the source doping level. Since most charge carriers contained in the source rely on the proportional modification of charge carriers that tunnel from source to channel, the I_{ON} (saturated I_{ds}) is not purposefully pretended by the N_C variation. The predominant majority charge carriers (carriers of source region which contributed to I_{ON} deviations) fractional modulation is not very noteworthy. However, when carrier density within the channel region changes because of doping concentration, the I_{ON} significantly varies with this. As a possible consequence, as shown in Figure 10.6, a noticeable N_C influence on the I_{ON} of the F-TFET can be seen. The maximum I_{ON} (1.08×10^{-4} A/μm) is achieved when N_C is set 10^{15} cm^{-3}. The transfer characteristics of F-TFET are shown in Figure 10.6 on both linear and semi-log axes.

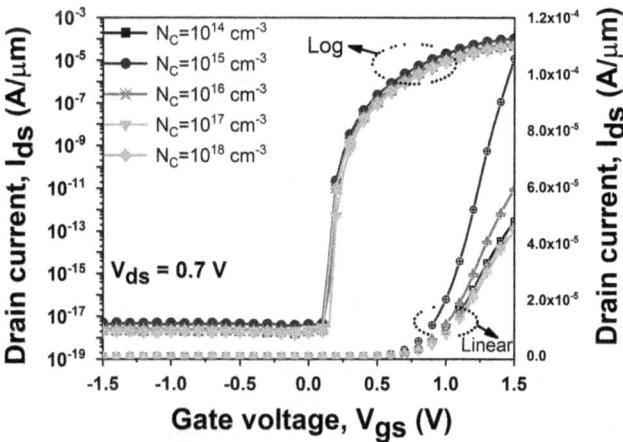

FIGURE 10.6 I_{ds}-V_{gs} plot to analyze the effect of channel doping deviations on transfer characteristics.

(a)

(b)

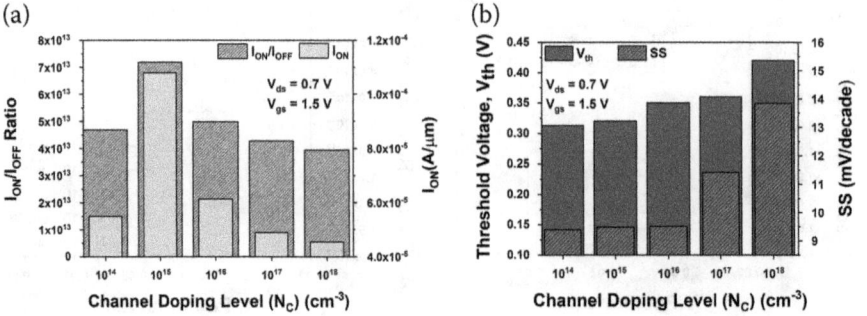

FIGURE 10.7 Variations in (a) I_{ON}/I_{OFF} ratio (Left Y-axis) and I_{ON} (Right Y-axis), (b) V_{th} (Left Y-axis), and SS (Right Y-axis) for different N_C.

$$SS_{avg} = \frac{V_{th} - V_{off}}{\log\left(\frac{I_{th}}{I_{off}}\right)} \tag{10.2}$$

The implications of N_C on the variation of ON-state current and ION/IOFF ratio are depicted in Figure 10.7(a). The higher ION of the simulated device is achieved for NC = 1015 cm^{-3} (Figure 10.7(a), lightly shaded plot) and the IOFF is not significantly affected by the N_C deviations. Hence, the I_{ON}/I_{OFF} ratio significantly varies ION accordingly. The maximum I_{ON}/I_{OFF} ratio achieved for the $N_C = 10^{15}$ cm^{-3} and the lowest for the $N_C = 10^{18}$ cm^{-3} (Figure 10.7(a), dark shaded plot). The slope of the $I_{ds} - V_{gs}$ plot in the subthreshold phase is inversely correlated to the SS value and its mathematical expression given by Equ. 10.2 [23]. Figure 10.7(b) shows that when N_C rises, the SS value rises (lined dark plot), which is not acceptable for the switching speed of the investigated device. On either side, the suggested device's V_{th} increases (black shaded plot, Figure 10.7(b)) as N_C increases, which is not favorable for applications requiring incredibly low power.

10.4 ANALYSIS OF ANALOG/RF FOMS

Parasitic capacitances like gate-drain and gate-source capacitance (C_{gd} and C_{gs}), transconductance (g_m), maximum cut-off frequency (f_t), Gain band with the product (GBP), and a few other crucial characteristics examined in this investigation aim to explain the effects of N_C variations on high efficiency of F-TFET. A high I_{ON} with lower V_{th} and superior SS values usually plays a vital role in the remarkable high-frequency performance of any FET device. Due to their simultaneous significance for parasitic oscillation across several frequency ranges, the C_{gd} and C_{gs} are essential to evaluating how well a device performs at high frequencies. For high-efficacy devices, C_{gd} and C_{gs} are being reviewed; the parasitic capacitances must be as minimal as feasible due to their prominence on device speed, resulting in a logical circuit latency [23,24].

The increment in dopants within the channel region helps to exacerbate the inversion layer through the presence of charge carriers within the channel

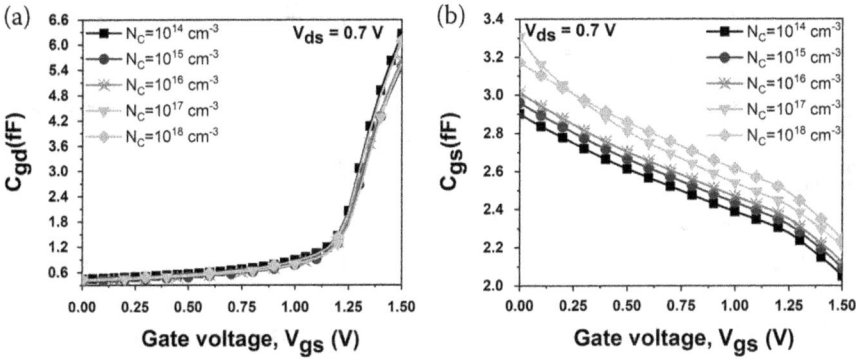

FIGURE 10.8 The plots of (a) C_{gd} and (b) C_{gs}.

region. However, the significant change in the inversion layer cannot be seen as N_C increases, so the C_{gd} is not changing significantly, as displayed in Figure 10.8(a). However, the presented potential barrier (reduced) at SCI_{int} varies according to the N_C (increase) values. Because of this, C_{gs} increase with increasing N_C, as described in Figure 10.8(b).

The findings show that, as V_{gs} increase from 0.0V to 1.50V, the F-TFET exhibits lower C_{gs} and C_{gd} at $N_C = 10^{15}$ cm^{-3} due to the increased carrier mobility across the channel and less inversion/accumulation process at both interfaces.

To analyze the device speed and switching response, we need to examine the gm parameters of the device. The g_m defines as the 1st derivative of I_{ds} to V_{gs}, and it should be high for improved device performances [25]. The g_m curves for various N_C concentrations are shown in Figure 10.9(a) to make it more convenient to analyze amplification or analyze the current device ability of F-TFET. The maxima of g_m plot are achieved for 10^{15} cm^{-3} doping of channel region because for this I_{ON} is higher. For other values of N_C, g_m starts decreasing as N_C increases from 10^{15} to 10^{18} cm^{-3}. By determining the ratio of the g_m to the $2\pi (C_{gd} + C_{gs})$, the value of the f_t can be obtained [26]. As shown in

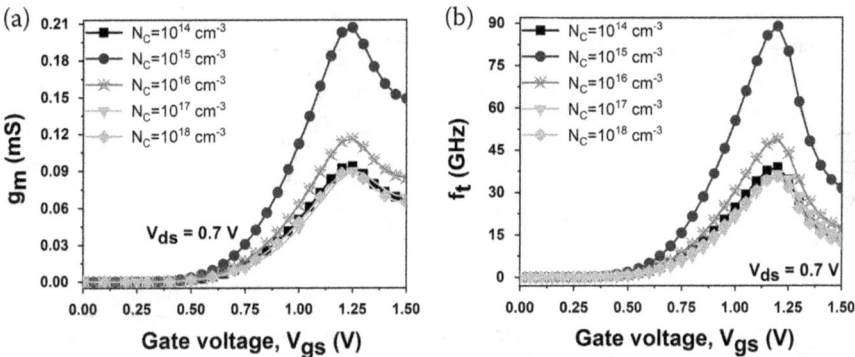

FIGURE 10.9 The curves of (a) g_m, (b) f_t under N_C deviations according to the gate voltage.

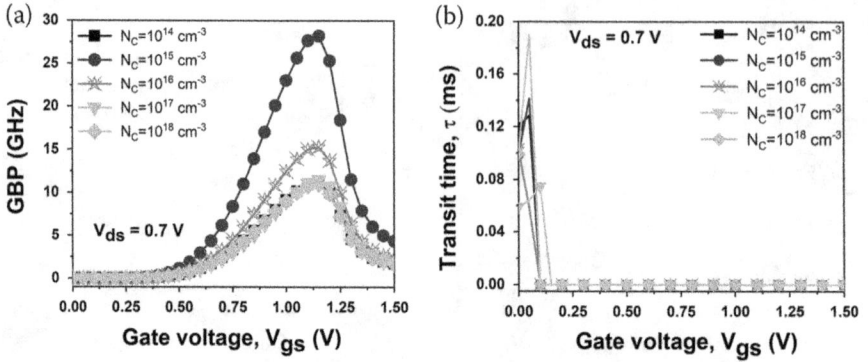

FIGURE 10.10   The variation in **(a)** GBP and **(b)** Transit time.

Figure 10.9(b), the significantly larger g_m and lower $(C_{gd} + C_{gs})$ for the $N_C = 10^{15}$ cm^{-3} with high V_{gs} allow for the attainment of the improved figure of f_t.

GBP is adversely correlated with the device's C_{gd} value and linearly proportionate to the g_m. To achieve superior high-frequency endurance, GBP must be large [27]. Although the g_m and C_{gd} both rise with N_C (10^{15} cm^{-3}), the very high gm causes GBP to improve with low $N_C = 10^{15}$ cm^{-3} and decline when $N_C > 10^{15}$ cm^{-3}. For N_C (10^{15} cm^{-3}), the percentage deviation of fluctuation in GBP is 4.56% (increase) with respect to $N_C > 10^{15}$ cm^{-3} (Figure 10.10(a)). Transit time (τ), another crucial factor, is used to examine the device response time and delay [28]. The considerable change in τ is seen when N_C varies; its minima (0.061 ms) are achieved when N_C is set to 10^{17} cm^{-3}, as depicted in Figure 10.10(b).

The study of TFP and TGF (Transconductance frequency and Generation factors) is important to demonstrate the device efficacy and offset among power dissipation and functioning bandwidth [26]. Figure 10.11(a) and (b) exhibit the effects of distinction N_C values on TFP and TGF. The TFP and TGF significantly decline with the higher value of N_C as the V_{gs} become high. For smaller V_{gs}, both the parameters start increasing and, after attaining its maxima, start decreasing because the device's maximum cut-off frequency is low (Figure 10.9(b)) and mobility saturation.

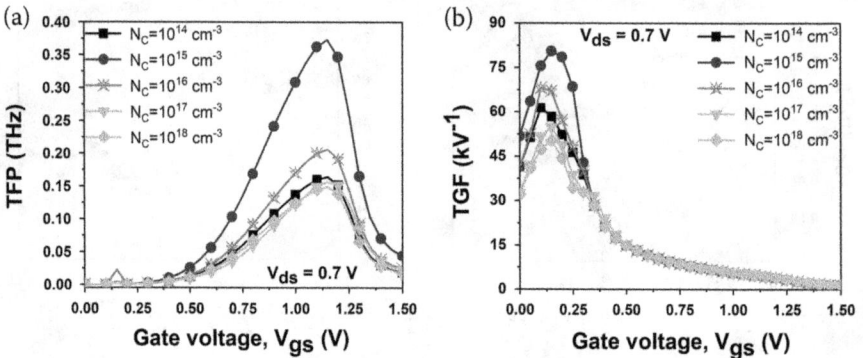

FIGURE 10.11   The variation in **(a)** TFP and **(b)** TGF.

TABLE 10.1

Overview of Critical Parameters for Different Doping Levels of Channel Region

Parameters	Channel Doping Level (N_C)				
	10^{14} cm^{-3}	10^{15} cm^{-3}	10^{16} cm^{-3}	10^{17} cm^{-3}	10^{18} cm^{-3}
I_{ON} (A/μm)	5.5×10^{-5}	1.08×10^{-4}	6.14×10^{-5}	4.89×10^{-5}	4.54×10^{-5}
I_{OFF} (A/μm)	1.17×10^{-18}	0.15×10^{-18}	2.33×10^{-18}	1.14×10^{-18}	1.15×10^{-18}
I_{ON}/I_{OFF}	4.68×10^{13}	7.18×10^{13}	4.98×10^{13}	4.28×10^{13}	3.94×10^{13}
I_{ambi} (A/μm)	2.17×10^{-18}	4.82×10^{-18}	2.63×10^{-18}	2.14×10^{-18}	2.00×10^{-18}
SS (mV/decade)	9.39	9.48	9.51	11.42	13.84
V_{th} (V)	0.313	0.32	0.35	0.36	0.42
C_{gd} (fF)	6.21	5.38	5.46	5.42	5.92
C_{gs} (fF)	2.05	2.092	2.18	2.25	2.34
g_m (mS)	0.081	0.208	0.118	0.091	0.86
f_t (GHz)	32.25	88.34	47.98	35.08	34.11
GBP (GHz)	12.84	27.89	16.21	14.56	13.02
TT (ms)	0.123	0.091	0.098	0.071	0.1001
TFP (THz)	0.18	0.38	0.201	0.173	0.168
TGF (kV^{-1})	42.05	51.21	43.1	33.81	33.53

The overall deviation in the F-TFET performance parameters with change in channel doping level is summarized in Table 10.1. From Table 10.1, we can clearly opt for the best-suited doping level for the simulated device structure channel region. Also, select the optimum frequency range within that device to work efficiently without any performance degradation.

10.5 CONCLUSION

In this chapter, the effect of channel doping on device efficacy is investigated to get the optimized N_C value. An improved SS and V_{th} values are observed if the NC is set between 1015 and 10^{16} cm^{-3}. Compared to lower N_C, there is a corresponding drop in I_{ON}/I_{OFF} ratio as lower I_{ON} is achieved at higher N_C, and ambipolar current is not much affected by N_C deviations. As the electric field and potential significantly lessen for increasing NC, the overall device performance degrades with this. As per consideration of outcomes, the peak value of I_{ON} (1.08×10^{-4} A/μm), lower SS (9.48 mV/decade) with optimum Vth (0.32 V) accomplished for $N_C = 10^{15}$ cm^{-3}.

Additionally, the functionality of the F-TFET is examined while considering the cumulative effects of N_C on ON-state amenities and RF efficacy aspects. Capacitances f_t and GBP are significantly impacted by higher N_C, which lowers the efficiency of the device in digital logic implementations. We found that RF limitations are more susceptible to higher N_C values, which does not allow us to assess the device's performance for high-frequency operations. The analysis suggests that the F-TFET might provide a viable option for enhanced analog/RF and ultra-low power applications when NC is between 10^{15} and 10^{16} cm^{-3}.

REFERENCES

[1] S.-W. Sun and P. G. Tsui, "Limitation of CMOS supply-voltage scaling by MOSFET threshold voltage variation," *IEEE J. Solid-state Circuits*, vol. 30, no. 8, pp. 947–949, 1995.

[2] S. M. Turkane and A. Kureshi, "Review of tunnel field-effect transistor (TFET)," *Int. J. Appl. Eng. Res.*, vol. 11, no. 7, pp. 4922–4929, 2016.

[3] U. E. Avci, D. H. Morris, and I. A. Young, "Tunnel field-effect transistors: Prospects and challenges," *IEEE J Electron Devices Soc*, vol. 3, no. 3, pp. 88–95, 2015.

[4] P. Singh, D. P. Samajdar, and D. S. Yadav, "Doping and dopingless tunnel field effect transistor," in 2021 6th International Conference for Convergence in Technology (I2CT), IEEE, pp. 1–7, 2021.

[5] W. Y. Choi, B.-G. Park, J. D. Lee, and T.-J.K. Liu, "Tunneling field effect transistors (tfets) with subthreshold swing (ss) less than 60 mv/dec," *IEEE Electron Device Lett*, vol. 28, no. 8, pp. 743–745, 2007.

[6] N. Kumar and A. Raman, "Prospective sensing applications of novel heteromaterial based dopingless nanowire-tfet at low operating voltage," *IEEE Trans Nanotechnol*, vol. 19, pp. 527–534, 2020.

[7] P. Singh and D. S. Yadav, "Design and investigation of f-shaped tunnel fet with enhanced analog/rf parameters," *Silicon*, pp. 1–16, 2021.

[8] D. S. Yadav, D. Sharma, R. Agrawal, G. Prajapati, S. Tirkey, B. R. Raad, and V. Bajaj, "Temperature based performance analysis of doping-less tunnel field effect transistor," in 2017 International Conference on Information, Communication, Instrumentation and Control (ICICIC), IEEE, pp. 1–6, 2017.

[9] N. Parmar, P. Singh, D. P. Samajdar, and D. S. Yadav, "Temperature impact on linearity and analog/rf performance metrics of a novel charge plasma tunnel fet," *Appl Phys A*, vol. 127, no. 4, pp. 1–9, 2021.

[10] P. G. Der Agopian, M. D. V. Martino, S. G. dos Santos Filho, J. A. Martino, R. Rooyackers, D. Leonelli, and C. Claeys, "Temperature impact on the tunnel fet off-state current components," *Solid-State Electron*, vol. 78, pp. 141–146, 2012.

[11] P. Singh and D. S. Yadav, "\Impactful study of f-shaped tunnel fet," *Silicon*, pp. 1–7, 2021.

[12] M. G. Upasana, R. Narang, and M. Saxena, "Impact of dielectric material and temperature variations on the performance of tfet with dielectric pocket," in 2016 IEEE Annual India Conference (INDICON), IEEE, pp. 1–4, 2016.

[13] S. Tirkey, D. Sharma, D. S. Yadav, and S. Yadav, "Analysis of a novel metal implant junctionless tunnel fet for better dc and analog/rf electrostatic parameters," *IEEE Trans Electron Devices*, vol. 64, no. 9, pp. 3943–3950, 2017.

[14] P. Singh, D. P. Samajdar, and D. S. Yadav, "A low power single gate l-shaped tfet for high frequency application," in 2021 6th International Conference for Convergence in Technology (I2CT), IEEE, pp. 1–6, 2021.

[15] S. B. Rahi, P. Asthana, and S. Gupta, ""Heterogate junctionless tunnel field-effect transistor: future of low-power devices," *Journal of Computational Electronics*, vol. 16, no. 1, pp. 30–38, 2017, IF:1.532,indexing:SCOPUS,SCI,ISSN:1569-8025. 10.1007/s10825-016-0936-9.

[16] P. Singh and D. S. Yadav, "Assessing the Impact of Drain Underlap Perspective Approach to Investigate DC/RF to Linearity Behavior of L-Shaped TFET," *Silicon*, 2022. 10.1007/s12633-022-01814-4.

[17] D. S. Yadav et al., "A Comparative Study of GaP/SiGe Hetero Junction Double Gate Tunnel Field Effect Transistor," 2017 IEEE International Symposium on Nanoelectronic and Information Systems (iNIS), pp. 195–199, 2017, 10.1109/iNIS. 2017.48.

[18] S. B.. Rahi, B. Ghosh, and B. Bishnoi, "Temperature Effect on Hetero Structure Junctionless Tunnel FET," *Journal of Semiconductors*, vol. 36, no. 3, pp. 034002_1–034002_5, 2015, indexing: Scopus, Web of Science. ISSN:2058-6140, 10.1088/1674- 4926/36/3/034002.

[19] S. Kumar and D. S. Yadav, "Assessment of Interface Trap Charges on Proposed TFET for Low Power High-Frequency Application," *Silicon*, 2022, 10.1007/s12633-021-01616-0.

[20] P. Singh and D. S. Yadav, "Impact of tunneling length on analog/RF performance of L-shaped TFET," 2021 First International Conference on Advances in Computing and Future Communication Technologies (ICACFCT), pp. 114–118, 2021, 10.1109/ICACFCT53978.2021.9837344.

[21] D. S. Yadav and M. Kamal, "Performance Analysis of Hetero Gate Oxide with Work Function Engineering Based SC-TFET with Impact of ITCs," *Silicon*, 2022. 10.1007/s12633-022-01792-7.

[22] S. B. Rahi and B. Ghosh, "High-k Double Gate Junctionless Tunnel FET with Tunable Bandgap," *RSC Advances*, vol. 5, no. 67, pp. 54544–54550, 2015, IF: 3.119, indexing: Web of Science. ISSN: 2046-2069, 10.1039/C5RA06954H.

[23] P. Singh and D. S. Yadav, "Performance analysis of ITCs on analog/RF, linearity and reliability performance metrics of tunnel FET with ultra-thin source region," *Appl. Phys. A*, vol. 128, p. 612, 2022, 10.1007/s00339-022-05741-4

[24] D. S. Yadav, D. Sharma, S. Tirkey, D. G. Sharma, S. Bajpai, D. Soni, S. Yadav, M. Aslam, and N. Sharma, "Hetero-material CPTFET with high-frequency and linearity analysis for ultra-low power applications," *Micro Nano Lett.*, vol. 13, pp. 1609–1614, 2018, 10.1049/mnl.2018.5075

[25] N. Kumar and A. Raman, "Low voltage charge-plasma based dopingless tunnel field effect transistor: analysis and optimization," *Microsyst Technol*, vol. 26, no. 4, pp. 1343–1350, 2020.

[26] P. Singh and D. S. Yadav, "Impact of temperature on analog/rf, linearity and reliability performance metrics of tunnel fet with ultra-thin source region," *Appl Phys A*, vol. 127, no. 9, pp. 1–15, 2021.

[27] M. Kamal and D. S. Yadav, "Effects of linearity and reliability analysis for hgo-dw-sctfet with temperature variation for high frequency application," *Silicon*, pp. 1–11, 2021.

[28] D. Kumar, S. B. Rahi, and P. Kuchhal, "Investigation of Analog Parameters and Miller Capacitance affecting the Circuit Performance of Double Gate Tunnel Field Effect Transistors (TFETs)," Int. Conference on Intelligent Communication, Control and Devices, 2020. Indexing: Scopus (Springer).

11 Design of Nanotube TFET Biosensor

*Anju, Bibhudendra Acharya, and
Guru Prasad Mishra*

CONTENTS

11.1 INTRODUCTION

Since a few decades ago, many viruses have been found which are extremely harmful to human life. As such, some time ago and still, we are going through a pandemic called coronavirus (COVID-19) [1]. A lot of research has found that these infections caused by new viruses spread very fast. So, there is the only way to break the cycle of these viruses, which is to detect them at early-stage [2]. Some symptoms are of a specific disease, so it is essential to keep looking at that disease and find its potential to progress. Due to all these things, biosensors have an important place in the field of medicine. In this, the biosensor examines the samples taken from the patient and detects diseases caused by various components.

Moreover, Biosensor is used in many fields like agriculture, environmental monitoring [3] forensics [4], drug discovery [5], etc., for their accurate results and short-time detection. Label detection methods like magnetic, fluorescent, and electrochemical techniques alter the natural quality of sample biomolecules [6–8]. Because of this, these techniques give incorrect results and take more time. As a result, many researchers are attracted by label-free detection techniques, which give accurate results, for further research [9]. In the past few years, biosensors based on FET have become very popular, as they operate on label-free detection. Detection speed and sensitivity are very important when designing any biosensor [10,11]. The literature on several types of research on biosensors includes ion-sensitive field-effect transistors (ISFETs) and dielectrically modulated FETs (DMFETs) label-free biosensors [10–14]. Reducing the dimensions of the device increases the sensing ability of the biosensor. But reducing the dimension of the MOSFET produces different short channel effects (SCEs), high power dissipations, low current ratios, etc., which affect the sensitivity of the biosensor. Additionally, the MOSFET device's subthreshold swing (SS) is equal to or greater than 60 mV/decade at room temperature. Due to this, it has a large response time that limits the highest sensitivity [15–18].

Therefore, the MOSFET should be changed by any semiconductor device that has the same structure and manufacturing process. In this regard, the TFET is a device to overcome these limitations because its working mechanism is based on quantum tunneling [19–22]. Because of its superior performance compared to MOSFET-based biosensors, researchers are more interested in it, especially in DM phenomena [23–26]. Here, sample biomolecules are sensed by variations in the various characteristics of the biosensor as the samples with various values of K are present in the nanogap cavity area. Furthermore, 3-D devices such as nanowires [27] and NT-TFET biosensors [28] improve the sensing ability in several ways. In addition, 3-D devices provide greater gate-to-channel controllability, are area-efficient, and perform better than planar devices [29–34].

Physically doped TFETs increase the complexity of the fabrication process, which needs high-cost thermal annealing techniques [35]. This causes a deviation in the V_{th} (threshold voltage), which reduces the device's performance. Doping-free phenomena come into the picture to overcome these concerns by using charge plasma (CP) phenomena for the formation of source and drain regions. The creation of abrupt junctions with high thermal cost can be minimized by using a charged plasma-based TFET [35–37].

In this chapter, we proposed a doping-free silicon nanotube TFET (DF-Si-NT-TFET) biosensor for sensing various sample (with and without charge) biomolecules. The source and drain regions are created by utilization of suitable metal electrode work function. Here, the cavity is created in the outer and inner sections of the device. Both the nanotube cavities effectively display the neutral and charged (DNA) biomolecules present. Also, the nanotube style provides improved sensing ability due to its inbuilt advantages. The availability of biomolecules in the nanogap cavity region enhances the capacitive coupling, resulting in improved device characteristics of the biosensor.

The rest of the chapter is structured as follows: Section 11.2 explains the proposed biosensor structure, dimensions, and models. Section 11.3 explains the results, including the basic electrical characteristics of neutral and charged (DNA) biomolecules in section 11.3.1. Sections 11.3.2 include the sensitivity analysis in terms of DC parameters. Section 11.3.3 consists of high-frequency parameters for biomolecules and section 11.3.4 has the sensitivity analysis in terms of high-frequency parameters. Section 11.3.5 is devoted to the optimization used for the proposed DF-Si-NT-TFET biosensor structure. Finally, 11.4 summarize the important points of the observations of this work.

11.2 BIOSENSOR STRUCTURE, DIMENSIONS, AND SIMULATION SETUP

Figure 11.1(a-b) shows the 3-D design and half of the 2-D cross-section design of the proposed DF-Si-NT-TFET biosensor. The biosensor is vertically designed with intrinsic silicon; here source gate and the drain region are created by using charge plasma phenomena. The proposed biosensor silicon and core gate diameter are taken as 10 nm. The other dimension of the device is considered as follows; drain, source, and channel region are 50, 50, and 20 nm. The silicon body's thickness is less than the Debye length, allowing charge to be induced by applying various metal electrodes to the silicon device.

The platinum and hafnium metal electrode with the work function of 5.93 and 3.9 eV is used for the creation of the CP source and CP drain region [23,35–37]. The cylindrical silicon device is covered with 1 nm oxide (SiO_2). The outer gate is separated by 2 nm from the CP drain and CP source electrodes. The core gate starts from the middle of the channel region and shifts towards the source region. The outer and core gate electrode metal work function is considered as 4.53 eV. The outer cavity region is designed in the channel region near the source and it is elevated 5 nm from the outer gate oxide. In addition, the inner cavity region is formed in the core section and beneath the source region of the device. This provides an equal allocation of the target biomolecules.

The details and uses of the biomolecules are mentioned in Table 11.1. The mass and thickness of most biomolecules are less than 5 and 2 nm, respectively. Therefore, these biomolecules can fit into the outer as well as the inner nanogap cavity, as the height of the outer cavity is 5 nm [38,39]. The proposed biosensor has the advantages of being doping-free and NT-TFET for the design of highly sensitive biosensors.

The simulation for the proposed DF-Si-NT-TFET biosensor is performed using 3-D Silvaco Atlas TCAD [40]. BBT.The Kane model is incorporated to consider three-dimensional tunneling in the cylindrical device. In addition, AUGER and Shockley Read Hall models are included for the computation of the recombination process in a semiconductor device. Further, CONMOB and FLDMOB are used to consider concentrate and field-dependent mobility. Fermi Dirac statistics are

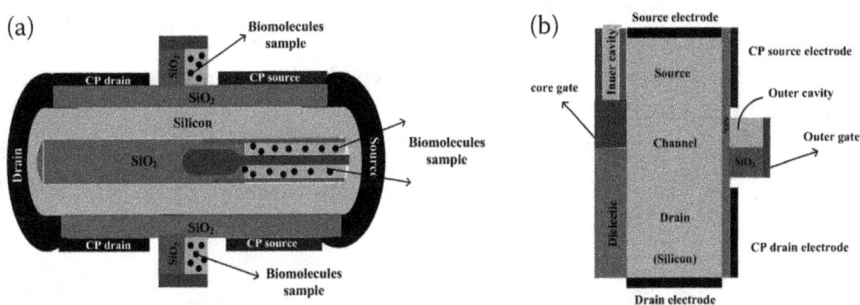

FIGURE 11.1 (a) 3-D design (b) half part of a 2-D vertical cross-section of DF-Si-NT-TFET biosensor.

TABLE 11.1

Dielectric Constant and Uses of Various Biomolecules

Name of Biomolecules	Dielectric Constant (K) [11,16,17,24]	Uses
Streptavidin	2.1	It is utilized for the identification of various nucleic acid proteins and lipids
Biotin	2.63	It is basically a vitamin B, it is available in edible things, supports to control blood sugar, and helps with the growth of hairs and nails
APTES	3.57	It is used for the process of silanization
Bacteriophage T7	6.3	It is used for killing bacteria
Protein	8	It is used for producing hormones and repairing tissue
DNA	6	It is used to track down blood relatives, identify bodies, and look for cures for various disease

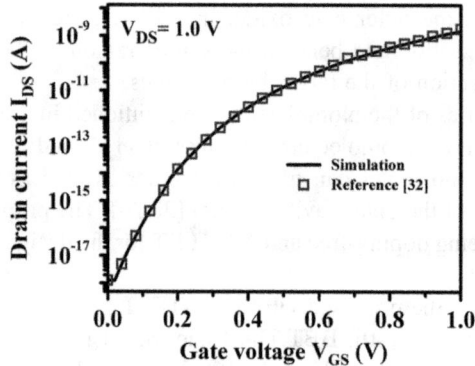

FIGURE 11.2   3-D calibrated result of [32].

utilized for modeling the band structure of the biosensor. Apart from these, mathematical calculations of the device physics are performed using the Gummel Newton Maxtrap approximation method.

For TCAD simulator calibration, we have kept the dimensions and bias voltages as specified in the reference [32]. The transfer characteristic is used to evaluate the simulation result. The simulation result of the nanotube TFET is approximately a replica of the reported results [32] and this is proved by Figure 11.2.

11.3   RESULT AND DISCUSSION

11.3.1   BASIC ELECTRICAL CHARACTERISTICS OF THE BIOSENSOR

The response of the DF-Si-NT-TFET biosensor to target biomolecules is investigated in this section with differences in some electrical characteristics. In this, the

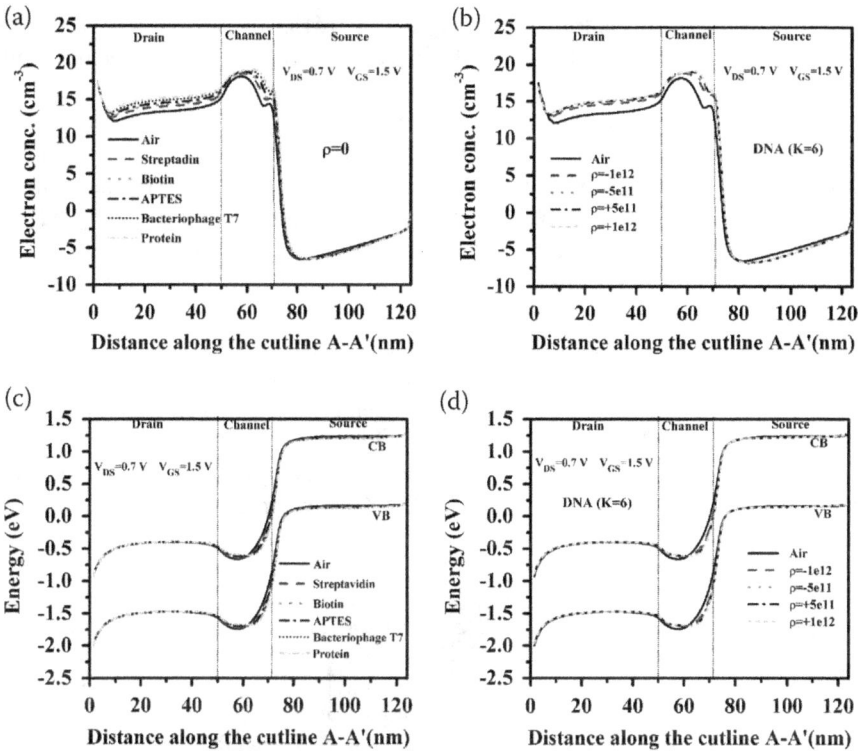

FIGURE 11.3 Impact of (a) neutrally charged (b) charged (DNA) biomolecules on e⁻ concentration and impact of (c) neutrally charged (d) charged (DNA) biomolecules on (EBD) energy band distribution.

variation in the electrostatic performance of the DF-Si-NT-TFET biosensor is accompanied. Figure 11.3 shows the difference in the electron concentration and energy band distribution (EBD) of the proposed DF-Si-NT-TFET biosensor in the presence of different neutral and charged (DNA) biomolecules. Figure 11.3(a) demonstrates that the increment in the value of K increases the capacitive coupling between the channel region and the gate electrode (outer and core), resulting in an increase in the electron concentration at the tunneling interface. Figure 11.3(b) depicts the difference in the e⁻ concentration in the presence of DNA biomolecules with different charge densities.

A positively/negatively charged biomolecule of DNA trapped in a nanogap cavity increases/decreases multiple electrons/holes at the tunneling interface, resulting in an increase/decrease in e⁻ concentration. The electron concentration is higher for the negative ($-1e12$ C/cm^{-2}) charge density than for air because the dielectric constant of DNA is higher than that of air.

Figure 11.3(c) shows the occupancy of biomolecules in the nanogap cavity by variation in the energy band distribution. As the value of K increases in the nanogap cavity region, the barrier width near tunneling junction decreases due to an enhancement in the electron tunneling rate. The increase in electron count in that region is verified by Figure 11.3(a). Figure 11.3(d) depicts the deviation in the energy band distribution (EBD) in the occupancy of DNA biomolecules with various charge densities.

As the higher positive charge ($+1e12$ C/cm^{-2}) is trapped in the nanogap cavity, the amount of electrons in the source and source/channel interface region is increased. Figure 11.3(b) is evident for the increase in electron for higher positive ($+1e12$ C/cm^{-2}) charge density. A reduction in the width of the tunneling barrier is observed due to this increase in electrons. An increase in electrons with K and ρ values in the channel and source region causes an increment in the peak value of the electric field at the tunneling interface.

The electric field deviation at the tunneling interface of the DF-Si-NT-TFET for neutral biomolecules is verified in Figure 11.4(a). The highest electric field peak of 3.53 MV/cm is observed for protein at the tunneling interface. The effect of various charge densities of DNA on the electric field is shown in Figure 11.4(a). The variation in an electric field is clearly visible for DNA in the cavity area compared to air. Since the deviation in the value of the K is much greater than the variation in the charge density.

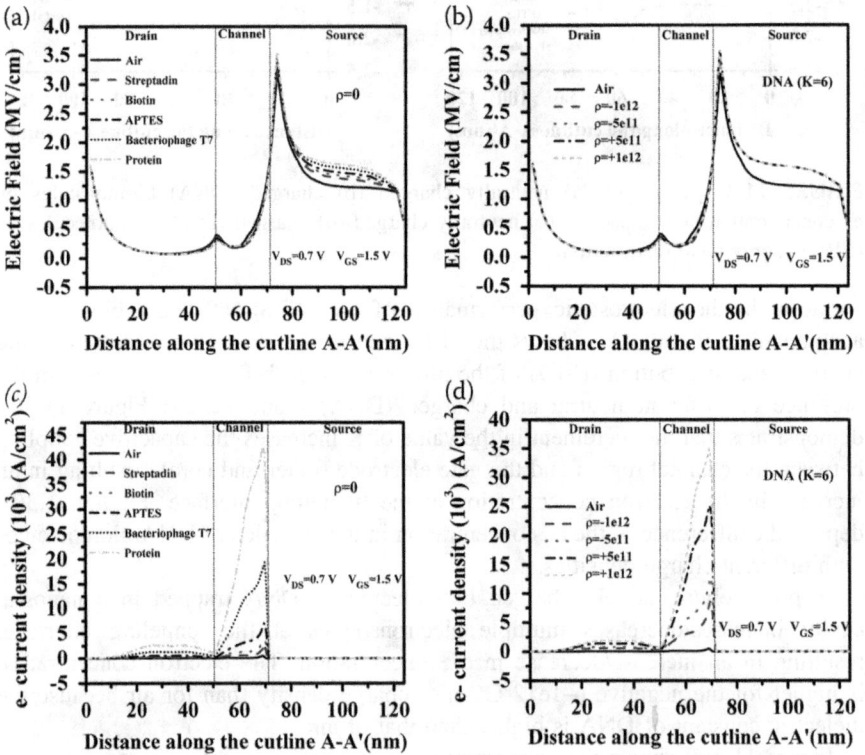

FIGURE 11.4 Impact of (a) neutrally charged (b) charged (DNA), biomolecules on electric field and impact of (c) neutrally charged (d) charged (DNA), biomolecules on electron current density.

The variation in the electron current density of DF-Si-NT-TFET for neutral and charge (DNA) biomolecules is depicted in Figure 11.4(c-d). Electron current density is defined as the number of electrons flowing per unit of time by a specific cross-sectional area of a semiconductor device.

The deviation of electron current density with the value of K for neutral biomolecules at the tunneling interface is shown in Figure 11.4(c). The lowest and highest peak of electron current density of 496.5 A/cm^2 and 42.5 × 10^3 A/cm^2 for air and protein are observed at tunneling junctions.

Similarly, the effect of various charge densities of DNA on electron current density is depicted in Figure 11.4(a). The electron current density at the tunneling interface changes with an enhancement in the value ρ from high negative (−1e12 C/cm^{-2}) to high positive (+1e12 C/cm^{-2}). The highest electron current density of 34.8 × 10^3 A/cm^2 is achieved for DNA with a charge density of +1e12 C/cm^{-2}.

Due to the target biomolecules in the inner and outer cavity, the electron density in the tunneling and CP source region increases, thereby reducing the tunneling barrier width of the proposed biosensor. This reduction in the width of the barrier initiates higher tunneling of electrons from CP source to the channel region.

This increase in electron tunneling is responsible for higher drain current in the DL-Si-NT-TFET. Figure 11.5(a) depicts the effect of different neutral biomolecules present in the cavity area of DF-Si-NT-TFET on I_{DS}-V_{GS} characteristics. The highest I_{DS} of 1.05 × 10^{6}A for the biosensor is achieved when protein is filled in both the nanogap cavity regions of biosensor. The high positive charge of DNA (+1e12 C/cm^{-2}) increases the amount of electrons in the CP source and tunneling region. So decreasing the tunneling width increases the probability of electron tunneling. This improves the I_{DS}, as seen in Figure 11.5(b).

Figure 11.6(a-b) shows the subthreshold swing for different target biomolecules. The value of V_{GS}, due to which the I_{DS} of a device increases by a decade or an order of magnitude, is known as SS.

The SS value is an essential parameter to classify it as a proficient biosensor, i.e., the biosensor which has a small value of SS rapidly detects various biomolecules. The accumulation of charge carriers increases due to the increment in the dielectric

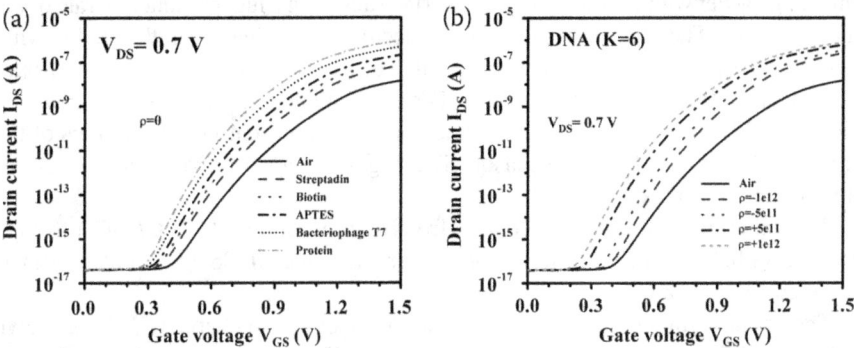

FIGURE 11.5 Impact of (a) neutrally charged (b) charged (DNA), biomolecules on transfer (I_{DS}-V_{GS}) characteristics.

(a)

(b)

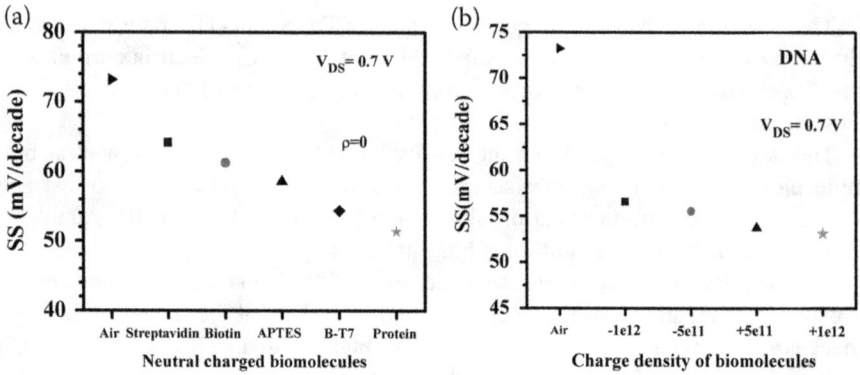

FIGURE 11.6 Impact of (a) neutrally charged (b) charged (DNA), biomolecules on subthreshold swing (SS).

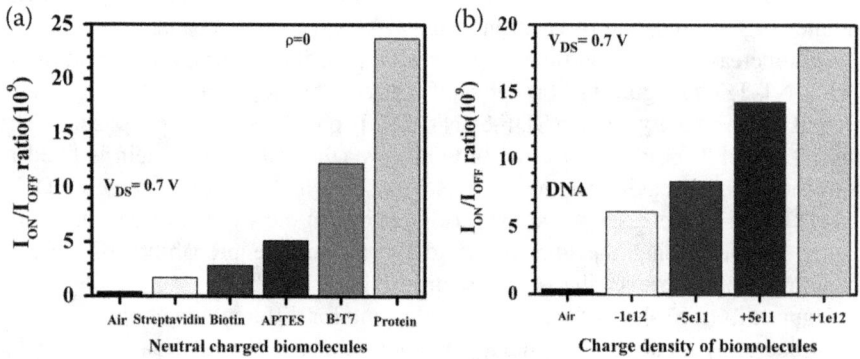

(a)

(b)

FIGURE 11.7 Impact of (a) neutrally charged (b) charged (DNA), biomolecules on I_{ON}/I_{OFF} ratio.

constant (K) value in the nanogap cavity region. This provides a decrement in the tunneling barrier width at the tunneling interface allowing a higher tunneling rate of the charge carrier. Therefore, for neutral biomolecules, the value of subthreshold swing (SS) decreases with the increment in the value of K, as represented in Figure 11.6(a).

Similarly, the effect of charged (DNA) biomolecules in SS is shown in Figure 11.7(b). The SS value is high when only air is present in both cavities of the proposed DF-Si-NT-TFET biosensor. The negative and positive charge densities are taken on the x-axis.

The value of SS for DNA increases from high negative (-1e12 C/cm^{-2}) to high positive (+1e12 C/cm^{-2}) values. The minimum value of SS (53 mV/decade) is obtained for +1e12 C/cm^{-2}.

The variation in the I_{ON}/I_{OFF} ratio of various neutral and charge (DNA) is shown in Figure 11.7(a-b). The increment in the value of K and ρ decreases the conduction band and valence band alignment at the tunneling interface, thereby improving the I_{DS} for the biosensor. Hence the I_{ON}/I_{OFF} ratio increases for the proposed biosensor.

The increment in the I_{ON}/I_{OFF} ratio for five different neutral biomolecules is represented in Figure 11.7(a). The protein-filled in the nanogap cavity shows a high ION/IOFF ratio value. The ION/IOFF ratio increase with the value of ρ is demonstrated in Figure 11.7(b). Lower and higher values of I_{ON}/I_{OFF} ratios of 6.15 × 10⁹ and 1.85 × 10¹⁰ have been detected for the DNA biomolecule, for the charge density of −1e12 C/cm⁻² and +1e12 C/cm⁻², respectively. This investigation shows that the I_{ON}/I_{OFF} ratio is an increasing function for the K and ρ.

11.3.2 SENSITIVITY ANALYSIS BY DC PARAMETERS

The sensitivity of any biosensor is a measure of its ability to detect. Therefore, every biosensor's sensing ability must be high to signify how proficiently it senses the target biomolecule. The sensing capability of DF-Si-NT-TFET biosensor is calculated in respect of air filled in the nanogap cavity area for the sample biomolecules, and the formula is given by

$$\text{Sensitivity} = \left| \frac{(S_1 - S_2)}{S_1} \right| \tag{11.1}$$

Here S_1 is the value when only Air is occupied in the cavity area, S_2 is the value when biomolecules are present in the nanogap cavity area; that value differs in magnitude. This section investigates the DF-Si-NT-TFET sensitivity in terms of I_{DS}, SS, and I_{ON}/I_{OFF} ratio sensitivity. Figure 11.8(a) depicts the I_{DS} sensitivity for different neutral biomolecules. In this, the IDS sensitivity increases as the value of K of the biomolecules increases. The lowest and highest I_{DS} sensitivity of 8.32 and 623.5 is detected for streptavidin and protein biomolecules. Similarly, with an enhancement in the value of ρ of DNA biomolecules from a high negative (+1e12 C/cm⁻²) to a high positive (+1e12 C/cm⁻²), I_{DS} sensitivity increases. The highest I_{DS} sensitivity is achieved for DNA with charge density +1e12 C/cm⁻², as shown in Figure 11.8(b). The best part of the I_{DS} sensitivity is that its peak is found at low V_{GS}.

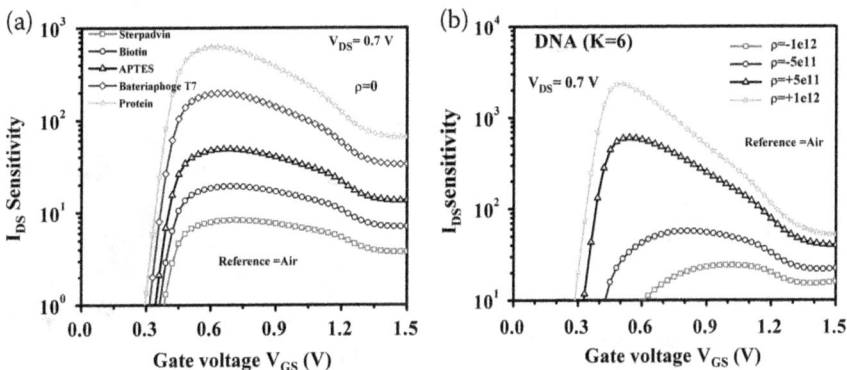

FIGURE 11.8 Impact of (a) neutrally charged (b) charged (DNA), biomolecules on (drain current) IDS sensitivity.

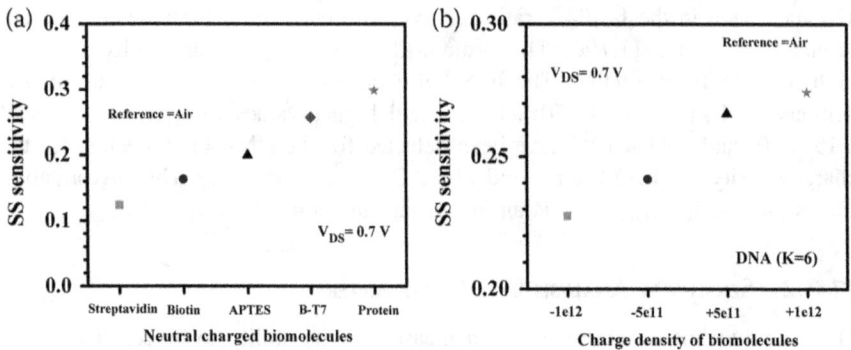

FIGURE 11.9 Impact of (a) neutrally charged (b) charged (DNA), biomolecules on subthreshold swing sensitivity.

The SS is a very important parameter for a biosensor in the process of detection because it defines the speed of the sensing, i.e., lower value of SS shows faster sensing ability. The SS sensitivity is calculated by the given formula in equation 11.1. Figure 11.9(a-b) shows the SS sensitivity of DF-Si-NT-TFET for various neutral and charged biomolecules. The steeps SS provide superior detection ability as well as improve the basic electric performance of the DF-Si-NT-TFET.

Figure 11.9(a) depicts the SS sensitivity of five different neutral biomolecules. Here, the lowest and highest SS sensitivity of 0.12 and 0.3 were obtained for streptavidin and protein biomolecules, respectively. Figure 11.9(b) demonstrates the SS sensitivity for DNA with four different negative and positive charge densities. From the results, we can say that SS sensitivity increases from negative ($-1e12$ C/cm^{-2}) to positive ($+1e12$ C/cm^{-2}) charge density. The highest SS sensitivity of 0.275 is achieved for DNA with a charge density of $+1e12$ C/cm^2.

The I_{ON}/I_{OFF} sensitivity is also determined using equation 11.1. Figure 11.10(a-b) shows the I_{ON}/I_{OFF} current ratio sensitivity for different target biomolecules. The I_{DS} increases with an increment in the value K which is due to the decrease in the barrier

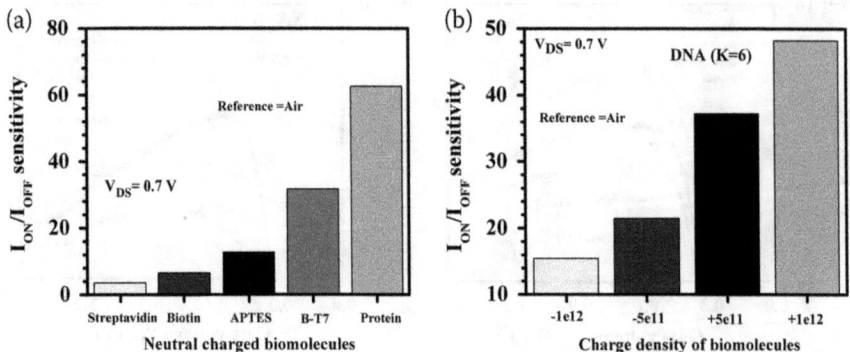

FIGURE 11.10 Impact of (a) neutrally charged and (b) charged (DNA) biomolecules on I_{ON}/I_{OFF} sensitivity.

width at tunneling interface. This results in an increase in the value of I_{ON}/I_{OFF} ratio. In Figure 11.10(a), the highest sensitivity of 64 is obtained for protein biomolecules with respect to air.

The I_{ON}/I_{OFF} current ratio enhances with an increment in the ρ from negative ($-1e12$ C/cm^{-2}) to positive ($+1e12$ C/cm^{-2}). Figure 11.10(b) demonstrated the I_{ON}/I_{OFF} sensitivity for DNA for four different charge densities. This is because the stabilization of the higher value of ρ ($-1e12$ C/cm^{-2}) inside both the cavity regions with the V_{GS} takes to a superior inversion below the cavity. Like the I_{DS} and SS sensitivity, the I_{ON}/I_{OFF} sensitivity is also proportional to the value of K and ρ.

11.3.3 HIGH-FREQUENCY PARAMETERS ANALYSIS OF NEUTRALLY CHARGED AND CHARGED BIOMOLECULES

High-frequency parameters are also used as sensing metrics for biosensor. So, some high-frequency parameters such as output conductance (g_{ds}), transconductance (g_m), cut-off frequency (f_T), and device efficiency (g_m/I_{DS}) [41–43] are calculated for further sensitivity analysis of DF-Si-NT-TFET biosensor.

The g_{ds} is a very important parameter of high-frequency parameters, which provides information about the dependence of I_{DS} in the drain voltage (V_{DS}). Figure 11.11 shows the impact on g_{ds} for various neutral biomolecules. It is observed that the curve of g_{ds} increases with drain voltage, but after a peak, it gets degraded. From the investigation, it is clear that g_{ds} increase with the value of K of the five different neutral biomolecules. The highest peak of 3.18 µS is found for protein at 0.5V of V_{DS} among all the biomolecules. Figure 11.11(b) demonstrates the g_{ds} with respect to V_{DS} for four different charge densities of DNA biomolecules. The gds curve increases from negative ($-1e12$ C/cm^{-2}) to positive ($+1e12$ C/cm^{-2}) charge densities of DNA biomolecules. The lowest and highest peaks of gds, 0.8 and 2.6 µS are found at 0.6 and 0.55 V_{DS}.

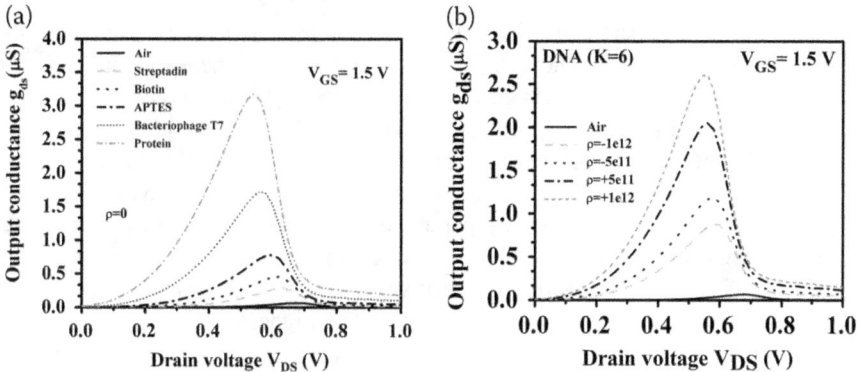

FIGURE 11.11 Impact of (a) neutrally charged (b) charged (DNA), biomolecules on output conductance (g_{ds}).

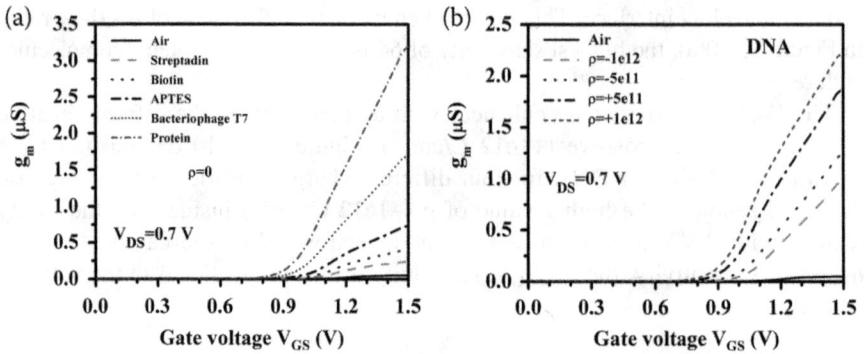

FIGURE 11.12 Impact of (a) neutrally charged (b) charged (DNA), biomoleuclues on transconductance (g_m).

g_m calculates the device's ability to convert V_{GS} to I_{DS}. The variation in gm for different neutral and charged (DNA) biomolecules is represented in Figure 11.12. The impact of the change of different values of K in the nanogap cavity is analyzed in Figure 11.12(a). Here the g_m enhances with the value of V_{GS} for all the biomolecules. The highest peak of g_m is achieved for protein biomoleucule.

Similarly, Figure 11.12(b) depicts the impact of deviation in the value of ρ in the nanogap cavity region for g_m. The higher positive ($+1e12$ C/cm^{-2}) charge density traps more electron tunneling interface, resulting in an increase in the I_{DS}. So the higher positive ($+1e12$ C/cm^{-2}) charge density of DNA has a great ability to convert the V_{GS} into I_{DS}.

f_T is the frequency where the current gain of the device becomes unity. f_T can be calculated by the formula given below:

$$f_T = \frac{g_m}{2\Pi\,(C_{gs} + C_{gd})} \tag{11.2}$$

The high-frequency parameter f_T is demonstrated as a sensitivity parameter for neutral and charge (DNA) biomolecules in Figure 11.13. The variation in f_T for different values of K is represented in Figure 11.13(a). Here, increasing the value of K from 1 to 8 increases f_T with respect to V_{GS}. The lowest and highest values of f_T are obtained for the air and protein, respectively. The maximum value of f_T, which is 0.82 GHz, is achieved in the presence of protein in the cavity area.

The deviation in high-frequency parameter f_T for DNA with different values of ρ is represented in Figure 11.13(b). f_T increases with the higher negative ($-1e12$ C/cm^{-2}) value of ρ to higher positive ($+1e12$ C/cm^{-2}) values of ρ. Due to the value of g_m increasing with the value of ρ. The highest value of f_T is obtained for a charge density of $+1e12$ C/cm^{-2}.

Device efficiency (g_m/I_{DS}) is the high-frequency parameter that is used to calculate the ability to convert I_{DS} into g_m. The biosensor's sensing ability is investigated in respect of device g_m/I_{DS} for the proposed biosensor and is represented in

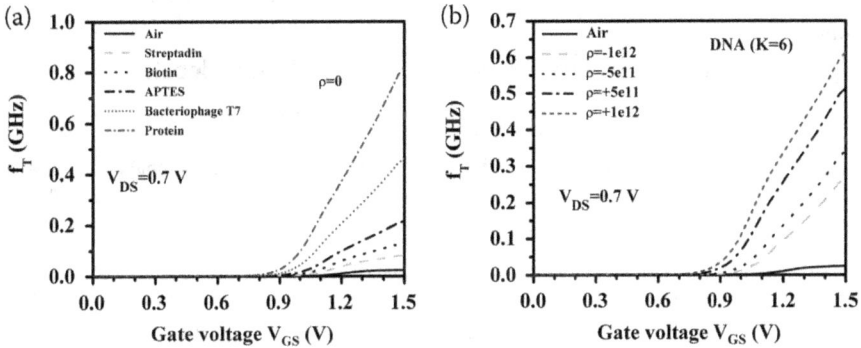

FIGURE 11.13 Impact of (a) neutrally charged (b) charged (DNA), biomolecules on cut-off frequency (f_T).

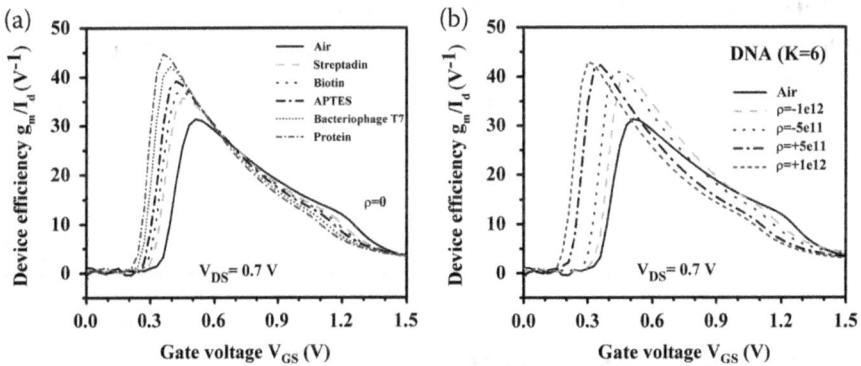

FIGURE 11.14 Impact of (a) neutrally charged (b) charged (DNA), biomoleuclues on g_m/I_{DS}.

Figure 11.14. The impact on the gm/I_{DS} for five neutral biomolecules is shown in Figure 11.14(a). The peak of g_m/I_{DS} enhances with an increment in the value of K of sample biomolecules. Also, the peak of g_m/I_{DS} is shifted to a lower value of V_{GS} with an increasing value of K. The difference between the V_{GS} for high and low peak values of gm/I_{DS} of protein and air is 0.2 V.

Similarly, the difference in the gm/I_{DS} in the presence of various charge densities of DNA present in the biomolecules is investigated in Figure 11.14(b). Here, the peak of g_m/I_{DS} shows a very slight increase with deviation in charge density. But shifting in V_{GS} for the particular peak of gm/I_{DS} is more. A difference of about 0.28V exists between air and high positive charge density (($+1e12$ C/cm^{-2}).

Therefore, the sensing parameter gm/I_{DS} is a very good candidate where a low supply voltage is required to sense any neutral and charged (DNA) biomolecules.

The values of the high-frequency parameters for various neutral and charged (DNA) biomolecules are presented in Table 11.2 and Table 11.3, respectively.

TABLE 11.2

High-Frequency Parameters for Neutrally Charged Biomolecules

High-Frequency Parameters	Neutrally Charged Biomolecules				
	Streptavidin (K = 2.1)	Biotin (K = 2.63)	APTES (K = 3.57)	Bacteriophage T7 (K = 6.3)	Protein (K = 8)
g_{ds} (µS)	0.27	0.45	0.78	1.72	3.18
g_m (µS)	0.23	0.4	0.72	1.7	3.21
f_T (GHz)	0.08	0.13	0.21	0.46	0.82
g_m/I_{DS} (V^{-1})	35.7	37.2	39.2	42.2	44.7

TABLE 11.3

High-Frequency Parameters for Charged Biomolecules

High-Frequency Parameters	Charged Biomolecules (DNA) (K = 6)			
	−1e12	−5e11	+5e11	+1e12
g_{ds} (µS)	0.88	1.19	2.05	2.6
g_m (µS)	0.98	1.25	1.88	2.23
f_T (GHz)	0.26	0.34	51.5	61.3
gm/I_{DS} (V^{-1})	40.5	40.9	42.3	43

11.3.4 SENSITIVITY ANALYSIS BY HIGH-FREQUENCY PARAMETERS

The sensitivity analysis by high-frequency parameters for various neutral and charged (DNA) biomolecules is performed using equation 11.1. The calculated sensitivity parameters for neutral and charged (DNA) biomolecules are specified in Table 11.4 and Table 11.5, respectively.

TABLE 11.4

High-Frequency Parameter Sensitivity for Neutral Biomolecules

High-Frequency Sensitivity Metrics	Neutrally Charged Biomolecules				
	Streptavidin (K = 2.1)	Biotin (K = 2.63)	APTES (K = 3.57)	Bacteriophage T7 (K = 6.3)	Protein (K = 8)
g_{ds} sensitivity	3.21	5.84	10.7	24.9	46.8
g_m sensitivity	3.92	7.37	14.1	34.2	65.4
f_T sensitivity	5.02	12.3	29.2	108.7	325.1
gm/I_{DS} sensitivity	0.14	0.19	0.25	0.35	0.43

TABLE 11.5

High-Frequency Parameter Sensitivity for Charged Biomolecules

High-Frequency Sensitivity Metrics	Charged Biomolecules (DNA)			
	−1e12	−5e11	+5e11	+1e12
gds sensitivity	12.3	16.9	29.9	38.1
gm sensitivity	19.4	24.8	37.8	45.2
f_T sensitivity	13	31.5	378.2	1974.8
g_m/I_{DS} sensitivity	0.30	0.31	0.35	0.38

The sensing ability is calculated by considering the peak value of each high-frequency parameter with the air as a reference. Here, like the high-frequency parameters, its sensing capability enhances with an increment in the value of K and ρ of the biomolecules.

11.3.5 OPTIMIZATION

Figure 11.15(a) shows the I_{DS} sensitivity for biotin with respect to the V_{GS}. It has been demonstrated that when an individual cavity is performed at a time, it gives a lower sensitivity than both cavities. The reason is that both cavities provide a large surface for immobilizing biomolecules, which increases the sensing capability of the biosensor. Figure 11.15(b) displays the I_{DS} sensitivity to demonstrate the importance of shifting the core gate towards the source region.

It can be clearly seen from the graph that the proposed biosensor with the shifted gate offers greater sensitivity for each biomolecule available in the cavity area. The core gate below the source region provides a sufficient number of electrons to decrease the tunneling barrier. This resulted in an improvement in the I_{DS} sensitivity of the proposed doping-free biosensor.

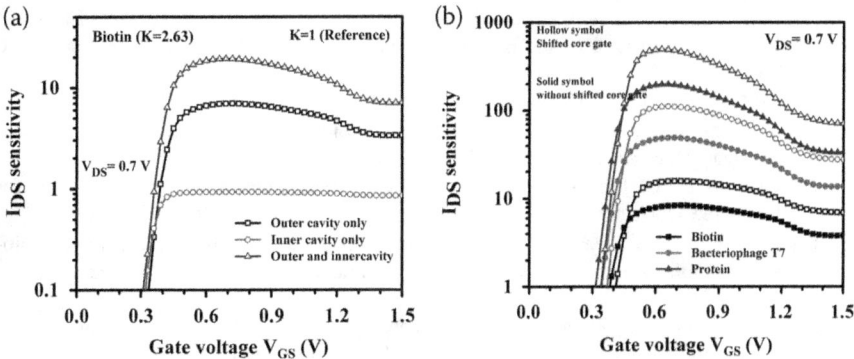

FIGURE 11.15 Impact of neutrally charged biomolecules on I_{DS} sensitivity for (a) cavity optimization (b) core gate position optimization.

11.4 CONCLUSION

This chapter suggests an insightful analysis of the utility of DC and high-frequency parameters as sensing metrics for doping-free silicon NT-TFET (DF-Si-NT-TFET) biosensors. The simulation results specify the variation in device characteristics when changing biomolecules in the nanogap cavities of the biosensor. The shifting of the core gate gives the biosensor greater sensitivity to different biomolecules. The collaborative working of both cavities increases the sensitivity of the biosensor. High-frequency parameters can also be a great option for measuring biosensors' sensing capability. Both DC and high-frequency parameters for sensitivity are an increasing function of K and ρ of biomolecules.

REFERENCES

[1] N. Jebril, "World Health Organization Declared a Pandemic Public Helth Menace: A Systematic review of the Coronavirus Disease 2019 'COVID-19'", *SSRN Electron J.*, April. 2020.

[2] J. Martin *et al.*, "Tracking SARS-CoV-2 in Sewage: Evidence of Changes in Virus Variant Predominance during COVID-19 Pandemic," *Viruses 2020*, vol. 12, p. 1144, Oct. 2020.

[3] V. T. Nguyen, Y. S. Kwon, and M. B. Gu, "Aptamer-based environmental biosensors for small molecule contaminants," *Current Opinion Biotechnol.*, vol. 45, pp. 15–23, Jun. 2017.

[4] J. Shin, S. Choi, J.-S. Yang, H.-I. Jung, and A. I. Jung, "Smart forensic phone: Colorimetric analysis of a bloodstain for age estimation using a smartphone," *Sens. Actuators B, Chem.*, vol. 243, pp. 221–225, May 2017.

[5] M. A. Cooper, "Optical biosensor in drug discovery", *Nature Review Drug Discovery*, vol. 1, pp. 515–528, Jul. 2002.

[6] G. Marrazza *et al.*, "Disposable DNA Electrochemical Sensor for Hybridization Detection," *Biosensors & Bioelectronics*, vol. 14, pp. 43–51, Jan. 1999.

[7] Bras *et al.*, "Control of Immobilization and Hybridization on DNA Chips by Fluorescence Spectroscopy," *J. Fluorescence*, vol. 10, pp. 247–253, Feb. 2000.

[8] M. M. Miller *et al.*, "A DNA Array Sensor Utilizing Magnetic Microbeads and Magnetoelectronic Detection," *J. Magnetism and Magnetic Materials*, vol. 225, pp. 138–144, Jan. 2001.

[9] Jang *et al.*, "Sublithographic Vertical Gold Nanogap for Label-free Electrical Detection of Protein-ligand Binding," *J. Vac. Sci.Technol. B.*, vol. 25, pp. 443–447, 2007.

[10] M. Barbaro, A. Bonfiglio, and L. Raffo, "A charge-modulated FET for detection of biomolecular processes: conception, modeling, and simulation," *IEEE Trans. Electron Devices*, vol. 53, pp. 158–166, 2006.

[11] C. H. Kim, C. Jung *et al.*, "Novel dielectric modulated field-effect transistor for label-free DNA detection," *Bio Chip Journal*, vol. 2, pp. 127–134, 2008.

[12] J. Y. Kim *et al.*, "An underlap channel-embedded-field-effect transistor for biosensor application in watery and dry environment," *IEEE Trans. Nanotechnol.*, vol. 11, pp. 390–394, Mar. 2012.

[13] X. P. A. Gao, G. Zheng, and C. M. Lieber, "Subthreshold regime has the optimal sensitivity for nanowire FET biosensors," *Nano Lett.*, vol. 10, pp. 547–552, 2010.

[14] N. Kannan and M. J. Kumar, "Dielectric-modulated impact-ionization MOS transistor as a label-free biosensor," *IEEE Electron Device Lett.*, vol. 34, pp. 1575–1577, Dec. 2013.

[15] R. Narang, M. Saxena, and M. Gupta, "Comparative analysis of dielectric-modulated FET and TFET-based biosensor," *IEEE Trans. Nanotechnol.*, vol. 14, pp. 427–435, May 2015.

[16] S. Kanungo *et al.*, "Study and analysis of the effects of SiGe source and pocket-doped channel on sensing performance of dielectrically modulated tunnel FET-based biosensors," *IEEE Trans. Electron Devices*, vol. 63, pp. 2589–2596, Jun. 2016.

[17] R. Narang *et al.*, "A dielectric-modulated tunnel-FET-based biosensor for label-free detection: Analytical modeling study and sensitivity analysis," *IEEE Trans. Electron Devices*, vol. 59, pp. 2809–2817, Oct. 2012.

[18] K-T Lam *et al.*, "A simulation study of graphene-nanoribbon tunneling FET with heterojunction channel," *IEEE Electron Device Lett.*, vol. 31, pp. 555–557, July 2010.

[19] A. C. Seabaugh and Q. Zhang, "Low-voltage tunnel transistors for beyond CMOS logic," *Proc. IEEE.*, vol. 98, pp. 2095–2110, Dec. 2010.

[20] W. Y. Choi *et al.*, "Tunneling field-effect transistors (TFETs) with subthreshold swing (SS) less than 60 mV/dec," *IEEE Electron Device Lett.*, vol. 28, pp. 743–745, July 2007.

[21] P. F. Wang *et al.*, "Complementary tunneling transistor for low power application," *Solid-State Electron.*, vol. 48, pp. 2281–2286, Dec. 2004.

[22] A. M. Ionescu and H. Riel, "Tunnel field-effect transistors as energy efficient electronic switches," *Nature*, vol. 479, pp. 329–337, Nov. 2011.

[23] D. Singh, S. Pandey, K. Nigam, D. Sharma, D. S. Yadav, and P. Kondekar, "A charge-plasma-based dielectric-modulated junctionless TFET for biosensor label-free detection," *IEEE Trans. Electron Devices*, vol. 64, pp. 271–278, Jan. 2017.

[24] M. Patil, A. Gedam, and G. P. Mishra, "Performance Assessment of a Cavity on Source ChargePlasmaTFET-Based Biosensor," *IEEE Sensors Journal*, vol. 21, pp. 2526–2532, Feb. 2021.

[25] M. Verma, S. Tirkey, S. Yadav, D. Sharma, and D. S. Yadav, "Performance Assessment of a Novel Vertical Dielectrically Modulated TFET-Based Biosensor," *IEEE Trans. Electron Devices*, vol. 64, pp. 3841–3848, Sep. 2017.

[26] Mahalaxmi, B. Acharya, and G. P. Mishra, "Design and analysis of Dual-Metal-Gate Double-cavity Charge- Plasma-TFET as a Label Free Biosensor," *IEEE Sensors Journal*, vol. 20, pp. 13969–13975, Dec. 2020.

[27] Ajay, R. Narang, M. Saxena, and M. Gupta, "Model of GaSb-InAs p-i-n Gate All Around BioTunnel FET," *IEEE Sensors Journal*, vol. 19, pp. 2605–2612.

[28] S. Shreya, A. H. Khan, N. Kumar, S. Intekhab Amin, and S. Anand, "Core-Shell Junctionless Nanotube Tunnel Field Effect Transistor: Design and Sensitivity Analysis for Biosensing Application," *IEEE Sensors Journal*, vol. 20, Jan. 2020.

[29] A. N. Hanna and M. M. Hussain, "Si/Ge hetero-structure nanotube tunnel field effect transistor," *J. Appl. Phys.*, vol. 117, Jul. 2015.

[30] A. N. Hanna, H. M. Fahad, and M. M. Hussain, "InAs/Si heterojunction nanotube tunnel transistors," *Sci. Rep.*, vol. 5, Apr. 2015.

[31] H. M. Fahad, C. E. Smith, J. P. Rojas, and M. M. Hussain, "Silicon nanotube field effect transistor with core–shell Gate stacks for enhanced high-performance operation and area scaling benefits," *Nano Lett.*, vol. 11, pp. 4393–4399, Sep. 2011.

[32] H. M. Fahad and M. M. Hussain, "High-Performance Silicon Nanotube Tunneling FET for Ultralow-Power Logic Applications," *IEEE Trans. Electron Devices*, vol. 60, pp. 1034–1039, Mar. 2013.

[33] G. Musalgaonkar, S. Sahay, R. S. Saxena, and M. J. Kumar, "Nanotube tunneling fet with a core source for ultra-steep subthreshold swing: A simulation study," *IEEE Transactions on Electron Devices*, vol. 66, pp. 4425–4432, Aug. 2019.

[34] G. Musalgaonkar, S. Sahay, R. S. Saxena, and M. J. Kumar, "A line tunneling field-effect transistor based on misaligned core-shell gate architecture in emerging nanotube FETs", *IEEE Transactions on Electron Devices*, vol. 66, pp. 2809–2816, Apr. 2019.

[35] C. Sahu and J. Singh, "Charge-plasma based process variation immune junctionless transistor," *IEEE Electron Device Lett.*, vol. 35, pp. 411–413, Mar. 2014.

[36] B. Ghosh and M. W. Akram, "Junctionless tunnel field effect transistor," *IEEE Electron Device Lett.*, vol. 34, pp. 584–586, May 2013.

[37] B. Venkata *et al.*, "Junctionless based dielectric modulated electrically doped tunnel FET based biosensor for label-free detection," *Micro Nano Lett.*, vol. 13, pp. 452–456, Dec. 2017.

[38] Kim *et al.*, "A transistor-based biosensor for the extraction of physical properties of from biomolecules," *Appl. Phys. Lett.*, vol. 101, p. 073703, Aug. 2012.

[39] N. Colloch *et al.*, "Crystal structure of the protein drug urate oxidase-inhibitor complex at 2.05 Å resolution," *Nature Struct. Mol. Biol.*, vol. 4, no. 514, pp. 947–952, Nov. 1997.

[40] ATLAS User's manual Device Simulation Software, Silvaco Int., Santa Clara, CA, USA, 2016.

[41] Anju, B. Acharya, and G. P. Mishra, "Junctionless Silicon Nanotube TFET for Improved DC and Radio Frequency Performance," *Silicon*, vol. 13, pp. 167–178, 2020.

[42] Anju, B. Acharya, and G. P. Mishra, "Interface Trap Charge Analysis of Charge-Plasma-Based Nanotube Tunnel FET for Improved Reliability," *Journal of Computational Electronics*, vol. 20, pp. 1157–1168, 2021.

[43] S. Yadav, D. Sharma, *et al.*, "A novel hetero-material gate-underlap electrically doped TFET for improving DC/RF and ambipolar behaviour," *Superlattices and Microstrcutures*, vol. 117, pp. 9–17, 2018.

12 TFET-based Memory Cell Design with Top-Down Approach

Young Suh Song, Youngjae Song,
T.S. Arun Samuel, P. Vimala, Shubham Tayal,
Ritam Dutta, Chandan Kumar Pandey,
Abhishek Kumar Upadhyay, Ilho Myeong, and
Shiromani Balmukund Rahi

CONTENTS

12.1 INTRODUCTION: DEMAND FOR LOW POWER DEVICES

Recently, the IT industry has been at the critical turning point [1–4]. As the demand for big data market and portable device market have been gradually expanded, the new portable electronic devices such as Apple watch, Samsung galaxy watch, and IC chips for self-driving car (eg. Tesla) are required to have '1) high data storage capacity' and '2) low power operation' [5–8]. In order to meet high data storage demand, lots of IC chip companies steadily continually design the integrated circuit (IC) chips with smaller transistor [9–11]. However, the steady miniaturization of transistors inevitably involves short channel effect (SCE) including increase of off-current (I_{off}), which makes hard to achieve "2) low power operation" [12–15].

DOI: 10.1201/9781003327035-12

According to mitigate the above issue, abundant research has been conducted to replace the conventional transistor structure, and it has been widely accepted that tunnel field-effect transistor (TFET, Tunnel FET) could be a promising candidate to replace the conventional transistor structure [16–19]. This is because TFET has strong immunity to SCE, and it also has low on/off current characteristics from the physical principal of its operation [20–23].

In this chapter, we are going to discuss what kinds of performances are required for designing next-generation memory devices, and what previous research has been done to accomplish each performance. On top of that, trade-off issues in designing the structure of memory device are also widely and specifically addressed. Then, this chapter will be finished by addressing the expected challenges for adopting TFET for memory chips.

12.2 FLOW CHART FOR EXPLAINING TFET-BASED MEMORY DEVICE

Figure 12.1 illustrates the flow chart for explaining TFET-based memory device [24]. In this chapter, TFET-based memory design will be addressed step-by-step in order to design ultra-low power memory device. Since the memory device has more complex structure compared to simple transistor (eg. MOSFET, TFET), this chapter will start from the physical structure of memory cell, and then explain the operation principle of memory device. Thereafter, finally, we discuss how to utilize TFET for memory design and some advantages of utilizing it.

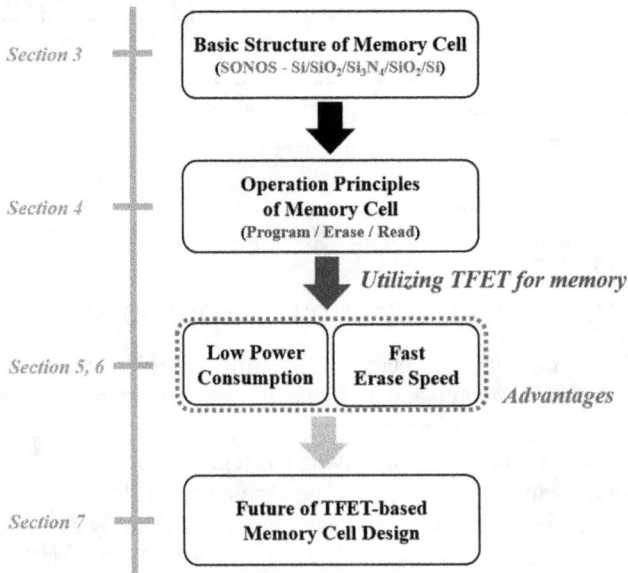

FIGURE 12.1   Flow chart with step-by-step approach to TFET-based memory cell.

12.3 BASIC STRUCTURE OF MEMORY CELL: SONOS STRUCTURE

In the previous chapters, the structure of metal-oxide-semiconductor field-effect transistor (MOSFET) and TFET have been explained. Compared to logic device (MOSFET, TFET), memory device has some additional oxide layers. For example, as shown in Figure 12.2, recent MOSFET (up-left) has the structure of Gate (metal) – gate dielectric (oxide) – channel (Si). Therefore, we can say the MOSFET has three layers (literally, Metal-Oxide-Silicon three layers, "MOS" in MOSFET).

On the other hand, memory cell basically has 5 layers, as shown in down side of Figure 12.2. The most representative memory structure is explained by "SONOS" structure with "Gate (poly-si) – blocking oxide (SiO_2) – charge trapping layer (Si_3N_4) - tunneling oxide (SiO_2) – channel (Si)". This "SONOS" structure was named after its materials poly-$Si/SiO_2/Si_3N_4/SiO_2/Si$ (= S/O/N/O/S).

Namely, MOSFET has three basic layers, and MOSFET-based memory cell has 5 basic layers with SONOS structure. Therefore, fabrication of memory cell is a little bit complicated compared to logic device (MOSFET, TFET), and understanding operation principle of memory cell is a little bit harder. However, fortunately, the structure of memory cell could be easily understood, if the readers already understand the operation principle of MOSFET.

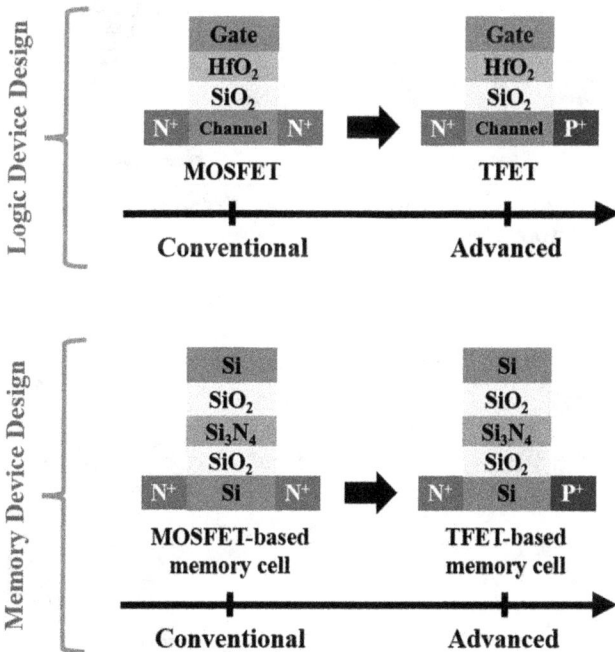

FIGURE 12.2 Schematic diagram illustrating the structure of MOSFET (up-left)/TFET (up-right)/MOSFET-based memory cell (down-left)/TFET-based memory cell (down-right).

12.4 OPERATION PRINCIPLE OF SONOS MEMORY STRUCTURE: PROGRAM/ERASE/READ

A memory cell has three types of operation (Program/Erase/Read) [Figure 12.3]. For example, as shown in left side of Figure 12.4, when there are lots of electron in charge trapping layer (CTL), we say "this memory cell is programmed". On the other hand, as shown in right side of Figure 12.4, when there is no electron in charge trapping layer (CTL), we usually say "this memory cell is erased". Namely, the binary information (0 or 1) is stored/erased through electrons.

Let's take a look further. When we want to program the memory cell (store electrons in CTL), we usually apply high voltage (16 ~ 20 V) to the gate. Then, electrons move from silicon channel to CTL (*This electron movement through oxide is called 'fowler nordheim (FN) tunneling'. If readers want to know more about this FN Tunneling, please kindly read another research after reading this chapter* [25]). Then, electrons are trapped in CTL, and no longer move to gate, because the oxide layer (SiO$_2$, blocking oxide) between CTL and gate is really

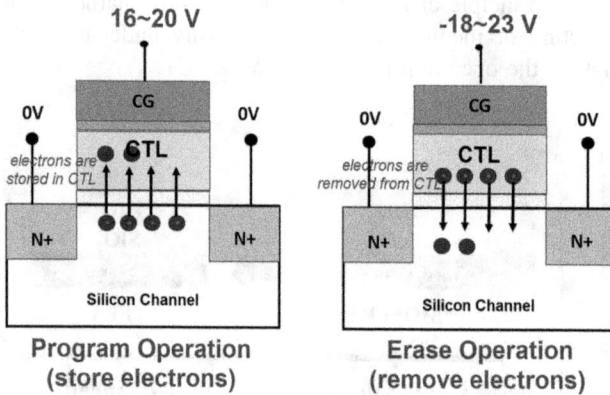

FIGURE 12.3 Program operation (left) and erase operation (right) in the memory cell.

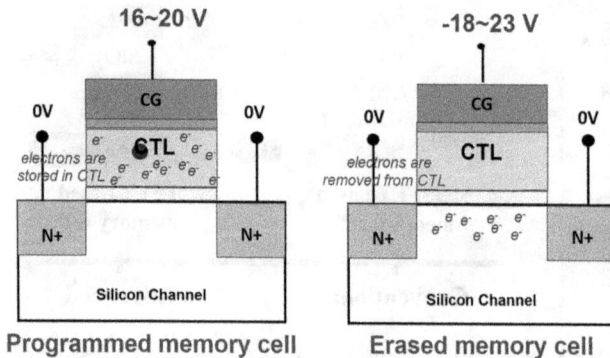

FIGURE 12.4 Schematic diagrams of programmed memory cell (left) and erased memory cell (right).

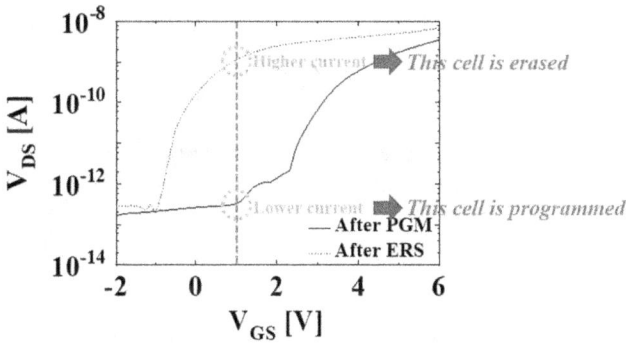

FIGURE 12.5 Method of determining whether certain cell is previously programmed or erased.

thick (~10 nm). Finally, CTL in the memory cell contains tons of electrons, and it is successfully programmed.

In contrast, when we want to erase the memory cell (remove electrons in CTL), we usually apply extreme negative voltage (−18 ~ −23 V) to the gate [26–28]. Then, electrons rather move from CTL to channel, due to electric field. After some erase operation with 40 ~ 400 ns, the memory cell is successfully erased.

Now, let's move on read operation. In read operation, the memory cell is analyzed with transfer characteristics. Specifically, by applying drain voltage (1 V) and gate voltage (from 0 V to 1V), the drain current is analyzed. By this read operation, we can determine whether the memory cell contains electrons or not. When electrons were previously stored (when memory cell was previously programmed), lower drain current might flow. This is because the electrons in CTL hinder the electric field caused by gate voltage (1 V). On the other hand, when electrons were previously removed (when memory cell was previously erased), higher drain current might flow. This is because the device operates without any hindrance from trapped electrons in CTL.

In sum, Figure 12.5 explains how to perform read operation and know whether the certain memory cell was previously programmed (1) or not (0). The programmed memory cell might have lower drain current, whereas the erased memory cell might have higher drain current. By comparing drain current to certain value, we can conclude the previous state of memory cell (1 = programmed or 0 = erased).

12.5 UTILIZING TFET FOR MEMORY DEVICE: TFET-BASED MEMORY FOR LOW POWER APPLICATION

In previous section, the basic operation of memory cell is explained, based on conventional MOSFET-based memory cell. In this section, TFET-based memory device will be discussed. Figure 12.6 shows the structure of TFET and TFET-based memory cell. Similar to MOSFET-based memory cell [Figure 12.2], TFET-based memory cell has 5 basic layers with SONOS structure as well.

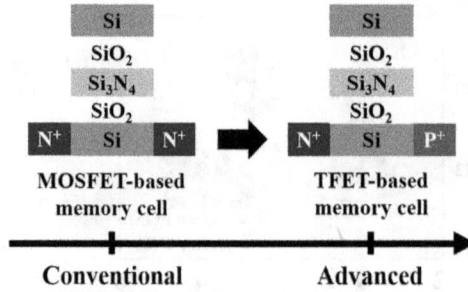

FIGURE 12.6 Schematic diagram illustrating basic structure of MOSFET-based memory cell (left) and TFET-based memory cell (right).

TABLE 12.1

Comparison Between MOSFET-Based Memory Cell and TFET-Based Memory Cell, in Terms of Power Consumption (Under Gate Voltage = Drain Voltage = 1 V) [29–34]

	MOSFET-based memory cell	TFET-based memory cell
Power (While reading programmed cell)	10,000 ~ 100,000 fW	1 ~ 10 fW
Power (While reading erased cell)	10,000 ~ 100,000 nW	10 ~ 100 nW

The program/erase operation of TFET-based memory cell is same with MOSFET-based memory cell. By applying high/low gate bias, the electrons are stored/removed in CTL, and the TFET-based memory cell is programmed/erased.

However, regarding read operation, TFET-based memory cell is different from MOSFET-based memory cell. Since TFET basically drives current by band-to-band tunneling (BTBT), different amount of current will flow in TFET-based memory cell, compared to MOSFET-based memory cell. Therefore, different amount of power might be consumed in MOSFET-based memory cell and TFET-based memory cell.

These values have remarkable meaning, especially for low-power applications. Table 12.1 summarizes the comparison between MOSFET-based memory cell and TFET-based memory cell. TFET-based memory cell usually shows 100 times lower on-current, and 1000 times lower off-current, so that significant power consumption could be saved. Therefore, this TFET-based memory cell structure is expected to be very strategic for future low-power applications, such as apple-watch, smart-watch, wearable device, smart phone, laptop computers.

12.6 ADDITIONAL ADVANTAGE OF TFET-BASED MEMORY DEVICE: FAST ERASE SPEED

In previous section, it has been discussed that TFET-based memory device has a cutting-edge advantage with 1/100 ~ 1/1000 lower power consumption. On top of that, TFET-based memory device has another advantage regarding erase speed [24]. Most readers might be familiar with this erase operation. For example, when we delete some documents in our PC or delete some music files in our smart phone, the semiconductor performs erase operation so that electrons in CTL of memory cell are removed. In this erase operation, TFET-based memory cell has remarkable advantage as well [24]. Especially, TFET-based memory cell has roughly 10,000 times higher erase speed, compared to conventional MOSFET-based memory cell [24].

Figure 12.7 shows schematic diagram of TFET-based memory cell, which is proposed by previous researcher [24]. In this structure [Figure 12.7], S/O/N/O stacks are located over the source so that off-current could be reduced (technically, for suppressing ambipolar current [24]).

Figure 12.8 describes the program/erase pulse for program/erase operation [24]. For analysis of programming, 16 V is applied to gate, and for analysis of erasing, $-16 \sim -20$ V is applied to gate [24]. As shown in Figure 12.9, regarding program speed, TFET-based memory cell has three times "lower" program speed, compared to the conventional MOSFET-based memory cell [24]. This is because, TFET-based memory cell is provided with electrons from only one side (n-type drain), whereas MOSFET-based memory cell is provided with electrons from two sides (n-type drain, n-type source) [24]. Because of this structure difference, TFET-based MOSFET has three times lower program speed.

Meanwhile, a TFET-based memory cell has 10,000 times faster program speed, compared to conventional MOSFET-based memory cell [24]. As shown in Figure 12.10, with broad range of erase voltage (V_{ERS}), TFET-based memory cell has excellent erase speed, compared to MOSFET-based memory cell [24]. This is because TFET-based memory cell has hole-supplier (p-type source), whereas conventional MOSFET-based memory cell doesn't have it [Figure 12.11] [24].

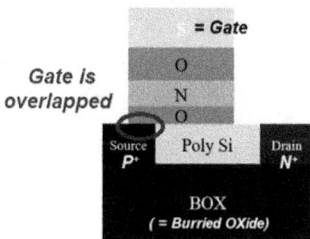

FIGURE 12.7 Schematic diagram explaining TFET-based memory cell.

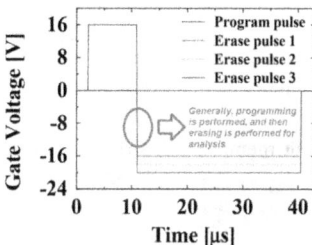

FIGURE 12.8 Gate voltage applied for analyzing performance of TFET-based memory cell.

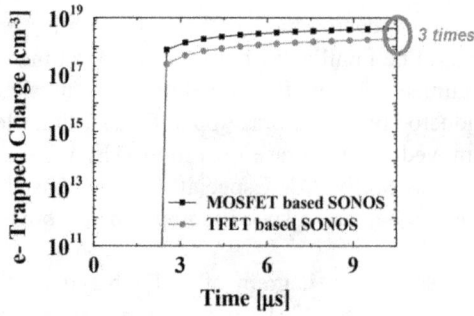

FIGURE 12.9   Comparison of program speed in 2 different memory cell structure.

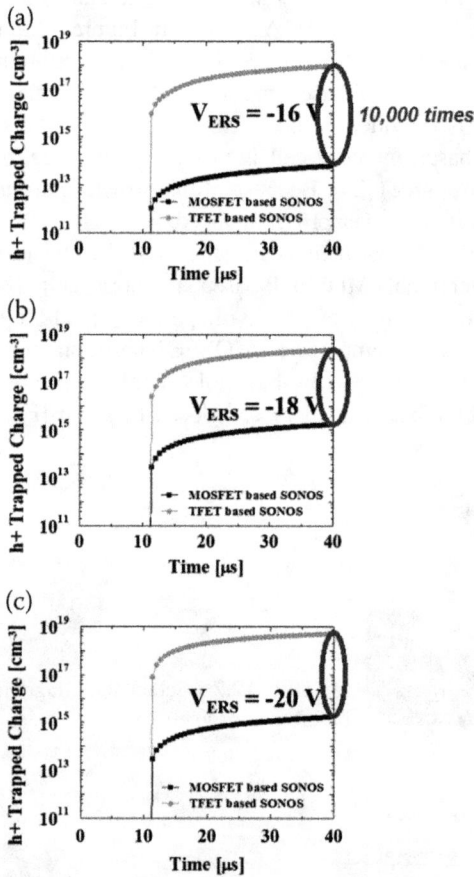

FIGURE 12.10   Erase speed comparison between 2 different memory cell structure under the gate voltage of (a) −16 V, (b) −18 V, and (c) −20 V.

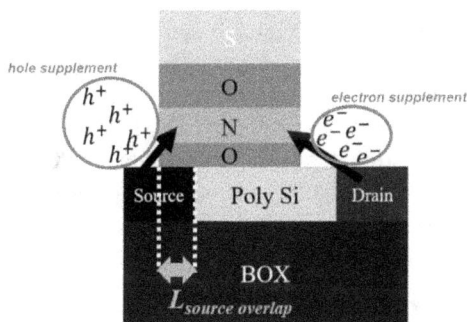

FIGURE 12.11 Dual supplement achieved by the structure of TFET. (especially, hole could be abundantly gained from source, and electron could be significantly gained from drain).

Namely, since TFET-based memory cell has different types of doping in source and drain, it can be either sufficiently provided with electrons and holes, at the same time. On the other hand, since conventional MOSFET-based memory cell has same types of doping in source and drain, unfortunately, conventional MOSFET-based memory cell might have problem of electron-supply or hole-supply. In this sense, the previous researchers have demonstrated that TFET-based memory cell has 10,000 times fast erase speed and three times lower program speed [24].

12.7 FUTURE OF TFET-BASED MEMORY CELL DESIGN: OPPORTUNITIES & CHALLENGES

So far, we have discussed design methodology by utilizing TFET for memory cell. It is clear and widely accepted that TFET-based memory cell is a promising candidate for future memory-semiconductor design. Especially, it has power advantage and erase speed at the same time, so that it might enable future wearable and portable devices with ultra-low battery consumption.

However, TFET-based memory design also has some challenges as well. For example, since TFET-based memory cell has different doping type in source and drain, more fabrication step is required for producing TFET-based memory. To be specific, compared to fabricating conventional MOSFET-based memory cell, TFET-based memory cell requires 5 more fabrication steps for one more doping (PR deposition – photolithography – PR removal – ion implantation (doping) – mask removal). Therefore, semiconductor designer should consider not only the performance boosting from TFET-based memory, but also increase in fabrication-cost. It is desirable to optimize this trade-off issue for TFET-based memory design, and we believe that future engineers could wisely optimize this trade-off relationship and make further improvements.

12.8 CONCLUSION

In this chapter, the step-by-step explanation has been carefully done for intuitive understanding. TFET-based memory cell has basically same structure (SONOS), compared to the conventional MOSFET-based memory. However, TFET-based memory cell has different type of dopant in source and drain, so that lower current could flow during operation. As a result, low power consumption could be achieved by this TFET-based memory cell. In addition, TFET-based memory cell has 10,000 times faster erase speed, so that data-erase-operation in PC or smart phone could be efficiently done. This TFET-based memory cell is expected to be very strategic for future semiconductor design with ultra-low power consumption.

ACKNOWLEDGMENTS

The chapter "TFET based memory cell design with Top-Down approach" has been written by utilizing some information obtained from various previous studies, and the sources of the information are indicated in a reference format. The author would like to express heartfelt thanks to previous researchers who have conducted various research and advanced technology for humanity. Especially, some figures in chapter "TFET based memory cell design with Top-Down approach" were written by re-editing and utilizing the figures from the open access journal, after obtaining permission from all authors of the corresponding manuscripts (Refs. 24). In addition, several plagiarism checks (Turnitin) have been conducted in writing this chapter. As a result, in this chapter with 2374 words, under 3 % similarity has been confirmed among all sentences of this chapter (except 'acknowledgment part' and 'reference list part'). I would like to express my gratitude to the previous studies and plagiarism screening program (Turnitin) for providing a variety of information.

REFERENCES

[1] J. T. C. Teng, V. Grover, and W. Guttler, "Information technology innovations: general diffusion patterns and its relationships to innovation characteristics," in *IEEE Transactions on Engineering Management*, vol. 49, no. 1, pp. 13–27, Feb. 2002, 10.1109/17.985744.

[2] Y. S. Song, S. Tayal, S. B. Rahi, J. H. Kim, A. K. Upadhyay, and B.-G. Park, "Thermal-Aware IC Chip Design by Combining High Thermal Conductivity Materials and GAA MOSFET," *2022 5th International Conference on Circuits, Systems and Simulation (ICCSS)*, 2022, pp. 135–140, 10.1109/ICCSS55260.2022.9802341.

[3] S. J. Kang, J. H. Kim, Y. S. Song, S. Go, and S. Kim, "Investigation of Self-Heating Effects in Vertically Stacked GAA MOSFET With Wrap-Around Contact," in *IEEE Transactions on Electron Devices*, vol. 69, no. 3, pp. 910–914, March 2022, 10.1109/TED.2022.3140283.

[4] P. Vimala, V. Singh, S. Gautam, T. Vijay, S. Singh, and T. S. Arunsamuel, "Performance and Characteristic Analysis of Graphene Field Effect Transistor with Different Channel Widths," *2021 IEEE Mysore Sub Section International Conference (MysuruCon)*, 2021, pp. 307–311, 10.1109/MysuruCon52639.2021.9641524.

[5] Z. Hao, W. Houjun, and Y. Peng, "Research on the Technology of High-Speed Large-Capacity Data Storage in DSO," *2007 8th International Conference on Electronic Measurement and Instruments*, pp. 4-882–4-885, 2007, 10.1109/ICEMI.2007.4351284.

[6] J. McHugh, P. E. Cuddihy, J. W. Williams, K. S. Aggour, V. S. Kumar, and V. Mulwad, "Integrated access to big data polystores through a knowledge-driven

framework," *2017 IEEE International Conference on Big Data (Big Data)*, 2017, pp. 1494–1503, 10.1109/BigData.2017.8258083.

[7] P. Vimala and N. B. Balamurugan, "New Analytical Model for Nanoscale Tri-Gate SOI MOSFETs Including Quantum Effects," in *IEEE Journal of the Electron Devices Society*, vol. 2, no. 1, pp. 1–7, Jan. 2014, 10.1109/JEDS.2014. 2298915.

[8] S. Tayal, B. Majumdar, S. Bhattacharya, and S. Kanungo, "Performance Analysis of the Dielectrically Modulated Junction-Less Nanotube Field Effect Transistor for Biomolecule Detection," in *IEEE Transactions on NanoBioscience*, 10.1109/TNB. 2022.3172702.

[9] A. Castro-Carranza *et al.*, "Effect of Density of States on Mobility in Small-Molecule n-Type Organic Thin-Film Transistors Based on a Perylene Diimide," in *IEEE Electron Device Letters*, vol. 33, no. 8, pp. 1201–1203, Aug. 2012, 10.1109/ LED.2012.2201441.

[10] D. Chakraborty, P. Raha, A. Bhattacharya, and R. Dutta, "Speed optimization of a FPGA based modified viterbi decoder," *2013 International Conference on Computer Communication and Informatics*, 2013, pp. 1–6, 10.1109/ICCCI.2013. 6466245.

[11] C. K. Pandey and S. Chaudhury, "Dual-Metal Graded-Channel Double-Gate Tunnel FETs for Reduction of Ambipolar Conduction," *2018 IEEE Electron Devices Kolkata Conference (EDKCON)*, 2018, pp. 572–576, 10.1109/EDKCON. 2018.8770449.

[12] A. K. Upadhyay, A. K. Kushwaha, and S. K. Vishvakarma, "A Unified Scalable Quasi-Ballistic Transport Model of GFET for Circuit Simulations," in *IEEE Transactions on Electron Devices*, vol. 65, no. 2, pp. 739–746, Feb. 2018, 10.1109/ TED.2017.2782658.

[13] N. Guenifi, S. B. Rahi, and M. Larbi, "Suppression of Ambipolar current and analysis of RF performance in double gate tunneling field effect transistors for low-power applications," *Int J nanoparticles nanotech*, vol. 6, p. 033, 2020.

[14] Y. S. Song, J. H. Kim, G. Kim, H. -M. Kim, S. Kim, and B. -G. Park, "Improvement in Self-Heating Characteristic by Incorporating Hetero-Gate-Dielectric in Gate-All-Around MOSFETs," in *IEEE Journal of the Electron Devices Society*, vol. 9, pp. 36–41, 2021, 10.1109/JEDS.2020.3038391.

[15] Y. S. Song, S. Kim, G. Kim, H. Kim, J.-H. Lee, J.H. Kim, and B.-G. Park, "Improvement of self-heating effect in Ge vertically stacked GAA nanowire pMOSFET by utilizing Al2O3 for high-performance logic device and electrical/ thermal co-design," in *Japanese Journal of Applied Physics*, vol. 60, pp. SCCE04, Mar. 2021, 10.35848/1347-4065/abec5c.

[16] W. Y. Choi, B. -G. Park, J. D. Lee, and T. -J. K. Liu, "Tunneling Field-Effect Transistors (TFETs) With Subthreshold Swing (SS) Less Than 60 mV/dec," in *IEEE Electron Device Letters*, vol. 28, no. 8, pp. 743–745, Aug. 2007, 10.1109/ LED.2007.901273.

[17] J. H. Kim, H. W. Kim, Y. S. Song, S. Kim, and G. Kim, "Analysis of current variation with work function variation in l-shaped tunnel-field effect transistor," *Micromachines*, vol. 11, no 8, pp. 780, Aug. 2020.

[18] T. S. Arun Samuel, and N. B. Balamurugan, "An analytical modeling and simulation of dual material double gate tunnel field effect transistor for low power applications," *Journal of electrical engineering and technology*, vol. 9, no. 1, pp. 247–253, 2014.

[19] P. Vimala, T. S. A. Samuel, D. Nirmal, and A. K. Panda, "Performance enhancement of triple material double gate TFET with heterojunction and heterodielectric," *Solid State Electronics Letters*, vol. 1, no. 2, pp. 64–72, 2019.

[20] G. Qi, W. Gan, L. Xu, J. Liu, Q. Yang, X. Zhu, J. Zhou, X. Ma, G. Hu, T. Chen, S. Yu, Z. Wu, H. Yin, and Y. Lu, "The Device and Circuit Level Benchmark of Si-Based Cold Source FETs for Future Logic Technology," *IEEE Transactions on Electron Devices*, vol. 69, no. 6, pp. 3483–3489, 2022.

[21] S. Tayal and A. Nandi, "Comparative analysis of High-K gate stack based Conventional & Junctionless FinFET," *2017 14th IEEE India Council International Conference (INDICON)*, 2017, pp. 1–4, 10.1109/INDICON.2017.8487675.

[22] R. Dutta, M. Rahaman, A. Guha, and N. Paitya, "Study of gate source-drain overlap/ gate-channel underlap in Heteojunction (50nm Ge channel) n-Double Gate TFET for different κ-spacer," *2019 International Conference on Smart Systems and Inventive Technology (ICSSIT)*, 2019, pp. 672–676, 10.1109/ICSSIT46314. 2019.8987903.

[23] C. K. Pandey, D. Dash, and S. Chaudhury, "Improvement in analog/RF performances of SOI TFET using dielectric pocket," *International Journal of Electronics*, vol. 107, no. 11, pp. 1844–1860, 2020.

[24] Y. S. Song, T. Jang, H.-M. Kim, J. -H. Lee, and B. -G. Park, "Erase Speed Enhancement with Low Power Operation by Incorporating Boron Doping," *Journal of Semiconductor Technology and Science*, vol. 21, no. 2, pp. 92–100, Apr. 2021.

[25] S. Venkatesan, and M. Aoulaiche, "Overview of 3D NAND Technologies and Outlook Invited Paper," *2018 Non-Volatile Memory Technology Symposium (NVMTS)*, 2018, pp. 1–5, 10.1109/NVMTS.2018.8603104.

[26] S.-J. Choi et al., "Enhancement of Program Speed in Dopant-Segregated Schottky-Barrier (DSSB) FinFET SONOS for NAND-Type Flash Memory," in *IEEE Electron Device Letters*, vol. 30, no. 1, pp. 78–81, Jan. 2009, 10.1109/LED.2008.2008667.

[27] M. Sathish kumar, T. A. Samuel, K. Ramkumar, I. V. Anand, and S. B. Rahi, "Performance evaluation of gate engineered InAs–Si heterojunction surrounding gate TFET," *Superlattices and Microstructures*, vol. 162, p. 107099, 2022.

[28] G. Naima and S. B. Rahi, "Design and Optimization of Heterostructure Double Gate Tunneling Field Effect Transistor for Ultra Low Power Circuit and System," *Electrical and Electronic Devices, Circuits, and Materials: Technological Challenges and Solutions*, pp. 19–36, 2021.

[29] F. Horst et al., "Static noise margin analysis of 8T TFET SRAM cells using a 2D compact model adapted to measurement data of fabricated TFET devices," *2017 Joint International EUROSOI Workshop and International Conference on Ultimate Integration on Silicon (EUROSOI-ULIS)*, 2017, pp. 39–42, 10.1109/ULIS.2017. 7962595.

[30] G. Naima, S. B. Rahi, and G. Boussahla, "Impact of Dielectric Engineering on Analog/RF and Linearity Performance of Double Gate Tunnel FET," *International Journal of Nanoelectronics & Materials*, vol. 14, no. 3, 2021.

[31] S. B. Rahi and B. Ghosh, "High-k double gate junctionless tunnel FET with a tunable bandgap," *RSC Advances*, vol. 5, no. 67, pp. 54544–54550, 2015.

[32] Y. S. Song, S. B. Rahi, S. Tayal, A. Upadhyay, and J. H. Kim, "Design Techniques for High Reliability FET by Incorporating New Materials and Electrical/thermal Co-optimization," in *Emerging Materials*, pp. 133–154. Singapore: Springer, 2022.

[33] T. Kim, K. Park, T. Jang, M. H. Baek, Y. S. Song, and B. G. Park, "Input-modulating adaptive neuron circuit employing asymmetric floating-gate MOSFET with two independent control gates," *Solid-State Electronics*, vol. 163, p. 107667, 2020.

[34] Y. Choi, K. Lee, K. Y. Kim, S. Kim, J. Lee, R. Lee, H. M. Kim, Y. S. Song, S. Kim, J.-H. Lee, and B.-G. Park, "Simulation of the effect of parasitic channel height on characteristics of stacked gate-all-around nanosheet FET," *Solid-State Electronics*, vol. 164, p. 107686, 2020.

13 Designing of Nonvolatile Memories Utilizing Tunnel Field Effect Transistor

Pramod Kumar, Neha Paras, and Manisha Bharti

CONTENTS

13.1 INTRODUCTION

Following the recent trends in semiconductor industries, low-power applications are the need as the devices are scaling down. Conventional MOSFET devices are prone to short-channel effects at sub-micron regions, and their use is limited; hence industries are moving towards other compatible devices. One of the most critical areas is nonvolatile memories, which need to be fast for faster and low-power operation. Due to the "Boltzman tyranny" [1], the subthreshold slope of conventional MOSFET limits to 60 mV / decade [2]. Tunnel field effect transistors, by far, are the most promising devices for nonvolatile memories owing to their better performance in the subthreshold region. The use of TFET having steep switching characteristics (SS < 60 mV/decade at 300 K) can provide superior performance, as

DOI: 10.1201/9781003327035-13

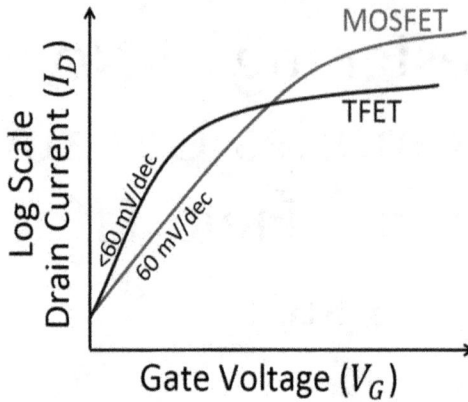

FIGURE 13.1 Comparison of subthreshold slope.

shown in Figure 13.1. This peculiar property of TFET helps us to scale down the power supply (V_{DD}) very aggressively, resulting in low power operation [2–10].

Quantum mechanical devices like Tunnel FET having band-to-band tunneling as transport phenomenon results in limited current. Unlike conventional, where the drift-diffusion phenomenon is responsible for the transportation of carriers, TFET optimizes the transport factor 'n' by using the tunneling transport phenomenon, which is temperature independent. TFET has an n+ source and p$^+$ drain and an intrinsic region between them forming n$^+$i-p^{+} as shown in Figure 13.2 (n tunnel FET). In the ON state, the current in tunnel devices is due to transport charges. In the ON state, the energy band alignment is such that it offers a minimum tunnel window for charge carriers [8–10].

Another emerging device is FeFET which integrates ferroelectric material inside FET's gate stack, which results in negative capacitance behavior. Figure 13.3 shows the structure of FeFET.

A steeper subthreshold slope exists in Negative capacitance in FeFET. As a result, vertical switching and a higher current ratio (I_{ON}/I_{OFF}) are the net results. The interaction of the negative capacitance of the ferroelectric layer with the positive capacitance underneath, properties of FeFET. The negative capacitance in FeFET arises due to the relation between the electric field (E) and polarization (P) of the ferroelectric material used. This interaction model is by the Landau-Khalatnikov equation (13.1).

$$E = \alpha\ P + \beta P^3 + \gamma P^5 + K_p dP/dt. \qquad (13.1)$$

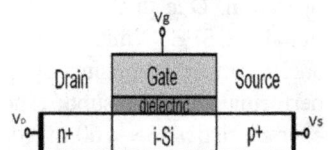

FIGURE 13.2   Tunnel FET.

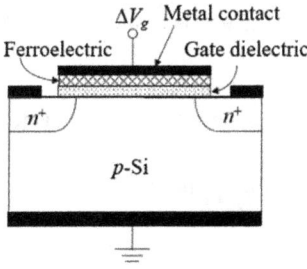

FIGURE 13.3 Ferro FET.

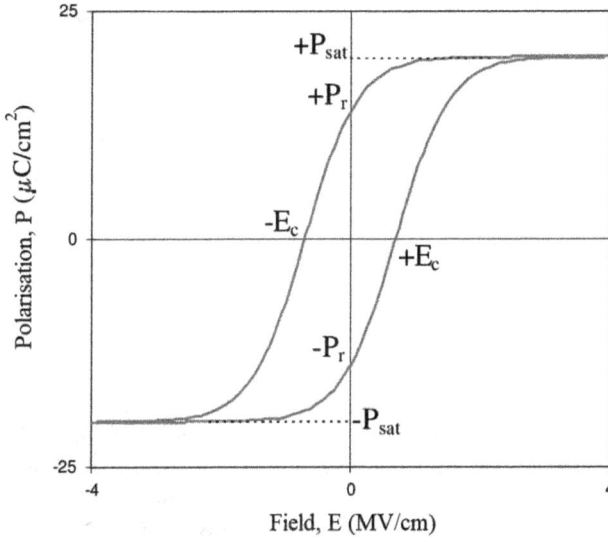

FIGURE 13.4 P–E curve of FeFET.

where α, β, and γ are static coefficients, and K_p is the kinetic coefficient.

For $K_p < 0$, there is a hysteresis in P-E characteristics shown below in Figure 13.4. The design of steep switching FeFETs to work in the negative capacitance region in the plot where the slope is negative.

This chapter uses an approach to the properties of FETs mentioned above to model a nonvolatile memory.

13.2 WORKFLOW AND SCOPE

13.2.1 WORKFLOW

TFET and FeFET emerged as one of the most promising devices in-memory applications. Nonvolatile memories on TFET and FeFET due to their robustness. For this chapter, we came across many techniques related to realizing these memories for low-power applications. Before using these types of memories, MOSFET-based SRAM cells existed, but the limitation of MOSFET is sub-threshold slope hindered their performance. MOSFET provides a weak sub-threshold slope, a major limiting factor for using MOSFET in the SRAM-based memory cells. With the introduction of the intrinsic junction instead of a doped junction and better insight into the sub-threshold region. The sub-threshold slope can be highly reduced (SS<60mV/decade at 300 K) with the introduction of TFET devices which in turn help in high-speed switching characteristics hence helping in the reduction of the access time in the SRAM cell [1].

The Ferroelectric material between the channel and gate gives us the negative capacitance domain in the FET, which also helps us reduce the sub-threshold slope. Although TFET and FeFET materials physics were too complicated to be applied to the FET devices, later the limitations, and now they are one of the materials to use for memories. Most researchers are using T-CAD models to simulate the new structures and make these devices more reliable, which is still going on. In the future, the technology may depend entirely on the TFET and FeFET memories [11]. Many TCAD simulations implement different device-related memory versions, which are helpful in-memory applications and decrease the memory area and power of the memories [11]. TCAD models introduce to cope with the ever-increasing need for memories such as FeFET memories, TFET memories, etc.

13.2.2 SCOPE

The scope of the ferroelectric memories is very vast as these memories provide a superior behavior in terms of area and power compared to other substitutes. Still, work is going on to perfect these memories. Ferroelectric-gate field effect transistors (FeFET) show excellent features as an integrated memory. Research works in [5,6,12,13] cover scalability, nonvolatile behavior, read-write speed, and higher temperature resistances. Even though FeFET has numerous advantages, memory retention time is still a significant concern in making it a practical device. In a Metal ferroelectric insulator semiconductor (MFIS) gate device, an insulator layer between the silicon and ferroelectric layers. Due to this arrangement, the interface damages occurring during the device's fabrication can restricts. Numerous experimental studies show charges trapped at the metal-ferroelectric and insulator-silicon junction. It causes the degradation of effective dielectric polarization and ultimately reduces the memory retention of the device. One way to increase the capacitive retention of the device is by improving the quality of the ferroelectric layer and interface, which achieves by treating it with thermal annealing and nitrogen radical. Moreover, much work has been done on MFIS FETs to improve their memory retention so that the devices can be practically realizable. These introductions to the FeFET devices are beneficial in applying nonvolatile memories.

Ferroelectric memories implemented on conventional MOSFET have attracted considerable attention for low-power nonvolatile memory applications, which may be helpful in upcoming times. Having superior features such as scalability, nonvolatility, and power-efficient switching readout is an overpowering feature of FeFET memories. They are highly distinguishable in two sectors as compared to memories, such as spin-based memories. They have high persistence compared to resistive RAMs, flash memories, and phase change memories. Based on the above reasons, it is an outstanding candidate for future memory technologies and nonvolatile memory applications.

13.3 ROADMAP FOR FUTURE MEMORY TECHNOLOGY FOR LOW POWER OPERATION

In recent years, the chip area has become a significant concern, along with the device's speed. The memories in the system need to be small in space and faster. One of the emerging areas in the respective field is ferroelectric memories, which are also faster (due to steep subthreshold slope) and precise. In recent years various concepts have emerged for memories shown in Figure 13.5.

The FeFET described by MISFET has ferroelectric oxide instead of SiO_x, $SiON$, or HfO_2 insulator. MSFET, which has epitaxial contact between ferroelectric to the semiconductor channel, in MFISFET, uses a buffer layer. With the floating gate between buffer and ferroelectric, MFMISFET obtains similar properties compared to other ferroelectric devices. Suppose the gate provides a sufficiently high voltage pulse in the FeFET, and the direction of polarization of ferroelectric material changes from inversion of the channel to accumulation mode, which results in a shift in the threshold voltage V_T that allows a read operation that is non-destructive in the memory [14]. There are some constraints in material and device requirements to make FeFET reliable and scalable. The main V_T shift in FeFET, as mentioned earlier, is the memory window. It determines through the voltage V_C which is the coercive voltage of the ferroelectric layer rather than the remnant polarization P_r [15–18]. It imposes a middle course between scaling and memory window. The dependence of coercive voltage on coercive field E_C and thickness d_{FE} of

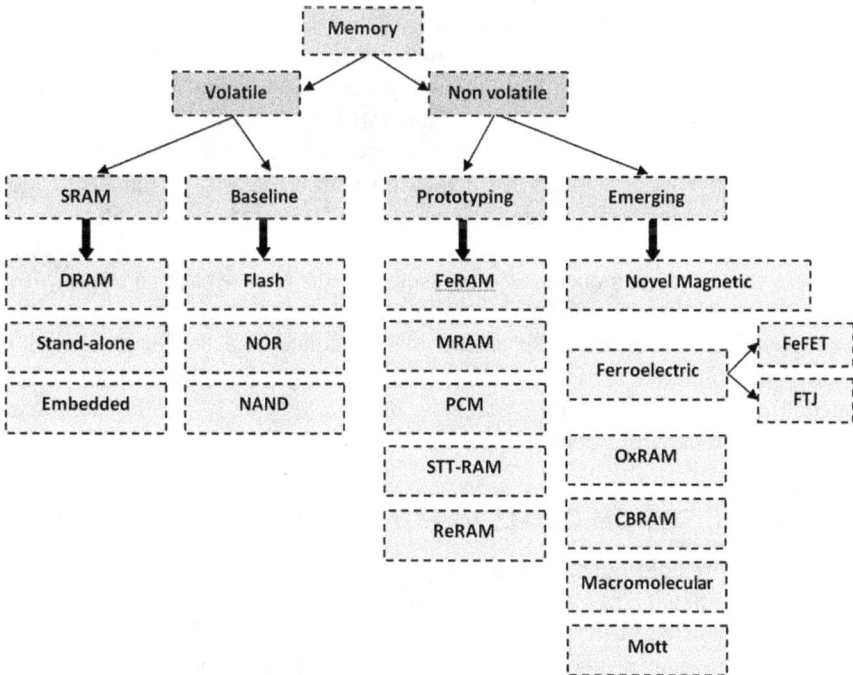

FIGURE 13.5 Taxonomy of various memory devices.

ferroelectric is the main factor that comes to mind for using FE. It is difficult for the commonly used perovskite-based FeFETs to scale down below 180nm because of thickness scaling and low E_C [19–26]. It hinders maintaining a reasonable memory window. For an affordable memory window, the thickness of the ferroelectric material should increase. FeFET issue rectifies by a highly coercive field and scalable Fe-HFO$_2$ [12]. HKMG technology is to scale down to 28nm, which was compatible with the CMOS technology and used for high-volume production. The technology mentioned above helps make the embedded memory have less mask count than the embedded flash memories.

Another essential property of the gate stack FeFET is its intrinsic capacitive voltage divider. The divided capacitive voltage results in a voltage drop in the Ferroelectric material and the nearby inherent regions. Because Of this action, there is a depolarization (inbuilt) field inside the FE material even in the no-bias condition, which opposes the applied electric field and ultimately results in the retention loss in standby mode and gate voltage distribution during a write operation. Rectification increases the insulator's capacitance as large as possible and ferroelectric capacitance as low as possible.

The ferroelectric tunnel junction is a design in which asymmetric electrodes interpose an ultra-thin ferroelectric film. It displays resistive switching, which is polarized and induced through a nonvolatile modulation of barrier height. The barrier height modifies the tunneling current, which is exponential, through which the orientation of the ferroelectric dipole for low or high resistance in ferroelectric tunnel junctions. The high or low resistance can read non-destructively, resulting in an electroresistance effect in ferroelectric tunnel junctions. It provides a 10 to 100 between HRS and LRS, which seems very good [27].

Further research shows a high tunneling electroresistance with ferroelectric tunnel junctions based on super tetragonal BiFeO$_3$. Another work is on the BaTiO$_3$ tunnel barrier, where one metal electrode replaces with a semiconducting electrode [28]. This new junction's design gives us variability in tunnel junction height and variable space charge regions in semiconductors and modifies barrier width.

These strategies mentioned above are helpful in the ferroelectric junctions where either the ferroelectric barrier with large polarization (BiFeO$_3$) is used or an electrode material of semiconductor nature that modulates the barrier height width through field-induced carrier depletion [28]. Although these ferroelectric tunnel junction memories seem very good, they are at a very early development stage and perovskite-based ferroelectrics for FeFET.

13.4   BASIC DESIGN GOALS IN FUTURE MEMORY DESIGN

- The future holds some new techniques related to FeFET.
- Electron detrapping method to enhance performance
- 1T nonvolatile memories.
- Digital coding states in memories by changing the polarization of FE material is already used nowadays in ferroelectric random access memory, which is capacitor-based.

But the identification of the memory state mainly depends upon the overall polarization charge stored in the capacitance-based ferroelectric material present in the gate stack configuration, which requires a destructive read operation. Alternatively, concepts such as FJT and FeFET allow nonharmful detection in the memory states and increase cell scalability. But their limitation of the FeFET between endurance and retention trade-off. HFO_2 FE has been demonstrated in some literature, giving an endurance range of 1012 switching cycles with ten years of retention data limit. Another concept is Ferroelectric tunnel junctions composed of Ferroelectric ultrathin film sandwiched by asymmetric electrodes. This polarization induces a resistive switching through a nonvolatile modulation barrier height—the non-destructive operation through orientation of Ferroelectric dipole codes a very low or high resistance path. The FJT-based structure may have a retention time of greater than ten years and endurance of 10^{14} cycles. The memories based on FJT are at a very early development stage, and still, the research is more focused on perovskite-based Ferroelectric. Moreover, for achieving the complex electrode ferroelectric system through perovskite based on the FJT concept, which relies on the throughput and CMOS compatible large substrates epitaxial growth techniques. If CMOS compatibility and thickness scalability of HfO_2-based Ferroelectric, then the FJT memories show considerable potential for the future.

The limitations in the VLSI industry lead to the use of memories based on DG MOSFET, FinFET, and GAA MOSFETs. Researchers are still working on multigate MOSFET structures for potential NVM applications (Nonvolatile Memories). SONOS (Si-Oxide-Nitride-Si) was the first for research in NVM for advanced MOSFET structures [14,19,28–37]. NVM based on TaN/Al2O3/HfO2 stack-based GAA nanowire structures were also reported earlier. One of the most conclusive works is from gate dielectric GAA Si Nanowire MOSFET, which embeds an nc-Si in GAA and has a superior performance concerning Nonvolatile memories. Other advancements such as GAA VNWFET (Gate All Around Vertical Nanowire Field Effect Transistors) emerged as suitable replacements for the conventional structures used for Nonvolatile memories. Due to the problem faced during the scaling of the flash memories beyond 22nm, GAA VNFET can be a possible replacement due to the crossbar arrangements leading to a high density [27,28].

These structures provide high current driving capability and scalability and are more feasible than other structures. The use of GAA VNFET as an access element in Nonvolatile memories is beneficial due to the device's high ION/IOFF ratio. Hence, access transistors with a minimal voltage and the access time are tiny, giving a far better result in the case of memories.

13.5   BASIC STRUCTURE OF MEMORY CELL

The use of portable devices in the modern world requires large battery life. The previously large hardware required a lot of energy to be driven. The RAM cell, which is one of the hardware architecture's main parts, needs to be small in size and drive much less power. SRAM cells retain the content present as long the power is ON. As SRAM prefers DRAM, it is a major contributor to the area consumption,

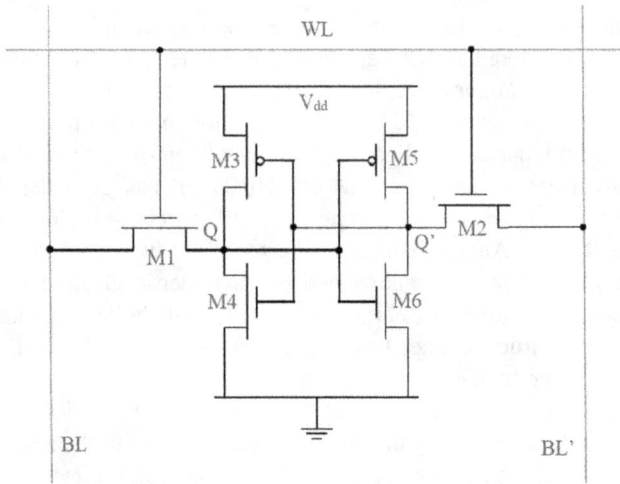

FIGURE 13.6 6T SRAM cell.

but the major concern for using the SRAM cells (CMOS Technology) is that due to scaling down the devices, the devices are more prone to leakages. Memory architecture uses a large number of transistors. The overall power reduces if there's a power reduction in the SRAM cell. Due to its simple design, the six transistor (6T) SRAM cell uses a memory cell. The delay and power of the 6T SRAM cell are also in balance. TFET is highly interested in its potential for steep subthreshold slope device operation. Due to this property, it is one of the most promising candidates for use in SRAM cells. In Figure 13.6, a TFET based 6T-SRAM cell is shown [9,10,38–44].

In the TFET based 6T SRAM cell, it includes inverter INV-1 (inverter 1) and INV-2 (inverter 2), which is in cross-coupling and access transistors M1 (NMOS) and M2 (NMOS). INV-1 has transistor M3, a pull-up transistor, and transistor M4 a pull-down transistor coupled to a positive voltage supply node V_{DD} and V_{SS}, the reference voltage. The second inverter also has a pull-up transistor M5 coupled to a supply voltage of V_{DD} and a pull-down transistor M6 related to reference supply V_{ss}. The INV-1 output is connected to the INV-2 input and vice-versa. These nodes from the result define Q and QB of the bit cell.

Node Q, a storage node, is coupled to the first bit line BL through M1 as an access transistor. Storage node Q connects to second-bit line BLB through second access transistor M2, where BL and BLB are complementary. Write line WL in bit cell control the access transistor M1 and M2 through the control signal. M1 and M2 access transistors configure to be in one direction only. When M1 is conducting, it allows the flow of current only from bit line BL to the first storage node Q and prevents current flow from storage node Q to bit line BL. If M2 is conducting, it only allows the flow of current from storage node Q to the bit line BLB and prevents current flow from storage node Q to the second line BLB. Thus we can say that in TFET, M1 uses in access configuration, which is inward & M2 in access configuration, which is outward.

13.6 PHYSICAL PRINCIPLE OF MEMORY CELL: PROGRAM/ERASE/ READ OPERATION

13.6.1 READ OPERATION

In the SRAM cell, an access transistor M1 performs the read operation, which allows higher RNM. First, the bit lines BLB and BL are precharged to V_{DD}. Then '1' is asserted to the write line WL to make the transistor M1 and M2 in enable mode. If bit '0' stores at node Q, then the bit line BL discharges through the access transistor M1 by producing a conducting path through transistors M1 and M4. Hence voltage (V_{DD}) at bit line BL is through M1 and M4. The sense amplifier senses the change and generates a signal indicating '0' stores in the SRAM bit cell. The bit line remains precharged to V_{DD}. If the node at Q is at '1', there is no need for charging or discharging. Figure 13.7 shows the read operation of the bit cell. A better read noise margin is due to the transistors M1 and M4; the M4 transistor is stronger than M1, obtained by increasing the width of transistor M4. The read operation depicts in Figure 13.7, where the conducting path to process more efficiently.

13.6.2 WRITE OPERATION

The write operation performs through M1 or M2 transistors depending on which data write on the bit cell. For writing '1' on node Q, which previously contained '0', the bit line BLB and BL are charged to V_{DD}. The write line WL gives '1' to enable the transistors M1 and M2. Simultaneously control logic sends a very minute positive voltage pulse to the V_{SS} node such that its magnitude is more significant than V_{SS} but less than V_{DD}. It reduces the gain of M4 temporarily. The QB node of inverter INV1 has '1' as stored. Here M4 remains conducting, but node Q of the first inverter raises above '0' for a short time. It reduces the positive feedback from INV-1 to INV-2 for writing '1'. The small voltage pulse given to V_{SS} makes it more convenient to write '1' at node Q as there is no direct access for the ground to node Q through M4. The node Q charge reaches the second inverter's trip point, and node QB becomes '0'. The voltage at V_{SS} can decrease to the same ground,

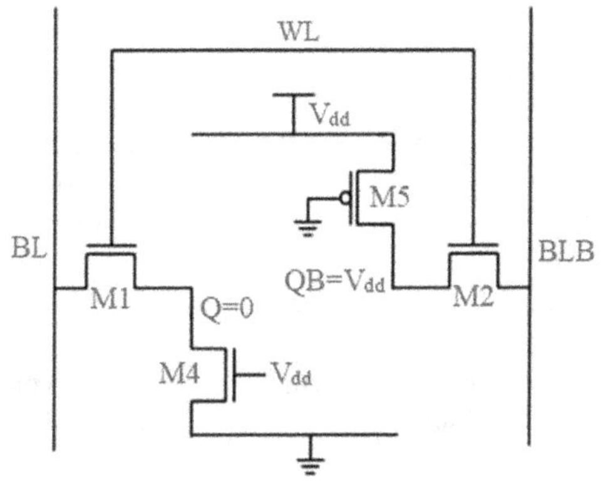

FIGURE 13.7 Read operation of 6T SRAM cell.

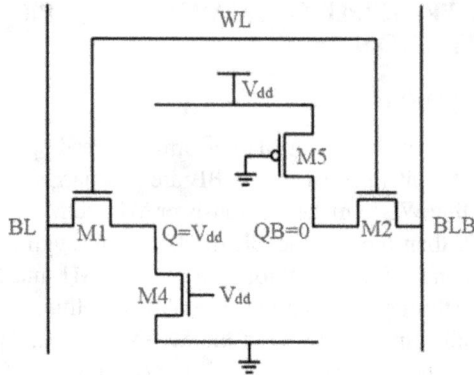

FIGURE 13.8 Write operation of 6T SRAM cell.

boosting positive feedback from INV1 to INV2 to maintain the new stored information on Q & QB for the SRAM cell. '1' is written through the M1 transistor. Though transistor M2 is also on, the voltage at BLB does not hinder the voltage at node Q or QB because the transistors make in such a way that they do outward conduction, and BLB is at V_{DD}. With these two conditions' help, node Q's discharging can be stopped during the write '1' operation. For flipping the bit content from '1' to '0', both BL and BLB are discharged to the ground. WL signal sets to '1' through control logic. The 6T SRAM bit cell's write operation is depicted in Figure 13.8.

13.7 DESIGN OPTIMIZATION WITH STEEP SUBTHRESHOLD SWING TRANSISTOR

Although the design of the previously discussed TFET-based SRAM cell is better than the conventional SRAM cells, there are also some limitations to TFET-based cell design. In this section, we will cover the regulations of TFET-based cell design and see what optimizations are in the literature.

13.7.1 FORWARD P-I-N CURRENT

The asymmetric and uncontrollable characteristic of current in TFET is a significant concern. Researchers have shown this effect using transmission gates [1]. If we talk about TFET-based transmission gates, the transistors cannot remove by the input and gate voltage, as shown in Figure 13.9. Even when C is '0', the p-i-n forward current still affects the TG_{out} voltage. The transmission gate suffers from the forward current limitation; thus, the SRAM cell containing transmission gate structure suffers from it. This problem discusses using the 6T SRAM cell. When the 6T-OA SRAM array goes into writing or read mode and node QB of an adjacent non-accessing cell is '0', the BLB is high, and the right AT (access transistor) generates forward p-i-n current though the WL is 0. It reduces the stability of the SRAM bit cell. Also, the HSNM (hold static noise margin) suffers, as shown in Figure 13.10.

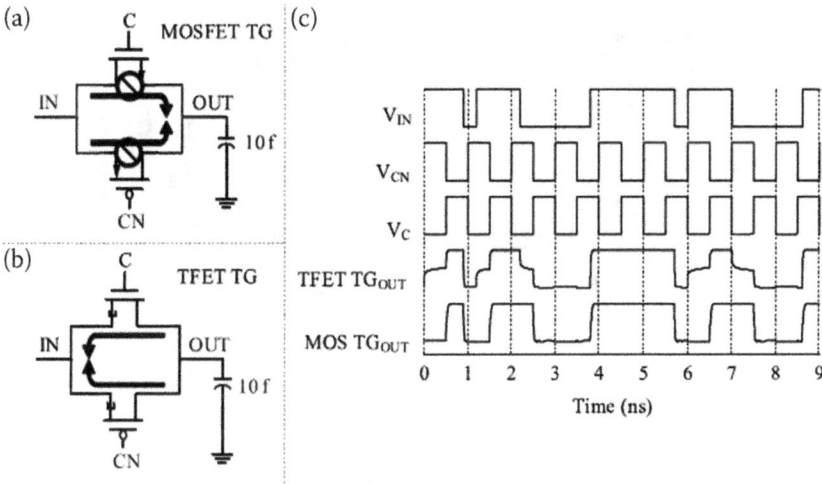

FIGURE 13.9 (a) MOSFET transmission Gate, (b) TFET transmission Gate, and (c) simulation of both [1].

The forward current does not only reduce HSNM, but due to the leakage current, it also increases static power dissipation, which generates by AT (access transistor), as shown in Figure 13.10b. Lastly, the bit line BL discharging may occur due to the p-i-n forward current, which can increase the read time and may also cause an error in reading the data stored at the node shown in Figure 13.10c. The above-discussed problem is encountered by making some changes in the SRAM cell. The authors have proposed that the p-i-n forward current eliminates the forward bias voltage effectively.

Hence a pro-11T cell is shown in Figure 13.11, which eliminates the write and read path forward bias voltage. For writing '1', the PTFET transistors AT4 and AT5 are connected in series. For writing '0', two NTFET transistors AT6 and AT7, are strings connected, such that the source of AT6 and drain of AT7 connects to ground and Q, respectively. For improving the write '1' ability, an NL2 transistor employs. The use of additional transistors increases the Hold Static Noise Margin (HSNM) and Read Static Noise Margin (RSNM). For writing '1', operations AT4 and AT5 are ON. The NL2 transistor interrupts the latch state, due to which the Q voltage pulls up through AT4 and AT5, and the write '1' capability is notably enhanced. For the write '0' function, AT6 and AT7 are ON, and the Q node discharges through these transistors.

13.7.2 Low Current of TFET

Another issue listed in the literature is the delay in the SRAM standard cell due to the low ON-current of some of the TFETs used. The issue of insufficient ON-current can mitigate by some of the device perspective changes, such as:

- Decreasing the adequate gate oxide thickness
- Changing the abruptness of the source-channel doping profile

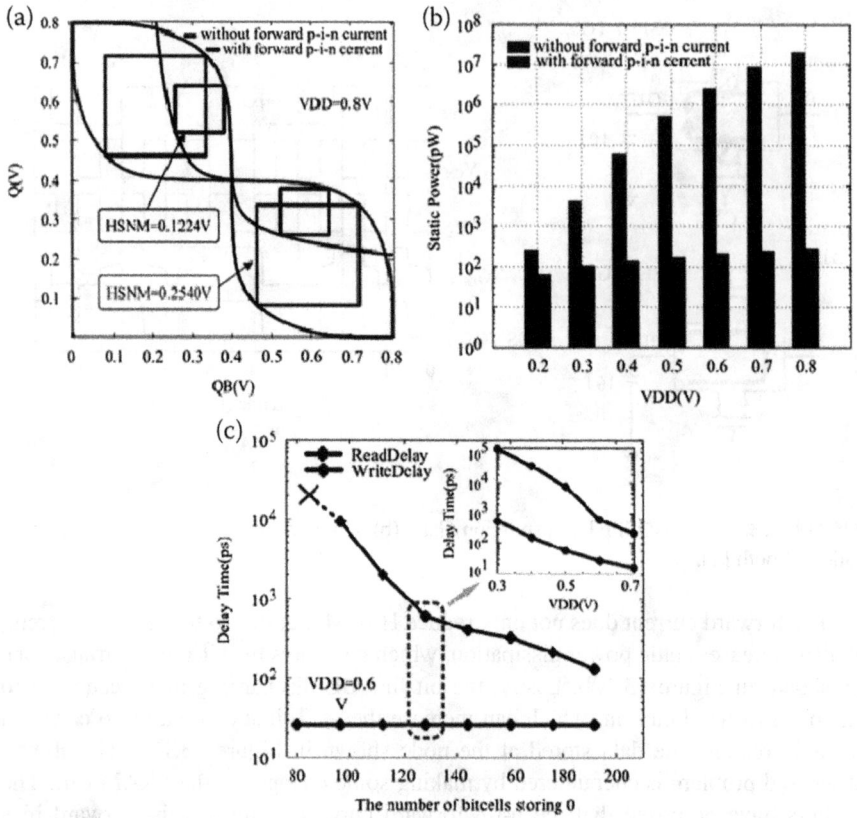

FIGURE 13.10 (a) Butterfly curve showing HSNM with and without p–i–n forward current, (b) static power dissipation comparison with and without forward p–i–n current, and (c) read-write delay in 6T-OA SRAM cell [1].

- Using a double-gate control mechanism
- Using high-k gate oxide
- Increasing the tunnel area

The problem mentioned was eliminated from a circuit perspective. The use of hybrid circuits can be a possible solution [1]. A MOSFET in the read path might enhance the speed of an SRAM bit cell. MOSFET introduces the leakage in way by using a control signal. A comparison of a TFET-based and hybrid-based SRAM cell shows a decrease in the delay of the SRAM cell.

13.8 CONCLUSION

In this chapter, we explored TFET and FeFET-based nonvolatile memories. We have seen the drawbacks of the conventional memories and how the tunnel transistor and FeFET overcome the limitations of the traditional devices and

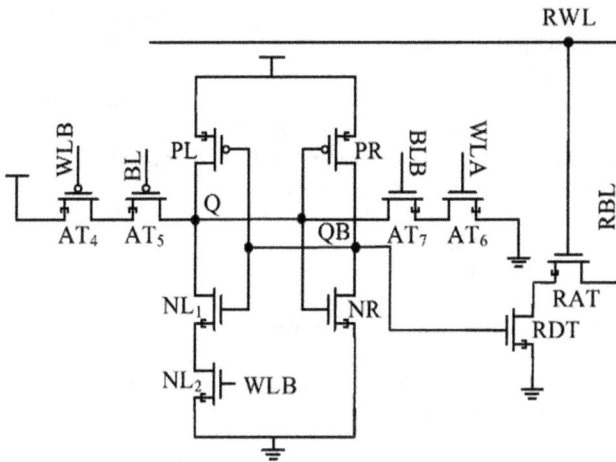

FIGURE 13.11 Pro 11-T SRAM cell [1].

design methodologies. We have also seen various problems associated with the TFET and how to overcome these problems. Given a brief overview of the future aspects of nonvolatile memories. An SRAM cell using TFET shows in this chapter. Through rigorously done research, the advantage of using TFET and FeFET and the disadvantages, such as the forward p-i-n current of TFET, which is asymmetric and uncontrollable. As present in the literature, the HSNM of TFET-based SRAM cells also discusses a significant concern for the memories based on TFET. We have also discussed above the ferroelectric tunnel junctions and how they are helpful in the making of FeFET. Researchers are further working with the ferroelectric tunnel junctions, which will be very helpful in making reliable FeFET-based SRAM cells.

REFERENCES

[1] Z. Lin, et al., "Challenges and Solutions of the TFET Circuit Design," in *IEEE Transactions on Circuits and Systems I: Regular Papers*, vol. 67, no. 12, pp. 4918–4931, December 2020, 10.1109/TCSI.2020.3010803.

[2] W. Young Choi, B.-G. Park, J. Duk Lee, and T.-J. King Liu, "Tunneling Field-Effect Transistors (TFETs) with Subthreshold Swing (SS) Less Than 60 mV/dec," *IEEE Electron Device Letters*, vol. 28, no. 8, pp. 743–745, August 2007, 10.1109/LED.2007.901273.

[3] H. Liu, et al., "ZrO_2 Ferroelectric FET for Nonvolatile Memory Application," in *IEEE Electron Device Letters*, vol. 40, no. 9, pp. 1419–1422, September 2019, 10.1109/LED.2019.2930458.

[4] A. Chen, "Emerging Nonvolatile Memory (NVM) Technologies," in Solid State Device Research Conference (ESSDERC), 2015 45th European, 2015, pp. 109–113, 10.1109/ESSDERC.2015.7324725.

[5] A. Sharma and K. Roy, "1T Non-Volatile Memory Design Using Sub-10nm Ferroelectric FETs," in *IEEE Electron Device Letters*, vol. 39, no. 3, pp. 359–362, March 2018, 10.1109/LED.2018.2797887.

[6] Y.-S. Liu and P. Su, "Improving the Scalability of Ferroelectric FET Nonvolatile Memories With High-k Spacers," in *IEEE Journal of the Electron Devices Society*, vol. 10, pp. 346–350, 2022, 10.1109/JEDS.2022.3169753.

[7] Rohit, G. Saini, "A Stable and Power Efficient SRAM Cell," in *IEEE International Conference on Computer, Communication, and Control*, 2015.

[8] M. Rahman, M. Li, J. Shi, S. Khasanvis, and C. A. Moritz, "A New Tunnel-FET Based RAM Concept for Ultra-Low Power Applications," *2014 IEEE/ACM International Symposium on Nanoscale Architectures (NANOARCH)*, 2014, pp. 57–58, 10.1109/NANOARCH.2014.6880505.

[9] S. Gupta, M. Steiner, A. Aziz, V. Narayanan, S. Datta, and S. K. Gupta, "Device-Circuit Analysis of Ferroelectric FETs for Low-Power Logic," in *IEEE Transactions on Electron Devices*, vol. 64, no. 8, pp. 3092–3100, August 2017, 10.1109/TED.2017.2717929.

[10] N. Paras and S. S. Chauhan, "Optimization of Design Parameters for Vertical Tunneling Based Dual Metal Dual Gate TFET," in 5th IEEE International Conference on Advances in Computing and Communication Engineering (ICACCE-2019), Tamil Nadu, India, 4–6th April 2019, ISBN: 978-1-7281-3250-1, 10.1109/ICACCE46606.2019.9079988.

[11] K. Lee, S. Kim, M. Kim, J. -H. Lee, D. Kwon, and B. -G. Park, "Comprehensive TCAD-Based Validation of Interface Trap-Assisted Ferroelectric Polarization in Ferroelectric-Gate Field-Effect Transistor Memory," in *IEEE Transactions on Electron Devices*, vol. 69, no. 3, pp. 1048–1053, March 2022, 10.1109/TED.2022.3144965.

[12] J. Mller, E. Yurchuk, T. Schlsser, J. Paul, R. Hoffmann, S. Mller, D. Martin, S. Slesazeck, P. Polakowski, J. Sundqvist, M. Czernohorsky, K. Seidel, P. Kcher, R. Boschke, M. Trentzsch, K. Gebauer, U. Schrder, and T. Mikolajick, "Ferroelectricity in HfO_2 nonvolatile data storage in 28 nm HkMg," in VLSI Technology (VLSIT), 2012 Symposium on, 2012, pp. 25–26, 10.1109/VLSIT.2012.6242443.

[13] J. R. Contreras, J. Schubert, H. Kohlstedt, and R. Waser, "Memory Device Based on a Ferroelectric Tunnel Junction," *60th DRC. Conference Digest Device Research Conference*, 2002, pp. 97–98, 10.1109/DRC.2002.1029532.

[14] J. Müller, T. S. Böscke, U. Schröder, S. Mueller, D. Bräuhaus, U. Böttger, L. Frey, and T. Mikolajick, "Ferroelectricity in Simple Binary ZrO_2 and HfO_2," *Nano Letters*, vol. 12, no. 8, pp. 4318–4323, 2012, 10.1021/nl302049k.

[15] N. Tasneem, *et al.*, "Efficiency of Ferroelectric Field-Effect Transistors: An Experimental Study," in *IEEE Transactions on Electron Devices*, vol. 69, no. 3, pp. 1568–1574, March 2022, 10.1109/TED.2022.3141988.

[16] T. Sandu, C. Tibeica, R. Plugaru, O. Nedelcu, and N. Plugaru, "Physical Modeling of Ferroelectric Tunnel Junctions," *2021 International Semiconductor Conference (CAS)*, 2021, pp. 85–88, 10.1109/CAS52836.2021.9604149.

[17] H. Duan, W. Fang, L. Liu, and W. Chen, "Theoretical Study of Bilayer Composite Barrier Based Ferroelectric Tunnel Junction Memory," *2020 IEEE MTT-S International Conference on Numerical Electromagnetic and Multiphysics Modeling and Optimization (NEMO)*, 2020, pp. 1–3, 10.1109/NEMO49486.2020.9343600.

[18] B. Max, M. Hoffmann, S. Slesazeck, and T. Mikolajick, "Direct Correlation of Ferroelectric Properties and Memory Characteristics in Ferroelectric Tunnel Junctions," in *IEEE Journal of the Electron Devices Society*, vol. 7, pp. 1175–1181, 2019, 10.1109/JEDS.2019.2932138.

[19] L. Van Hai, M. Takahashi, and S. Sakai, "Downsizing of Ferroelectric Gate Field-Effect-Transistors for Ferroelectric-NAND Flash Memory Cells," in *Proceedings of 3rd IEEE International Memory Workshop (IMW)*, May 2011, pp. 1–4, 10.1109/IMW.2011.5873239.

[20] U. Schroeder, S. Mueller, J. Mueller, E. Yurchuk, D. Martin, C. Adelmann, T. Schloesser, R. van Bentum, and T. Mikolajick, "Hafnium Oxide Based CMOS Compatible Ferroelectric Materials," *ECS Journal of Solid State Science and Technology*, vol. 2, no. 4, pp. N69–N72, 2013, 10.1149/2.010304jss.

[21] A. Sheikholeslami and P. G. Gulak, "A Survey of Circuit Innovations in Ferroelectric Random-Access Memories," *Proceedings of the IEEE*, vol. 88, no. 5, pp. 667–689, 2000, 10.1109/5.849164.

[22] M. H. Lee, S. T. Fan, C. H. Tang, P. G. Chen, Y. C. Chou, H. H. Chen, J. Y. Kuo, M. J. Xie, S. N. Liu, M. H. Liao, C. A. Jong, K. S. Li, M. C. Chen, and C. W. Liu, "Physical Thickness 1. x nm ferroelectric HfZrOx Negative Capacitance FETs," in Electron Devices Meeting (IEDM), 2016 IEEE International, 2016, pp. 12–1, 10.1109/IEDM.2016.7838400.

[23] H.-T. Lue, *et al.*, "3D AND: A 3D Stackable Flash Memory Architecture to Realize High-Density and Fast-Read 3D NOR Flash and Storage-Class Memory," *2020 IEEE International Electron Devices Meeting (IEDM)*, 2020, pp. 6.4.1–6.4.4, 10.1109/IEDM13553.2020.9372101.

[24] N. Ronchi, *et al.*, "A Comprehensive Variability Study of Doped HfO_2 FeFET for Memory Applications," *2022 IEEE International Memory Workshop (IMW)*, 2022, pp. 1–4, DOI: 10.1109/IMW52921.2022.9779294.

[25] "Ferroelectric Field-effect Transistors as High-density, Ultra-fast, Embedded Non-volatile Memories," *2022 International Symposium on VLSI Technology, Systems and Applications (VLSI-TSA)*, 2022, pp. i, 10.1109/VLSI-TSA54299.2022.9770965.

[26] A. Saeidi, F. Jazaeri, I. Stolichnov, and A. M. Ionescu, "Double-Gate Negative-Capacitance MOSFET With PZT Gate-Stack on Ultra Thin Body SOI: An Experimentally Calibrated Simulation Study of Device Performance," in *IEEE Transactions on Electron Devices*, vol. 63, no. 12, pp. 4678–4684, December 2016, 10.1109/TED.2016.2616035.

[27] T. K. Agarwal, O. Badami, S. Ganguly, S. Mahapatra, and D. Saha, "Design Optimization of Gate-All-Around Vertical Nanowire Transistors for Future Memory Applications," *2013 IEEE International Conference of Electron Devices and Solid-state Circuits*, 2013, pp. 1–2, 10.1109/EDSSC.2013.6628113.

[28] A. Sengupta and C. K. Sarkar, "Comparative Study on Nanocrystal Embedded Gate Dielectric and Oxide-Nitride Oxide Stack Dielectric GAA MOSFET Nonvolatile Memory Devices," *2012 International Conference on Informatics, Electronics & Vision (ICIEV)*, 2012, pp. 511–515, 10.1109/ICIEV.2012.6317402.

[29] A. G. Maslovskaya, L. I. Moroz, A. Yu Chebotarev, and A. E. Kovtanyuk, "Theoretical and Numerical Analysis of the Landau–Khalatnikov Model of Ferroelectric Hysteresis," *Communications in Nonlinear Science and Numerical Simulation*, vol. 93, p. 105524, 2021, ISSN 1007-5704, 10.1016/j.cnsns.2020.105524.

[30] A. S. Starkov and I. Starkov, "Asymptotic Description of the Time and Temperature Hysteresis in the Landau–Khalatnikov Equation Framework," *Ferroelectrics*, vol. 461, no. 1, pp. 50–60, 2014, 10.1080/00150193.2014.889544.

[31] L. Lu, D. Mohata, and S. Datta, "Scaling Length Theory of Double-Gate Interband Tunnel Field-Effect Transistors," *IEEE Transactions on Electron Devices*, vol. 59, no. 4, pp. 902–908, 2012.

[32] J. S. Meena, S. M. Sze, U. Chand, *et al.* "Overview of Emerging Nonvolatile Memory Technologies," *Nanoscale Research Letters*, vol. 9, p. 526, 2014, 10.1186/1556-276X-9-526.

[33] H. Takasu, Y. Fujimori, and T. Nakamura, "The Ferroelectric Memory Technologies and Its Applications," in Meeting Abstract of The 1999 Joint International Meeting, Honolulu, 1999, p. 1033.

[34] S. Ahmad, S. A. Ahmad, M. Muqeem, N. Alam, and M. Hasan, "TFET-Based Robust 7T SRAM Cell for Low Power Application," in *IEEE Transactions on Electron Devices*, vol. 66, no. 9, pp. 3834–3840, September 2019, 10.1109/TED.2019. 2931567.

[35] N. Hossain, A. Iqbal, H. Shishupal, and M. H. Chowdhury, "Tunneling Transistor Based 6T SRAM-Bit Cell Circuit Design in Sub-10 nm Domain," *2017 IEEE 60th International Midwest Symposium on Circuits and Systems (MWSCAS)*, 2017, pp. 1485–1488, 10.1109/MWSCAS.2017.8053215.

[36] Y. Sun, *et al.*, "Junction-Less Stackable SONOS Memory Realized on Vertical-Si-Nanowire for 3-D Application," *2011 3rd IEEE International Memory Workshop (IMW)*, 2011, pp. 1–4, 10.1109/IMW.2011.5873187.

[37] P. Kumar Sahu, Sunny, Y. Kumar, and V. N. Mishra, "Design and Simulation of Low Leakage SRAM Cell," *Third International Conference on Devices, Circuits and Systems (ICDCS)*, 2016, pp. 73–77.

[38] A. Agal, Pardeep, and B. Krishnan, "6T SRAM Cell: Design and Analysis," *International Journal of Engineering Research and Applications*, vol. 4, no. 3, pp. 574–577, March 2014.

[39] A. Bhaskar, "Design and Analysis of Low Power SRAM Cells," 2017 Innovations in Power and Advanced Computing Technologies (i-PACT), 2017, pp. 1–5, 10.11 09/IPACT.2017.8244888.

[40] N. Sah and N. Goyal, "Analysis of Leakage Power Reduction in 6T SRAM Cell," *International Journal of Advance Engineering Research and Technology*, vol. 3, no. 6, pp. 196–201, June 2015.

[41] A. C. Seabaugh and Q. Zhang, "Low-Voltage Tunnel Transistors for Beyond CMOS Logic," in *Proceedings of the IEEE*, vol. 98, no. 12, pp. 2095–2110, December 2010, 10.1109/JPROC.2010.2070470.

[42] J. Singh, R. Krishnan, S. Mookerjea, S. Datta, and V. Narayanan, "TFET based 6T SRAM cell," Patent: US 20120106236 A1.

[43] N. Paras and S. S. Chauhan, "Vertical Tunneling Based Tunnel Field Effect Transistor with Workfunction Engineered Hetero-Gate to Enhance DC Characteristics," *Journal of Nanoelectronics and Optoelectronics*, vol. 14, no. 1, American Scientific Publishers.(ISSN: 1555-1318), 2019, 10.1166/jno.2019.2427.

[44] Y. Lee, et al., "Low-Power Circuit Analysis and Design Based on Heterojunction Tunneling Transistors (HETTs)," *IEEE Transactions of Very Large Scale Integration (VLSI) System*, vol. 21, no. 9, pp. 1632–1643, September 2013, 10.1109/ TVLSI.2012.2213103.

14 TFET-based Universal Filter for Low-Power Applications

Mohd Yasir and Naushad Alam

CONTENTS

14.1 INTRODUCTION

TFET has a SS of less than 60 mV/dec at room temperature. Because the current transmission in TFET is on Band-to-band-tunnelling (BTBT), it has a lower leakage current than planar CMOS. Consequently, low power and low voltage operation of ICs is possible using TFET [1,2]. The low voltage and low power operation of TFET allow for applications like the Internet of Things (IoT), portable electronics, etc. Also, BTBT provides a low-temperature variation on TFET's characteristics [3,4]. This property of low-temperature variation will be helpful for its use in applications where temperature variation is of great concern, e.g., satellite communication, military, medical sector, aerospace, etc. However, the problem of lower ON current is associated with TFET [5–9]. The TFET model used here is the 20 nm planar InAs double gate Verilog-A SPICE model [10–12]. There are many other models of TFET available in the literature [13–15].

Active filters are the most commonly used in many signal processing applications, communication circuits, biomedical circuits, instrumentation, control systems like FM demodulators, PLLs, touch-tone telephone systems, speaker systems, etc. Because they consume much lower power than passive filters, active filters prefer different ICs. Voltage mode Multi-Input Single-Output (MISO) active filters are most common in different ICs. The option of having all the filter responses from a single topology makes them versatile and flexible for IC implementation. The Current Mode (CM) functional building blocks are extensively used to implement universal filters as they provide a higher linearity range, dynamic range, Bandwidth (BW), and low power consumption. There are various active filters implemented using different building blocks. Like Voltage Differencing Transconductance

DOI: 10.1201/9781003327035-14

251

Amplifier (VDTA), Voltage Differencing Differential Amplifier (VDDA), Voltage Differencing Buffered Amplifier (VDBA), Second-Generation Current Conveyor (CCII), Operational Transconductance Amplifier (OTA), Fully Differential Second-Generation Current Conveyor (FDCCII), Differential Voltage Current Conveyor (DVCC), Differential Difference Current Conveyor (DDCC), Current Differencing Transconductance Amplifier (CDTA), etc. [16–25].

This chapter implements a TFET-based DO-CCII and its application, i.e., a universal bi-quadratic filter. The circuits were analyzed using HSPICE simulations, providing low power and low voltage operation. The courses are free from physical resistors and use voltage-controlled TFETs. The capacitance values are also of the order of pF, which can easily implement in ICs. Therefore, circuits can integrate into any IC. The temperature variation analysis is also done on the courses from −50°C to 150°C. The temperature analysis shows a negligible effect of temperature variation on the circuits' performance.

The other sections of this chapter are as follows:

The overview of TFET and DO-CCII discusses in Sections 14.2 and 14.3. Section 14.3 also presents the performance-related aspects of DO-CCII and temperature variation analysis. The performance investigation of the voltage mode MISO Universal Biquadratic Filter is done in Section 14.4 using HSPICE, followed by conclusions drawn in Section 14.5.

14.2   OVERVIEW OF TFET

TFET is a beyond CMOS technology introduced to minimize the Short Channel Effects (SCEs) like Gate Induced Drain Leakage (GIDL), Drain Induced Barrier Lowering (DIBL), etc., which are the main hurdles in CMOS to follow Moore's law in the nanotechnology regime. Also, TFET provides low power and low voltage operation of ICs as it uses BTBT as a conduction mechanism. The low power and low voltage operation of TFET allow applications like the Internet of Things (IoT), portable electronics, etc. BTBT conduction mechanism provides low leakage; thereby, low and low power operation is possible [1,2]. Also, the BTBT mechanism is less affected by temperature variation. Therefore, TFET-based circuits are less affected by temperature variation [3,4]. This property of low-temperature variation will be helpful for its use in applications where temperature variation is of great concern, e.g., satellite communication, military, medical sector, aerospace, etc. The TFET is a new technology. Hence, its SPICE models are not mature enough. Most of them are based on Verilog-A [10,11,26,27]. This chapter shows a Verilog-A-based 20 nm double gate InAs TFET model. Its schematic is in Figure 14.1a. Here gmV_{GS} is modeling the drain current, and C_{GS} and C_{GD} are modeling the capacitance effect. Figure 14.1b shows the cross-section of the TFET used in this chapter. Here it has three terminals, namely Drain, Gate, and Source. The Gate terminal uses the governing terminal like CMOS. Therefore, TFET is compatible with the CMOS process [12].

TFET provides various advantages over planar CMOS in temperature constant low voltage and low power applications [28,29]. In TFET, the I_{ON} is mainly due to BTBT and has a smaller Subthreshold Slope (SS). The steep SS characteristics

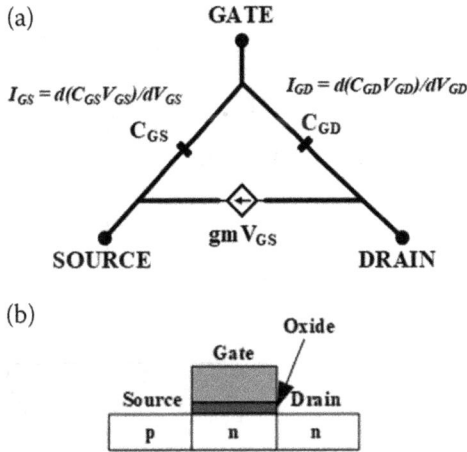

(a) GATE

$I_{GS} = d(C_{GS}V_{GS})/dV_{GS}$

C_{GS}

$I_{GD} = d(C_{GD}V_{GD})/dV_{GD}$

C_{GD}

gm V_{GS}

SOURCE   DRAIN

(b)

Oxide

Gate

Source   Drain

| p | n | n |

FIGURE 14.1 (a) Representation of Verilog-A model and (b) representation of cross-section of an n-channel TFET [12].

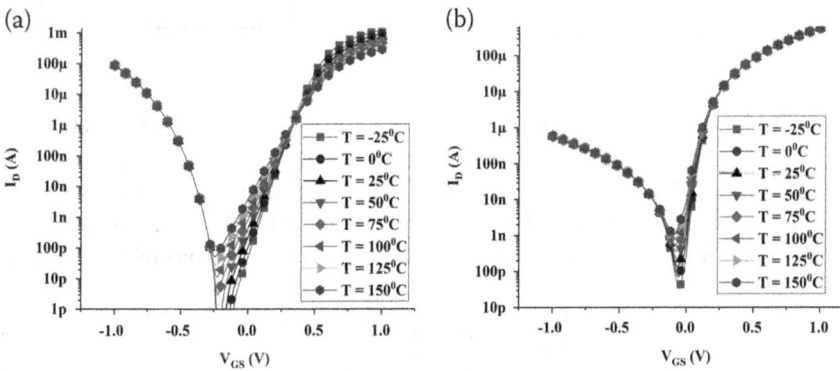

FIGURE 14.2 I_D–V_{GS} characteristics of (a) CMOS and (b) TFET for varying temperatures.

of TFET make its gm/I_D much more prominent in the subthreshold region. Therefore, TFET provides low voltage and low power for low-frequency applications. The BTBT mechanism's temperature dependence is negligible. Thus, TFET shows low-temperature support for its characteristics, as shown in Figure 14.2. Also, TFET has little effect on channel length modulation as it does not have a p-n connection. Because of this, TFET has an exceptionally high r_0. This high r_0 and high gm in the subthreshold region offer higher gain (gmr_0).

14.3 PERFORMANCE INVESTIGATION OF DO-CCII

A Dual Output Second Generation Current Conveyor (DO-CCII) is a CM functional building block. It provides various advantages over Op-Amp, like a higher linearity range, dynamic range, BW, and low power consumption. It consists of four ports,

(a)

(b)

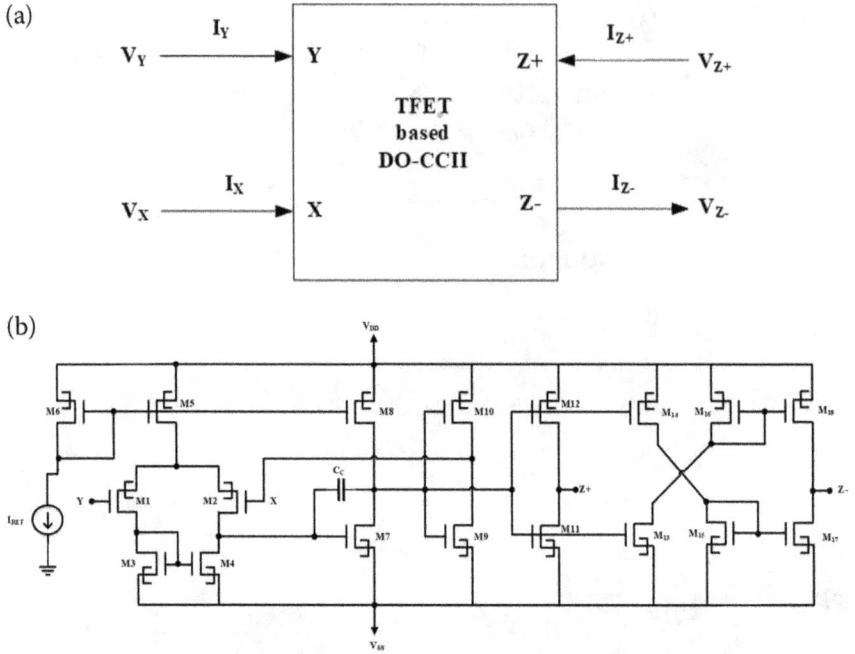

FIGURE 14.3   (a) Symbol and (b) transistor-level diagram of TFET-based DO-CCII.

namely X, Y, Z+, and Z−, respectively. The symbolic representation and its transistor-level schematic are shown in Figures 14.3a and 14.3b, respectively.

The following equation gives the relationship among various ports of the DO-CCII.

$$\begin{bmatrix} I_Y \\ V_X \\ I_{Z\pm} \end{bmatrix} = \begin{bmatrix} 0 & 0 & 0 \\ 1 & 0 & 0 \\ 0 & \pm 1 & 0 \end{bmatrix} \begin{bmatrix} V_Y \\ I_X \\ V_{Z\pm} \end{bmatrix} \tag{14.1}$$

where
   I_Y is the input current at the Y terminal,
   V_X is the voltage at the X terminal, and
   $I_{Z\pm}$ is the output currents at the Z± ports,
   V_Y is the input voltage at the Y terminal,
   I_X is the input current at the X terminal, and
   $V_{Z\pm}$ is the voltages at Z± ports, respectively.

All the TFETs used in the DO-CCII have lengths equal to 20 nm and widths, as shown in Table 14.1. The simulation uses supply voltages of ±0.5 V. The reference current shown in Figure 14.3b, i.e., I_{REF} used, is 1.5 μA. The average power consumed by the DO-CCII is 47.6 μW [30]. The temperature variation analysis for different parameters of DO-CCII is done from −50°C to 150°C and shows negligible effect.

TABLE 14.1

Summary of DO-CCII Transistor Sizing (L = 20 nm for all Transistors)

Transistors	M_1, M_2	M_3, M_4	M_5, M_6	M_7, M_8	M_9–M_{18}
Width (μm)	8	0.2	1, 0.5	6, 15	2

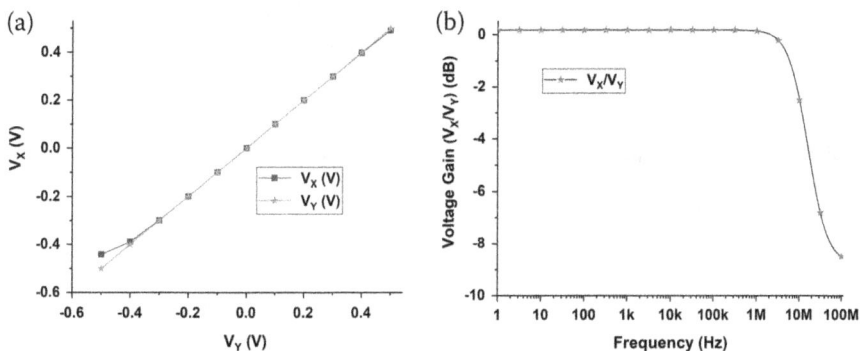

FIGURE 14.4 (a) Voltage transfer features between Y and X terminals and (b) frequency response of voltage gain (V_X/V_Y).

The voltage transfer features between Y and X ports are in Figure 14.4a. The linear range for voltage transfer characteristics calculates by applying −500 mV to 500 mV at the Y port. The signal at X port follows Y port, and the linear range is −330 mV to 410 mV. The frequency response of voltage gain (V_X/V_Y) plots in Figure 14.4b. The 3 dB Bandwidth and voltage gain (β) at lower frequencies are 11.5 MHz and 179 MDB, respectively.

The current transfer features among X, Z+, and Z− terminals are in Figure 14.5a. The linear range for contemporary transfer characteristics calculates by applying −300 μA to 300 μA at the X port. The signals at Z+ and Z− ports are followed by the sign at X port, and the linear range is −250 μA to 250 μA. The frequency response of current gains (I_{Z+}/I_X) and (I_{Z-}/I_X) is in Figure 14.5b. The 3 dB Bandwidth and current gain (α) at lower frequencies for (I_{Z+}/I_X) are 11.5 MHz and −119 μdB, respectively. The 3 dB Bandwidth and current gain (α) at a lower frequency for (I_{Z-}/I_X) are 11.5 MHz and 88 MDB, respectively.

The transient response of voltage transfer and current transfer among DO-CCII's ports are plotted in Figures 14.6a and 14.6b, respectively. For the quick response of voltage transfer between Y and X terminals, a sine wave of ±250 mV of amplitude and 1 MHz of frequency applies at the Y terminal. The X terminal's signal follows the movement of the Y port according to the eq. (14.1). Similarly, for the transient response of current transfer among X, Z+, and Z− terminals, a sine wave of ±100 μA of amplitude and 1 MHz of frequency is applied at the X terminal. According to the eq, the signal at the X terminal follows the signs at Z+ and Z− ports. (14.1).

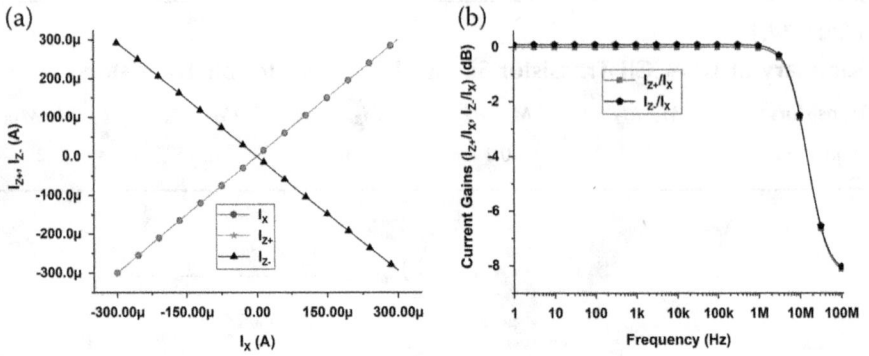

FIGURE 14.5 (a) Current transfer features among X, Z+, and Z− terminals and (b) frequency response of current gains (I_{Z+}/I_X) (I_{Z-}/I_X).

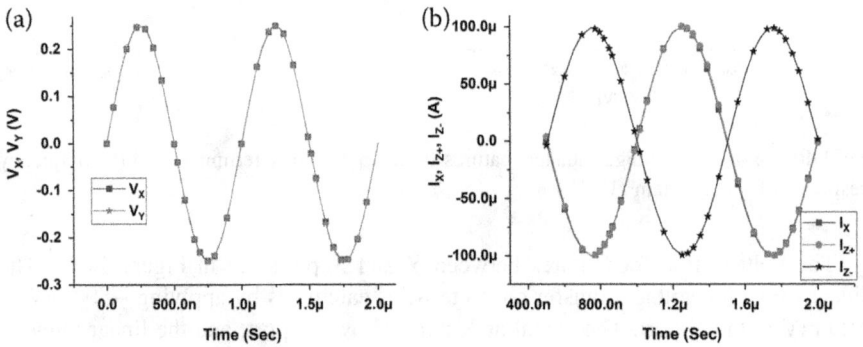

FIGURE 14.6 Transient response of (a) voltage transfer features between Y and X terminals and (b) current transfer among X, Z+, and Z− terminals.

The temperature variation analysis on the DC voltage transfer features between Y and X ports (Figure 14.7a). Signal at the Y port by the signal at X port and shows negligible effect. The temperature variation on the DC transfer features among X, Z+, and Z− terminals (Figure 14.7b). The signals at Z+ and Z− ports are followed by the sign at X port and show negligible effect.

The temperature variation analysis on the frequency response of voltage gain (V_X/V_Y) and current increases (I_{Z+}/I_X) and (I_{Z-}/I_X) in Figures 14.8a–14.8c. The analysis shows that at lower frequencies and up to 3 dB frequency, variation in voltage gain (β) and current increases (I_{Z+}/I_X) and (I_{Z-}/I_X) is negligible. After 3 dB frequency, variation is there but of low value.

The temperature variation analysis on the transient response of voltage transfer and current transfer among DO-CCII's ports are plotted in Figures 14.9a and 14.9b, respectively. The X port's signal at the Y port has approximately no variation. On the other hand, alerts at Z+ and Z− ports follow the movement at X port with low variation.

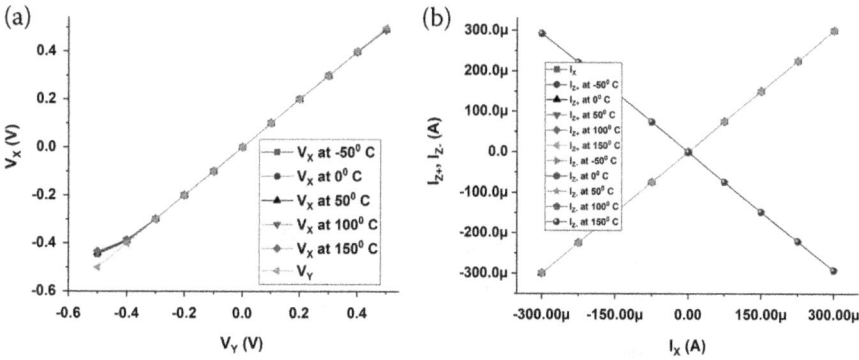

FIGURE 14.7 (a) Voltage transfer features between Y and X terminals and (b) current transfer features among X, Z+, and Z− terminals at different temperatures (−50°C to 150°C).

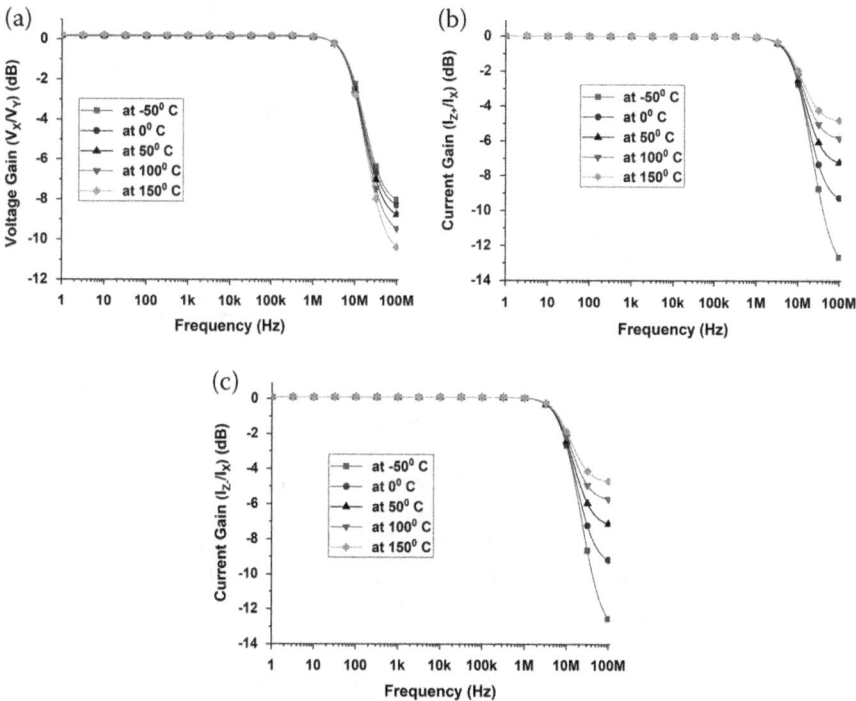

FIGURE 14.8 Frequency response of (a) voltage gain (V_X/V_Y), current gains, (b) (I_{Z+}/I_X), and (c) (I_{Z-}/I_X) at different temperatures (−50°C to 150°C).

14.4 DO-CCII-BASED UNIVERSAL BIQUADRATIC FILTER

The block diagram of the DO-CCII-based voltage-mode MISO universal biquadratic filter is in Figure 14.10a. It contains two DO-CCIIs, two resistors, and two capacitors. The resistor implementation in an IC consumes a large amount of

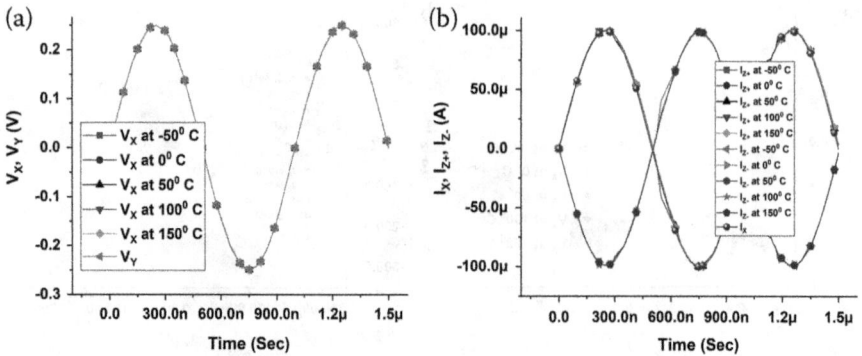

FIGURE 14.9 Transient response of (a) voltage follower between Y and X terminals and (b) current transfer among X, Z+, and Z− terminals at different temperatures (−50°C to 150°C).

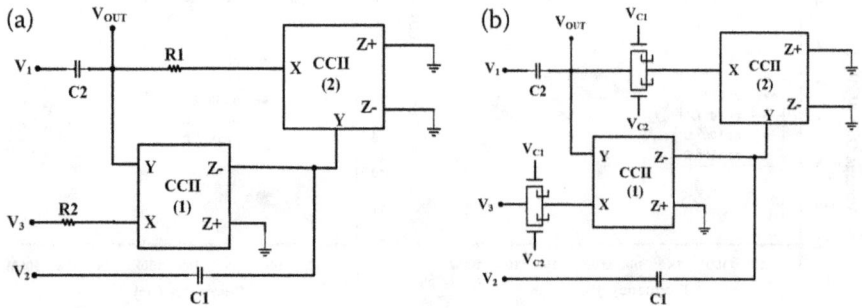

FIGURE 14.10 (a) DO-CCII-based Universal Filter presented in [32] and (b) its Implementation using voltage-controlled transistors.

Si area. Therefore, in Figure 14.10b, the performance of this filter is shown using voltage-controlled transistors to implement resistor behavior [31]. The filter transfer function (TF) for the voltage mode MISO filter can easily be calculated using DO-CCII's characteristics and applying a routine analysis in Figure 14.10a.

The output voltage (V_{OUT}) in terms of V_1, V_2, and V_3 shows in eq. (14.2).

$$V_{OUT} = \frac{s^2 R_1 R_2 C_1 C_2 V_1 + s R_2 C_1 V_2 + V_3}{s^2 R_1 R_2 C_1 C_2 + s R_2 C_1 + 1} \tag{14.2}$$

Transfer functions for different filters by applying various combinations of inputs in eq. (14.2) as shown in Table 14.2. If $V_1 = V_2 = 0$ and $V_3 = V_{IN}$ is applied in eq. (14.2) LPF response is obtained as shown below.

$$\frac{V_{LP}}{V_{IN}} = \frac{1}{s^2 R_1 R_2 C_1 C_2 + s R_2 C_1 + 1} \tag{14.3}$$

TABLE 14.2

Transfer Function of the Universal Biquadratic Filter

Filter	LPF	HPF	BPF	BRF	APF
V_1	0	1	0	1	1
V_2	0	0	1	0	−1
V_3	1	0	0	1	1

Similarly, if $V_2 = V_3 = 0$ and $V_1 = V_{IN}$ is applied in eq. (14.2), HPF response is obtained as shown below.

$$\frac{V_{HP}}{V_{IN}} = \frac{s^2 R_1 R_2 C_1 C_2}{s^2 R_1 R_2 C_1 C_2 + s R_2 C_1 + 1} \tag{14.4}$$

Similarly, if $V_1 = V_3 = 0$ and $V_2 = V_{IN}$ is applied in eq. (14.2), BPF response is obtained as shown below.

$$\frac{V_{BP}}{V_{IN}} = \frac{s R_2 C_1}{s^2 R_1 R_2 C_1 C_2 + s R_2 C_1 + 1} \tag{14.5}$$

For BRF, $V_1 = V_3 = V_{IN}$ and $V_2 = 0$ is applied in eq. (14.2).

$$\frac{V_{BR}}{V_{IN}} = \frac{s^2 R_1 R_2 C_1 C_2 + 1}{s^2 R_1 R_2 C_1 C_2 + s R_2 C_1 + 1} \tag{14.6}$$

The equation for zero frequency of BRF can be given as:

$$\omega_Z = \frac{1}{(R_1 R_2 C_1 C_2)^{\frac{1}{2}}} \tag{14.7}$$

For APF, $V_1 = -V_2 = V_3 = V_{IN}$ is applied in eq. (14.2).

$$\frac{V_{AP}}{V_{IN}} = \frac{s^2 R_1 R_2 C_1 C_2 - s R_2 C_1 + 1}{s^2 R_1 R_2 C_1 C_2 + s R_2 C_1 + 1} \tag{14.8}$$

The equations for natural frequency (ω_0) and quality factor (Q) can easily be obtained from the eq. (14.2) using simple mathematical analysis.

$$\omega_0 = \frac{1}{(R_1 R_2 C_1 C_2)^{\frac{1}{2}}} \tag{14.9}$$

$$Q = \left(\frac{R_1 C_2}{R_2 C_1}\right)^{\frac{1}{2}}$$ (14.10)

The ω_0 can easily be controlled using R_1 or R_2, while Q can be controlled using ratios R_1/R_2 or C_2/C_1. The passive sensitivities of ω_0 and Q can easily be obtained using routine mathematical analysis.

$$S_{C_1}^{\omega_0} = S_{C_2}^{\omega_0} = S_{R_1}^{\omega_0} = S_{R_2}^{\omega_0} = -\frac{1}{2}$$ (14.11)

$$S_{C_2}^{Q} = S_{R_1}^{Q} = -S_{C_1}^{Q} = -S_{R_2}^{Q} = \frac{1}{2}$$ (14.12)

The passive sensitivities obtained are of smaller values for ω_0 and Q.

All the simulations used HSPICE software with a 20 nm TFET model [12]. The supplies used are ±0.5 V, and the power used by the universal biquadratic filter is 93.8 μW. The control voltages V_{C1} and V_{C2} for voltage-controlled transistors simulating the behavior of resistors are 0.5 V for NTFET transistors and −0.5 V for PTFET transistors, respectively. The values of C_1 and C_2 pF. With the combination of control voltages simulating the behavior of resistors and values of capacitors, the values achieved for f_0 and Q are 155 MHz and 1, respectively.

Figure 14.11b shows the frequency response of HPF gain, while Figures 14.11a, 14.12a, 14.12b, and 14.12c show the frequency response of LPF, BPF, BRF, and APF gain and Phase, respectively. The collective frequency response of all the filters (Figure 14.12d) for the same control voltages and capacitor values. The joint frequency response of all filters shows that the f_0 for all the filter responses is 155 MHz – the study of process variation using Monte Carlo simulations. The analysis uses a 10% tolerance in capacitor values. There are 100 runs in this study. Figures 14.13a and 14.13b show the Monte Carlo analysis of BPF frequency response of gain and phase responses. The study shows lesser variations.

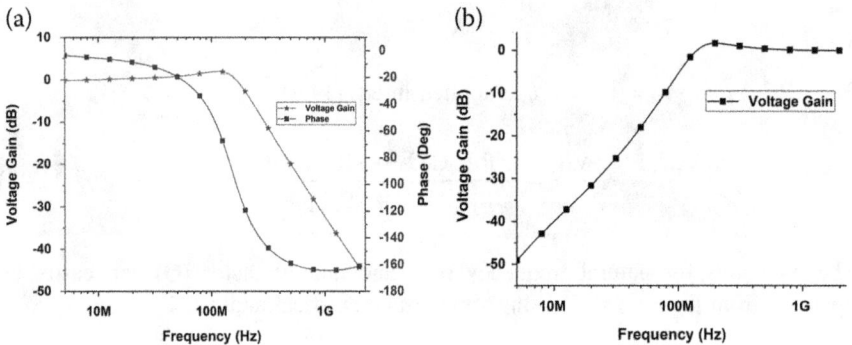

FIGURE 14.11   Universal Filter frequency response of (a) LPF gain and phase and (b) HPF gain.

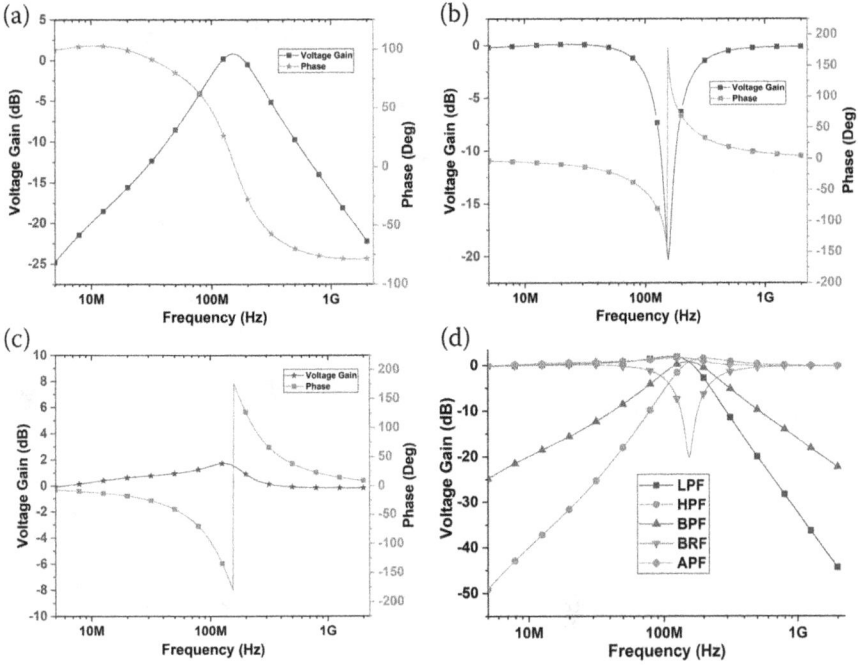

FIGURE 14.12 Universal Filter frequency response of (a) BPF gain and phase, (b) BRF gain and phase, (c) APF gain and Phase, and (d) voltage gain frequency responses of the universal biquadratic filter.

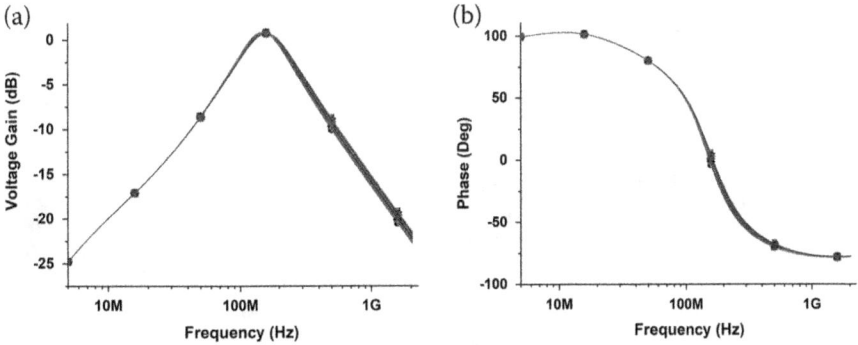

FIGURE 14.13 Monte Carlo simulations of BPF for (a) Voltage gain (b) Phase.

The temperature variation on the frequency response of gains of LPF, HPF, BPF, BRF, and APF (Figures 14.14a, 14.14b, 14.15a, 14.15b, and 14.15c, respectively). The analysis shows that variation in gain-frequency responses of different filters is negligible.

A comparison of the VM universal biquadratic filter with previously reported circuits is shown in Table 14.3.

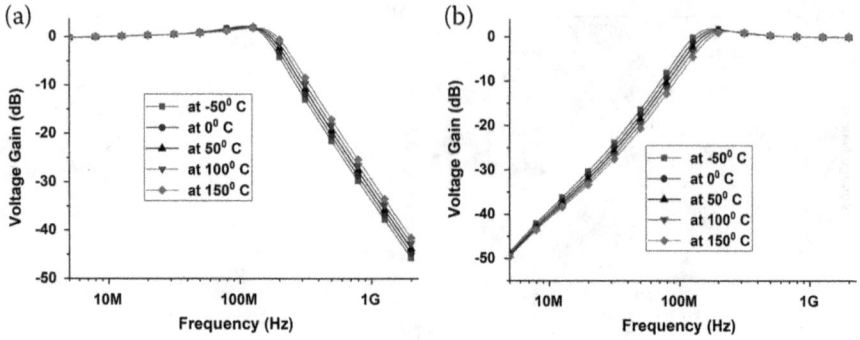

FIGURE 14.14 Universal filter frequency response of (a) LPF gain and (b) HPF gain at different temperatures (−50°C to 150°C).

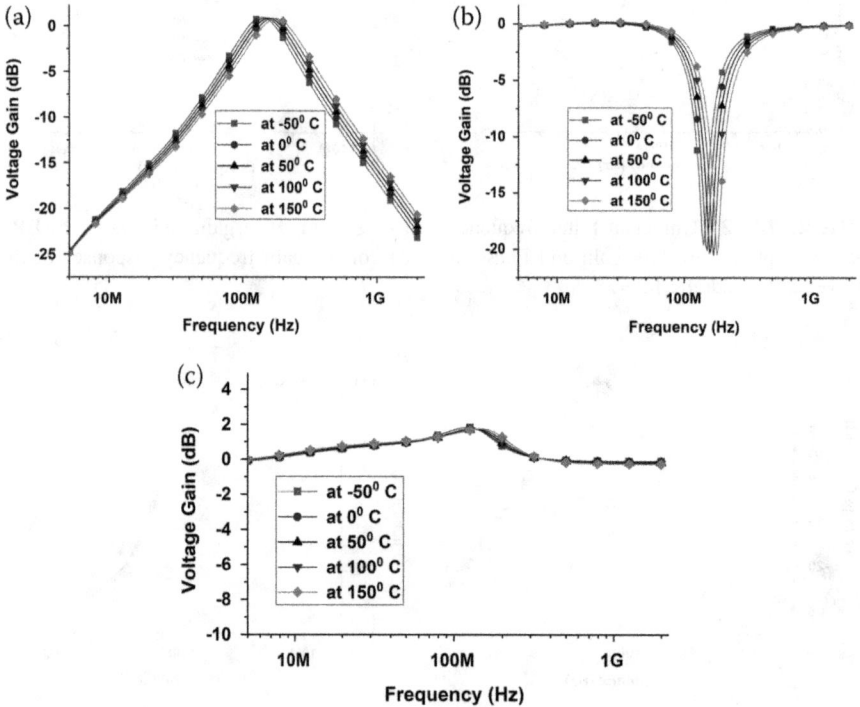

FIGURE 14.15 Universal filter frequency response of (a) BPF gain, (b) BRF gain, and (c) APF gain at different temperatures (−50°C to 150°C).

The 32 nm CNTFET is used in Reference [33]. Its limitation is that it does not provide all five standard filter functions. The 180 nm CMOS is used in Reference [18], and BJT in [16]. They both have a limitation: they do not provide resistorless topology as the physical resistor consumes a large amount of Si area. Therefore, the IC integration of these circuits is not good enough. The 450 nm

TABLE 14.3

Comparison of the VM Universal Biquadratic Filter with Previously Reported Circuits

Specifications	This Work	[33]	[18]	[17]	[16]
Process (nm)	20	32	180	450	–
Technology	TFET	CNTFET	CMOS	CMOS	BJT
Active and passive elements	2 DO-CCII, 2 C, 4 Switches	2 OTA, 2 C	3 DDCC, 2 C, 5 R	1 VDVTA, 2 C	3 VDDDA, 2 C, 1 R
Resistorless topology	Yes	Yes	No	Yes	No
Offer 5 Standard Filter Functions	Yes	No	Yes	Yes	Yes
Supply Voltages (V)	±0.5	±0.9	±1.8	±1	±5
Power Dissipation (μW)	93.8	–	–	368	–

CMOS is used in Reference [17]. It has a limitation because it consumes much power compared to this work.

14.5 CONCLUSION

The chapter presents a TFET-based DO-CCII. A voltage mode MISO universal biquadratic filter using two DO-CCII blocks, two capacitors, and four voltage-controlled transistors. The 3 dB Bandwidth of DO-CCII for voltage gain and current gains is 11.5 MHz. The average power consumption of DO-CCII is 47.6 μW. The DC voltage and current range DO-CCII are −330 mV to 410 mV and −250 μA to 250 μA, respectively. As the universal biquadratic filter is of MISO type, it provides all five filter responses from the same topology. It has achieved the values of 155 MHz and 1 for f_0 and Q, respectively. Since the circuit is not having any physical resistors, it is suitable for IC integration. The simulations use a 20 nm TFET model with the HSPICE software.

REFERENCES

[1] P. K. Asthana, Y. Goswami, S. Basak, S. B. Rahi, and B. Ghosh, "Improved Performance of a Junctionless Tunnel Field Effect Transistor with a Si and SiGe Heterostructure for Ultra-low Power Applications," *RSC Advances*, vol. 5, no. 60, pp. 48779–48785, 2015.

[2] S. B. Rahi, S. Tayal, and A. K. Upadhyay, "A Review on Emerging Negative Capacitance field effect transistor for low power electronics," *Microelectronics Journal*, vol. 116, p. 105242, 2021.

[3] S. B. Rahi, B. Ghosh, and B. Bishnoi, "Temperature Effect on Hetero Structure Junctionless Tunnel FET," *Journal of Semiconductors*, vol. 36, no. 3, p. 034002, 2015.

[4] G. Naima and S. B. Rahi, "Low Power Circuit and System Design Hierarchy and Thermal Reliability of Tunnel Field Effect Transistor," *Silicon*, pp. 1–11, 2021.

[5] A. C. Seabaugh and Q. Zhang, "Low-Voltage Tunnel Transistors for Beyond CMOS Logic," *Proceedings of the IEEE*, vol. 98, no. 12, pp. 2095–2110, 2010.

[6] H. Lu and A. Seabaugh, "Tunnel Field-Effect Transistors: State-of-the-Art," *IEEE Journal of the Electron Devices Society*, vol. 2, no. 4, pp. 44–49, 2014.

[7] U. E. Avci, D. H. Morris, and I. A. Young, "Tunnel Field-Effect Transistors: Prospects and Challenges," *IEEE Journal of the Electron Devices Society*, vol. 3, no. 3, pp. 88–95, 2015.

[8] Y.-N. Chen, M.-L. Fan, V. P.-H. Hu, P. Su, and C.-T. Chuang, "Evaluation of Sub-0.2 V High-Speed Low-power Circuits Using Hetero-channel MOSFET and Tunneling FET Devices," *IEEE Transactions on Circuits and Systems I: Regular Papers*, vol. 61, no. 12, pp. 3339–3347, 2014.

[9] K. Boucart and A. M. Ionescu, "Length Scaling of the Double Gate Tunnel FET with a High-k Gate Dielectric," *Solid-State Electronics*, vol. 51, no. 11–12, pp. 1500–1507, 2007.

[10] L. Barboni, M. Siniscalchi, and B. Sensale-Rodriguez, "TFET-Based Circuit Design Using the Transconductance Generation Efficiency gm/Id Method," *IEEE Journal of the Electron Devices Society*, vol. 3, no. 3, pp. 208–216, 2015.

[11] G. B. Beneventi, E. Gnani, A. Gnudi, S. Reggiani, and G. Baccarani, "InAs TFET Optimized Using TCAD to Meet all the ITRS Specs at Vdd= 0.5 V," in *Proceedings of International Semicondonductors Device Research Symposium (ISDRS)*, pp. 1–2, 2013.

[12] T. Y. Hao Lu and A. Seabaugh, "Universal TFET Model," Jan 2015. Accessed 26 December 2019.

[13] S. B. Rahi, P. Asthana, and S. Gupta, "Heterogate Junctionless Tunnel Field-Effect Transistor: Future of Low-Power Devices," *Journal of Computational Electronics*, vol. 16, no. 1, pp. 30–38, 2017.

[14] N. Guenifi, S. Rahi, and T. Ghodbane, "Rigorous Study of Double Gate-tunneling Field Effect Transistor Structure Based on Silicon," *Materials Focus*, vol. 7, no. 6, pp. 866–872, 2018.

[15] B. Samani, S. B. Rahi, and S. Labiod, "Analytical Compact Model of Nanowire Junctionless Gate-All-Around MOSFET Implemented in Verilog-A for Circuit Simulation," *Silicon*, pp. 1–10, 2022.

[16] P. Supavarasuwat, M. Kumngern, S. Sangyaem, W. Jaikla, and F. Khateb, "Cascadable independently and electronically tunable voltage-mode universal filter with grounded passive components," *AEU-International Journal of Electronics and Communications*, vol. 84, pp. 290–299, 2018.

[17] V. Kumar, R. Mehra, and A. Islam, "Design and Analysis of MISO Bi-Quad Active Filter," *International Journal of Electronics*, vol. 106, no. 2, pp. 287–304, 2019.

[18] C.-N. Lee, "Independently Tunable Plus-Type DDCC-Based Voltage-Mode Universal Biquad Filter with MISO and SIMO Types," *Microelectronics Journal*, vol. 67, pp. 71–81, 2017.

[19] M. Yasir and M. S. Ansari, "Performance Investigation of Fin-Shaped Field Effect Transistor Based Multi Output Differential Voltage Current Conveyor and Its Application in Balanced Modulator," *Journal of Nanoelectronics and Optoelectronics*, vol. 14, no. 5, pp. 705–715, 2019.

[20] S. Kumari and M. Gupta, "Design and analysis of tunable voltage differencing inverting buffered amplifier (VDIBA) with enhanced performance and its application in filters," *Wireless Personal Communications*, vol. 100, no. 3, pp. 877–894, 2018.

[21] B. Singh, A. K. Singh, and R. Senani, "A New Universal Biquad Filter Using Differential Difference Amplifiers and its Practical Realization," *Analog Integrated Circuits and Signal Processing*, vol. 75, no. 2, pp. 293–297, 2013.

[22] F. Kacar, A. Yesil, and A. Noori, "New CMOS Realization of Voltage Differencing Buffered Amplifier and Its Biquad Filter Applications," *Radioengineering*, vol. 21, no. 1, pp. 333–339, 2012.

[23] F. Kaçar and A. Yeşil, "Voltage Mode Universal Filters Employing Single FDCCII," *Analog Integrated Circuits and Signal Processing*, vol. 63, no. 1, pp. 137–142, 2010.

[24] H.-P. Chen, "Single CCII-Based Voltage-Mode Universal Filter," *Analog Integrated Circuits and Signal Processing*, vol. 62, no. 2, pp. 259–262, 2010.

[25] D. Prasad, D. Bhaskar, and A. Singh, "Multi-Function Biquad Using Single Current Differencing Transconductance Amplifier," *Analog Integrated Circuits and Signal Processing*, vol. 61, no. 3, pp. 309–313, 2009.

[26] K. K. Bhuwalka, J. Schulze, and I. Eisele, "A Simulation Approach to Optimize the Electrical Parameters of a Vertical Tunnel FET," *IEEE Transactions on Electron Devices*, vol. 52, no. 7, pp. 1541–1547, 2005.

[27] W. G. Vandenberghe, A. S. Verhulst, G. Groeseneken, B. Soree, and W. Magnus, "Analytical Model for a Tunnel Field-Effect Transistor," in *MELECON 2008-The 14th IEEE Mediterranean Electrotechnical Conference*, pp. 923–928, IEEE MELECON 2008-The 14th IEEE Mediterranean Electrotechnical Conference, 2008.

[28] S. Strangio, F. Settino, P. Palestri, M. Lanuzza, F. Crupi, D. Esseni, and L. Selmi, "Digital and Analog TFET Circuits: Design and Benchmark," *Solid-State Electronics*, vol. 146, pp. 50–65, 2018.

[29] B. Sedighi, X. S. Hu, H. Liu, J. J. Nahas, and M. Niemier, "Analog Circuit Design Using Tunnel-FETs," *IEEE Transactions on Circuits and Systems I: Regular Papers*, vol. 62, no. 1, pp. 39–48, 2014.

[30] M. Yasir and N. Alam, "Design of CNTFET-Based CCII Using gm/Id Technique for Low-Voltage and Low-Power Applications," *Journal of Circuits, Systems, and Computers*, vol. 29, no. 09, p. 2050143, 2020.

[31] H. Çiçekli, I. Karacan, and A. Gökçen, "Current Operational Amplifier Based Voltage Mode MOS-C All-Pass Filter and Its Application," *Politeknik Dergisi*, vol. 23, no. 2, pp. 409–414, 2020.

[32] J.-W. Horng, M.-H. Lee, H.-C. Cheng, and C.-W. Chang, "New CCII-Based Voltage-Mode Universal Biquadratic Filter," *International Journal of Electronics*, vol. 82, no. 2, pp. 151–156, 1997.

[33] M. Cen, S. Song, and C. Cai, "A High-Performance CNFET-Based Operational Transconductance Amplifier and Its Applications," *Analog Integrated Circuits and Signal Processing*, vol. 91, no. 3, pp. 463–472, 2017.

15 TFET-based Level Shifter Circuits for Low-Power Applications

Ashish Kumar Sharma and Naushad Alam

CONTENTS

DOI: 10.1201/9781003327035-15

15.1   INTRODUCTION

15.1.1   MOTIVATION

Different system modules on chip (SoC), such as digital, analog, and passive components, are manufactured on a single chip and are optimized to work with multiple supply voltages to decrease the overall power dissipation. In a VLSI system, a level shifter cell converts the voltage levels from one voltage domain to another. Using different supply voltages in the same system is one of the valuable ways to reduce leakage and dynamic power. Voltage level shifters are also helpful when the chips are in between the core voltage and analog front-end circuitry (such as physical interphase of DDR) voltage. For example, if the core works with a smaller voltage in the range of 1.2 V, the analog frontend circuit works with a higher voltage of 3.3 V. The difference in a voltage domain may cause the problem with stably working the circuitries.

In Figure 15.1, signals from 3.3 V domains drive a module that works with 5 V. The lower range of voltage signal might not achieve the higher voltage if the voltage difference is significant. A voltage Level Shifter (LS) cell guarantees that

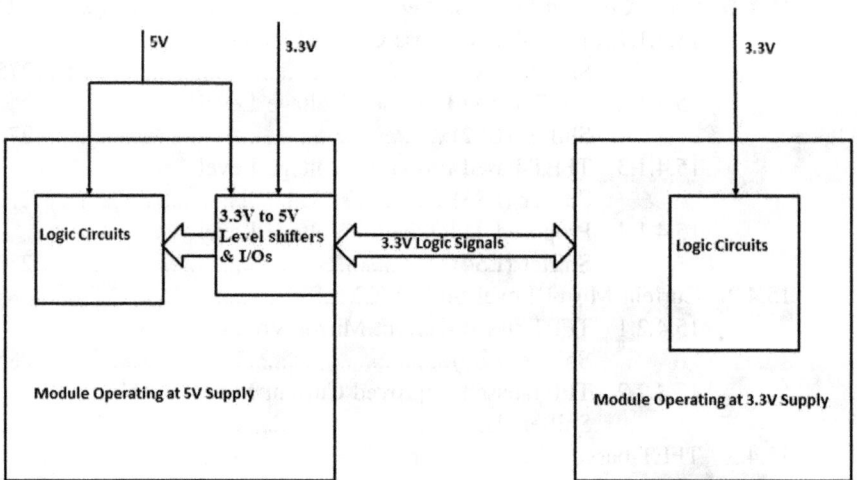

FIGURE 15.1   SOC block diagram with different modules and level shifter.

the multi-voltage domain modules function correctly. There are two possible scenarios: one in which the source signal voltage is low and the other in which the source signal voltage is high. When the source voltage is low, we need the high voltage at the input to drive the circuitries. In that case, we will use low-voltage translators and vice versa.

Nowadays, voltage reduction decreases the power consumption of systems. The power dissipation of circuits is quadratic about the supply voltage. If we reduce the supply voltage, the power dissipation of the courses reduces. For example, a 45 nm IA CPU [1] has 48 cores and eight voltage domains; for proper operation of the CPU, we require a voltage level shifter between modules that work with different supply voltages. As a result, TFET-based voltage Level shifters (LS) have become very promising, especially in systems with aggressive voltage scaling. Modules working on the different voltages require TFET-based voltage Level Shifters for voltage to go from one module voltage domain to another module. If the voltage coming at the node is higher than the module voltage level, then the system is not correctly working. Level shifters are used in the System on Chip to ensure that the modules operating at different voltages function correctly. As signals transition from one voltage domain to another, Level Shifters must provide correct driving strength and accurate timing. There are also some issues with level shifter designs with low conversion ranges.

In VLSI circuits, supply voltage and parasitic capacitances decide the power of the integrated circuits and systems. When the supply voltage lowers that caused, low voltage swings, insufficient noise margins, and leakage currents occur [2]. As technology has advanced to the deep submicron level, circuits operating at lower voltages have leakage power challenges. Also, at a deep submicron level, the leakage power is the dominating fraction of the total power component. If size and supply voltage scaling trends are to be maintained, the static power component of power consumption considers.

The primary concern while designing a Level Shifter circuit is the power consumption, Delay, and the area utilized. The purpose is to use TFET in voltage level shifter to increase conversion range and decrease Power & Delay. The purpose of inserting the MUX in a cross-coupled voltage Level Shifter (LS3) is to speed up the level shifter. The purpose of using two cross-coupled stages in the proposed voltage Level Shifter (LS4) is to improve both Power & Delay which implies lower PDP.

15.1.2 TFET Overview

TFET is an emerging transistor with a similar structure to a conventional MOSFET with the difference that its source & Drain are not interchangeable. The conduction mechanism in a TFET is different, making this Device suitable for low-power applications. Current conduction in TFET involves the quantum tunneling phenomena rather than carrier transfer over the barrier by applying sufficient gate voltage. In CMOS devices, the minimum value of subthreshold swing is 60 mV per decade, so we require a minimum of 60 mV for one decade of current change. But in TFET, the drain current is not limited by the subthreshold swing. A minimum subthreshold swing of 33 mV per decade. In TFET, Current conduction on band-to-band tunneling (BTBT) defies the Boltzmann-limited 60 mV/decade subthreshold swing.

15.1.3 WORKING OF TFET

As shown in Figure 15.2, a TFET gate controls BTBT across a p-n junction and depicts the cross-sectional view and band diagrams of an n-type Tunnel Field Effect Transistor (TFET) at different gate voltages. Usually, the Device turns off. When the gate is biased with zero voltage, the conduction band of the Channel is at higher energy than the valance band energy of the source, which prevents band-to-band tunneling. When we apply some positive gate bias voltage, the channel conduction band shifts below the source valence band. The TFET transistor is turned on, and electrons in the valence band have enough energy to tunnel into vacant states in the conduction band of the Channel.

The p-type TFET works the same way as the n-type TFET, but the source and drain doping differ. In the NTFET, the current profile also decreases when we decrease the Vgs voltage. Whenever Vgs is positive, or Vgs is higher, the electron will tunnel from source to channel and deposit by the drain side, and we can say the tunnel window is open. When we apply the negative voltage at the gate, the valence band of the Drain can be shifted over the conduction band, resulting in electron tunneling from the Channel into the Drain. TFET can switch on at the channel drain junction, and the tunneling window reopens as a result, with the tunnel junction carrier moved from the source to the drain side. When this happens, the channel conduction switches from one carrier type to another, resulting in an ambipolar transfer characteristic. Across all TFET geometries, this behavior is consistent. When the gate bias is positive, and the drain bias is negative, the TFET functions as an Esaki diode, and the output characteristics show NDR behavior.

An n-type TFET's intrinsic region enables by adding gate bias at the Channel. The channel band bending is steeper, and BTBT happens when the Intrinsic region's conduction band is at the same level as the valence band of the P region at a suitable gate voltage. Current can flow across the Device because electrons from the p-type region's valence band tunnel into the intrinsic region's conduction band. As the potential gate decreases, the energy band diagram shows the shift and the current on the drain side decrease, even with the more shift current being zero at the drain side.

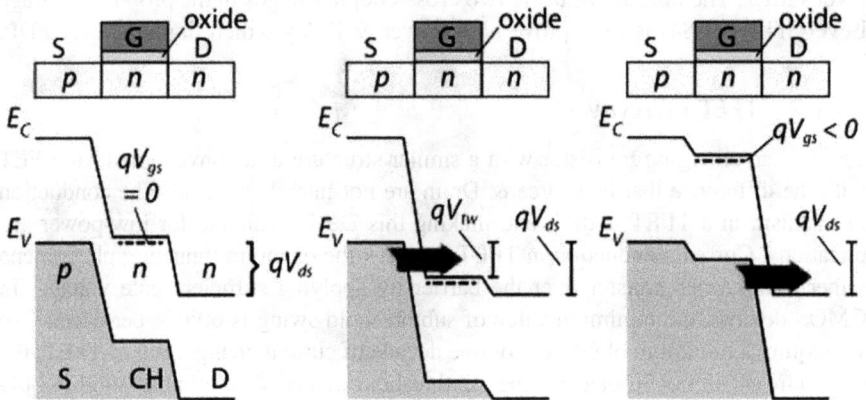

FIGURE 15.2   Band diagram of Tunnel Field Effect Transistor with the bias condition [3].

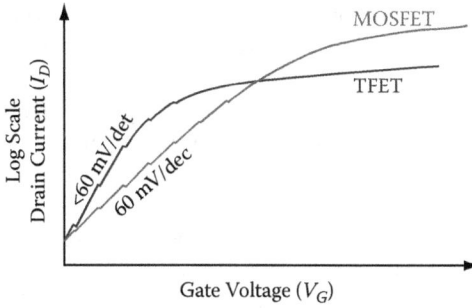

FIGURE 15.3 I_D vs. V_Gs characteristic of TFET.

Energy band diagram of n-type TFET biases in

a. OFF state,
b. ON state, and
c. Ambipolar state

In Energy Band Diagram, Ec stands for conduction band, Ev stands for valence band, Vgs stands for gate-source voltage, Vds stands for drain-source voltage, and Vtw stands for tunneling window.

The 60 mV/decade subthreshold limit is a fundamental limitation in traditional MOSFET devices. The ratio between the threshold voltage and the subthreshold slope determines the percentage of ON-current to OFF-current (Figure 15.3).

The ON-current of a transistor represents its speed. The higher the ON current value, the faster the Transistor can charge/discharge capacitive load. The subthreshold slope determines the minimum threshold voltage for a particular transistor and speed and the maximum subthreshold leakage that can tolerate. The constant field scaling principle necessitates reducing the threshold voltage. Key technology developers have been unable to scale supply voltage since 2003 and rely nearly entirely on threshold voltage scaling.

As a result, CPU speeds have not advanced at the same rate as before 2003. The industry will be able to maintain the scaling trends that began in the 1990s, when processor frequency doubled every three years, with the introduction of a TFET with a subthreshold slope of less than 60 mV/decade.

15.2 LITERATURE REVIEW

15.2.1 LITERATURE REVIEW

The components of integrated circuits, such as digital, analog, and passive components, are fabricated on a single chip and require different voltages to operate correctly. A Level Shifter (LS) is used in a VLSI system to change the voltage level from one voltage domain to another [4].

Analog and digital IC designers are emphasizing improved energy efficiency. Dynamic power depends on supply voltage, so a reduction in the supply voltage of modules is an efficient way to decrease overall power.

Nowadays, sensor nodes in the Internet of Things [1] have systems with many voltage domains working on different voltages to operate systems properly. We require TFET-based voltage Level Shifters between modules. The Level Shifter uses a split type of Inverter [5] to reduce power dissipation. In this work, the split type of Inverter is used in the proposed cross-coupled Level Shifter (LS2) & also in improved CMLS to decrease the power.

This proposed Level Shifter (LS3) uses a multiplexer [6] to speed up the operation, showing significant Delay reductions. This proposed Level Shifter (LS4) uses a cross-coupled and split type of Inverter [5,7–14] which shows a significant reduction of power dissipation, Delay, and power-delay-product in all topologies.

This proposed Level Shifter (LS3) uses a voltage divider network [6] to generate the V_{DDM} voltage at the input of PMOS. So the speed of the PMOS & NMOS is comparable, and the TFET-based proposed voltage Level Shifter (LS3) works faster and has a minimum delay among all topologies.

15.3 LEVEL SHIFTER CIRCUITS

15.3.1 LEVEL SHIFTER

We have different modules in integrated circuit design on a single chip. Level Shifters require signals that move from one module voltage level to another in other supply voltage modules. Calls that cross the one-module voltage levels cannot record without level shifters (Figure 15.4).

In the System on Chip, LS ensures that blocks with differing voltages work together appropriately. As a signal changes from one module voltage level to another, level shifters must provide correct driving strength and accurate timing. LS can employ during the synthesis and implementation phases.

We can use a voltage level shifter in both cases when voltage translation from a high to a low voltage domain and the voltage translation from a low voltage to a high voltage domain. In the case of high to standard translation voltage, we use two series of inverters and provide a more insufficient VDDL supply to the Inverter that helps to charge the parasitic node to VDDL potential.

The power domain boundaries are close to the level shifters. On the other hand, level shifters have two power boundaries

1. The primary power boundaries are on the voltage level shifter's top and bottom borders.
2. The level shifter's horizontal center line serves as the secondary power boundary.

The voltage that the primary power boundaries match determines the Level shifter's power domain. If the direct power boundaries of the level shifter are 0.8V, it should

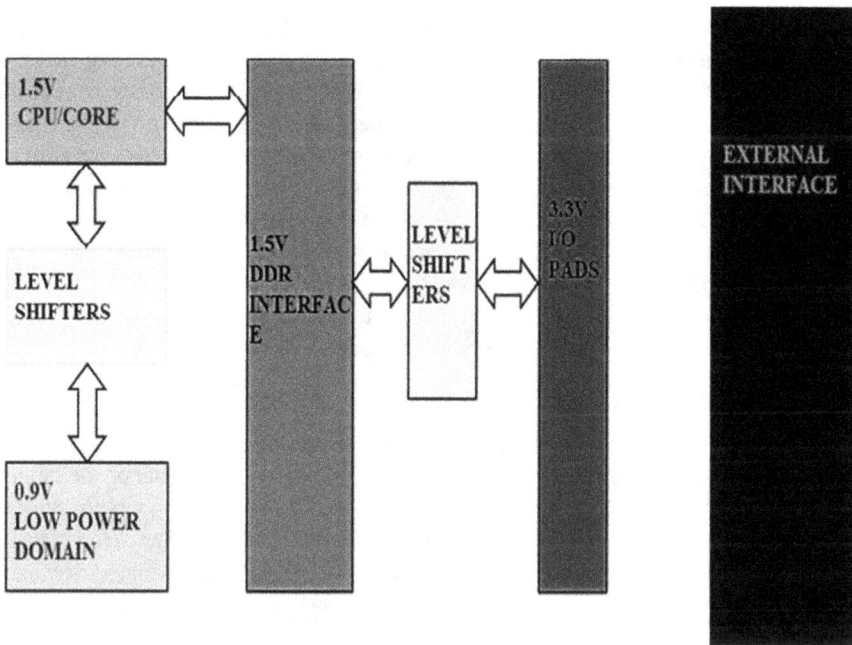

FIGURE 15.4 SOC block diagram of level shifter 1.

place in the 0.8V power domain. As a result, determining which power domain, the Level shifter should be placed in necessitates some library knowledge.

15.3.2 MEASUREMENT OF PERFORMANCE METRICS OF LEVEL SHIFTER

Level shifters are assessed based on various performance metrics. Level shifter design for targeted applications uses trade-offs among these metrics. This section discusses multiple figures of merit (FoM) of Level shifters and their measurement methodologies.

15.3.2.1 Frequency of Operation

The Frequency of operation of the Level shifter gives information about at which Frequency the output is close to 99% of the VDDH value means the level shifter properly works. TFET-based Level shifter work on the MHz range. In the case of TFET-based voltage level shifters, output resistance is higher, leading to a higher frequency of operation in TFET-based voltage level shifters compared to CMOS-based voltage level shifters.

15.3.2.2 Power Consumption

Identifying such components is a critical issue in minimizing the overall power consumption of TFET-based voltage Level shifters. The power consumption of the Level shifter comprises two major components: static or leakage power and dynamic power. Static power is the power drawn from the Level shifters' supply.

In the static case, we apply the zero at the input and calculate power due to V_{DDH} and V_{DDL} at the input and calculate power due to V_{DDH}. And figure avg of both power give information about static power. The dynamic power consumption during the operation of a Level shifter can be measured by adding up the emotional power drawn by different capacitive loads that are charged and discharged during the operations. In level shifter, design power consumption depends on the supply voltage V_{DDH}, V_{DDL}. In the dynamic case, when the input changes from 0 to V_{DDL}, the parasitic output capacitance is charge and discharge, which decides the fiery power consumption for the voltage level shifter.

So in the voltage level shifter, we calculate power due to V_{DDH} and V_{DDL}. The sum of these powers gives information about the total power.

15.3.2.3 Level Shifting Range

Level Shifting Range in TFET-based voltage Level shifters provides information on the range of operation. The lower conversion range of the voltage level shifter on the supply voltage of the V_{DDL} type inverter. The decrease in the supply of the V_{DDL} type inverter means we reduce the lower conversion range of the voltage level shifter. If the voltage level shifter works between 0.1 and 1.2 volts, it can translate voltage between 0.1 and 1.2 volts. This level shifter can raise the voltage of a 0.1V module to 1.2V.

15.3.2.4 Delay of Level Shifters

Delay of the voltage Level Shifter gives information about how fast the voltage Level shifter can operate. Mainly delay depends upon how quickly the Transistor can switch means how fast transistors are ON/OFF. TFET-based devices have a higher ON/OFF current ratio that gives information about transistors changing very fast deciding the Device's Delay. In TFET, appliances have a higher current at a lower voltage. This higher current helps to charge & discharge the parasitic node very fast, that's by TFET-based voltage level shifter have a lesser delay. TFET device is operating at more secondary Voltage (Threshold of the Device is more deficient as compared to CMOS). We are using a 20nm Double gate TFET Model. This device threshold is 0.145V less than CMOS.

15.3.2.5 Power Delay Product

The power–delay product (PDP) is an energy-efficiency merit figure. Power consumption P_{avg} (averaged over a switching time) multiplied by the input-output Delay or duration of the switching event to calculate switching energy. It calculates the energy utilization for each switching occurrence and has an energy dimension.

The Power delay product is the product of Power and Delay that give information about the combined effect of both Power & Delay.

15.4 CIRCUIT TOPOLOGIES OF LEVEL SHIFTERS

15.4.1 Cross Coupled Level Shifter

Cross-coupled voltage Level shifters have positive feedback in the pull-up network that helps to off another side of the cross-coupled Transistor so that less grow-bar current flows from VDDH to GND. Cross-coupled level shifters have unequal strength of pull-up & Pull-down transistors. module1 (Which is working on lower voltage) voltage goes to the driving NTFET type transistor, and module2 (which is working on higher voltage) voltage goes across the PTFET that creates the contention in the Pullup and pull-down network.

15.4.1.1 TFET-based Cross Coupled Voltage Level Shifter (LS1)

TFET-based Cross coupled voltage Level shifter converts the lower voltage even less than the Device's threshold to the higher module voltages. TFET-based cross-coupled voltage Level shifter encounters severe difficulty in the extreme conflict between the solid upper Pull-up and weak Lower pull-down networks, which leads to failure due to different strengths of pull-up & pull-down. With the help of two series stages with a lower supply voltage in the first stage to limit the unequal Power of PTFET & NTFET. With wide-range voltage conversion, the cross-coupled LS in [15] provides robust voltage shifting from 0.188V to 1.2V. However, it causes a significant delay and a penalty in terms of area. By successfully regulating the contention, the LS in [1] accomplishes a stable and quick voltage translation from the lower sub-threshold to the above threshold region. [7] The LS uses a self-adapting pull-up structure to convert from very low input voltages to the nominal voltage of 1.8V while consuming very little static power. Still, these circuits have a trade-off between the Transistor and conversion speed.

In Cross coupled voltage Level shifter Driving Transistor is weak because the input to the driving transistor is low (0 to VDDL) voltage; because of the Low voltage at the information, the strength of the Transistor is low. For proper operation of the cross-coupled voltage level shifter, we increase the Width of the driving NTFET transistor is 350 nm at the 20 nm TFET technology node because the strength of the Level shifter depends upon the gate voltage & Width of the Transistor (Figure 15.5).

In the TFET-based cross-coupled voltage level shifter, crowbar current flows when the input changes from VDDL to GND or V_{DDL}. Because at the edge, when information changes, PTFET of cross-coupled pull-up transistor require some time to off the PTFET of the opposite side, That's by a concise amount of grow bar current flows from V_{DDH} to GND. By seeing the waveform, we observe that TFET-based cross-coupled voltage level shifter (LS1) has more leakage current at the edge because TFET has a higher gate to drain capacitance at the border (Figure 15.6).

The above waveform shows that the TFET-based cross-coupled voltage level shifter translates voltage from 250 mv to 1.2 V. The waveform result currently due to V_{DDH} and V_{DDL} gives the information about the power.

Table 15.1 compares the CMOS and TFET implementation of cross-coupled Voltage Level Shifters. From this table, we observe that the TFET-based cross-coupled voltage level shifter has a better level shifting range, lower power dissipation, and lower delay as compared to TFET-based cross-coupled voltage level shifter.

CMOS-based cross-coupled voltage level shifter gives a better response at 1.2 MHz frequency, and TFET-based voltage level shifter gives a better response at 20 MHz.

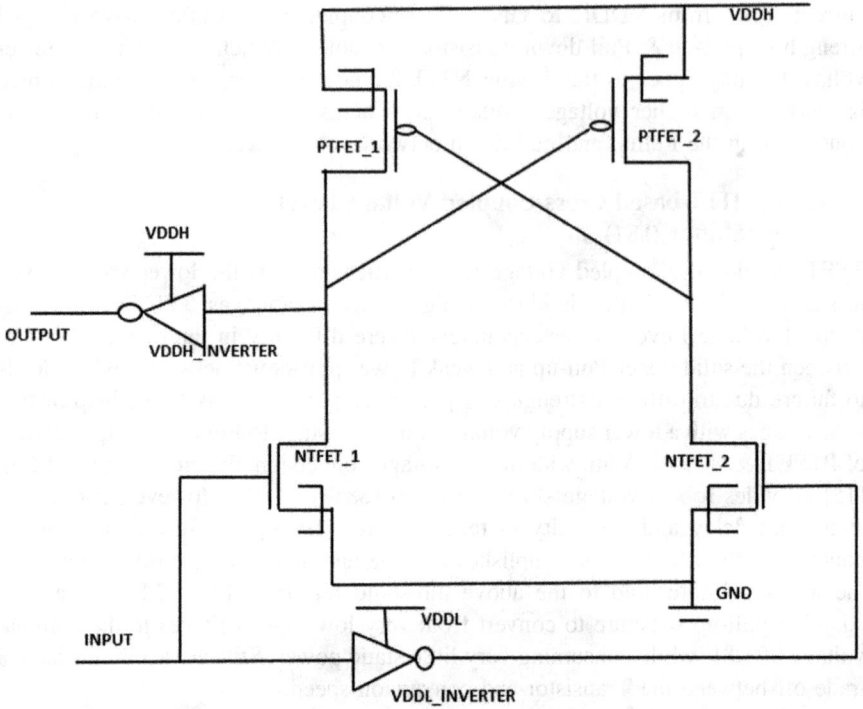

FIGURE 15.5   TFET-based cross-coupled voltage level shifter.

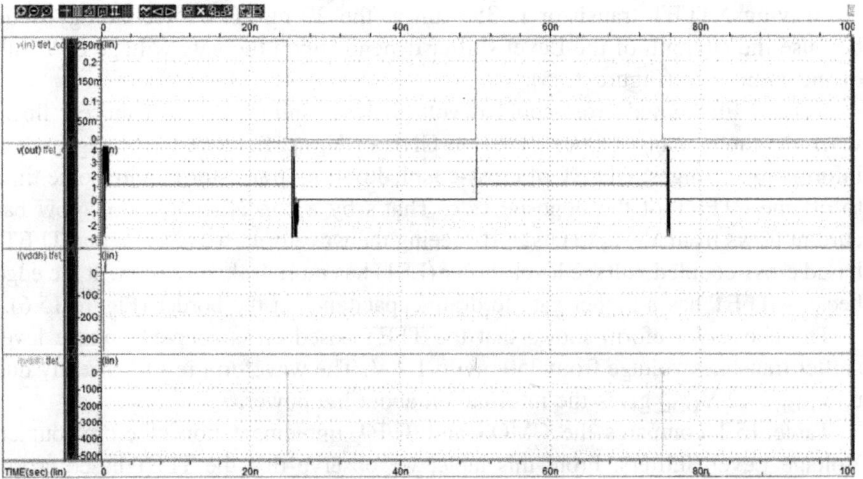

FIGURE 15.6   Waveform of V_{IN}, V_{OUT}, IV_{DDH}, and IV_{DD} of TFET-based cross-coupled level shifter.

TABLE 15.1

Comparison Table of CMOS-Based Cross-Coupled Voltage Level Shifter and TFET-Based Cross-Coupled Voltage Level Shifter

Parameter	CMOS-Based Cross-Coupled Level Shifter	TFET-Based Cross-Coupled Level Shifter (LS1)
Frequency of operation	1.2 MHz	20 MHz
Level shifting Range	0.3–1.2 V	0.25–1.2 V
Total Power	16.54 μW	556.23 pW
Delay	31.7 nS	71.389 pS
Power Delay Product (PDP)	524.413×10^{-15} J	39.708×10^{-21} J

In a CMOS-based cross-coupled voltage level shifter (LS1), we obtained a level shifting range from 0.3 to 1.2 V, and in TFET-based cross-coupled voltage level shifter (LS2), we received a level shifting range from 0.25 to 1.2 V. In CMOS-based voltage level shifter, we obtained power of 16.54 μW, but in the TFET-based cross-coupled voltage level shifter (LS1), we received Power of 556.23 pW and Power due to VDDH being 549.961 pW & VDDL is 6.273 pW. TFET-based Circuit is faster as compared to CMOS-based cross-coupled level shifter Circuit.

In the TFET-based voltage Level shifter circuit, we obtained Delay in the range of pS. But in the CMOS-based voltage level shifter (LS1) we obtained a delay of 31.7 nS but In the TFET-based cross-coupled voltage level shifter, we received a postponement of 71.389 pS.

The waveform result currently due to V_{DDH} & V_{DDL} gives the information about the power. We obtained leakage power when input varies from GND to V_{DDL} or vice versa.

15.4.1.2 TFET-based Proposed Voltage Level Shifter (LS2)

To improve the power of the cross-coupled voltage level shifter (LS1), we add the current limiting Transistor between the driving and cross-coupled (Positive feedback) transistor and connect the splitting type inverter at the output. These current-limiting Diode Connected transistors decrease the current from V_{DDH} to GND of the Circuit. For proper level shifting operation, we adjust the Width of the splitting type inverter (Figure 15.7).

From the below waveform, we observe that grow bar current that flows from V_{DDH} to GND is less in the TFET-based proposed voltage level shifter (LS2). It is a decrease due to the current limiting Transistor that leads to less power dissipation in the proposed voltage level shifter (LS2) compared to the TFET-based cross-coupled voltage level shifter (LS1) (Figure 15.8).

Table 15.2 compares the CMOS and TFET implementation of the proposed voltage Level Shifter (LS2). The Frequency of operation of the TFET-based

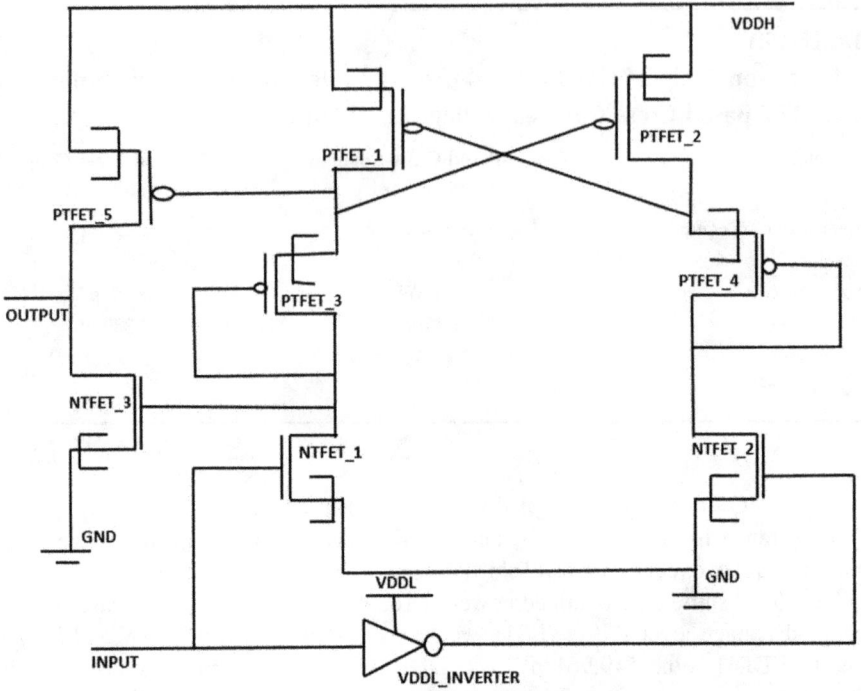

FIGURE 15.7 TFET-based improved cross-coupled level shifter.

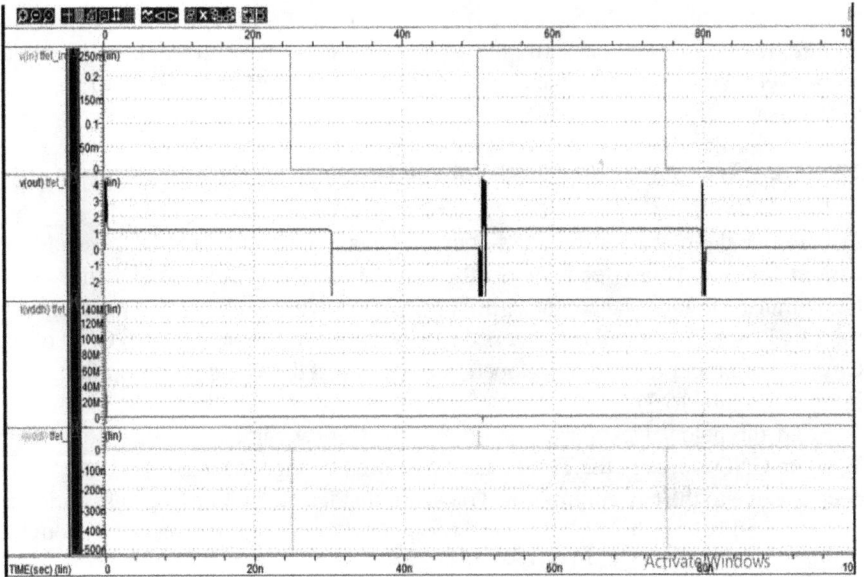

FIGURE 15.8 Waveform of V_{IN}, V_{OUT}, IV_{DDH}, and IV_{DD} of TFET-based proposed cross-coupled level shifter (LS2).

TABLE 15.2

Comparison Table of CMOS-Based Proposed Voltage Level Shifter (LS2) and TFET-Based Suggested Voltage Level Shifter (LS2)

Parameter	CMOS-Based Proposed Level Shifter (LS2)	TFET-Based Proposed Level Shifter (LS2)
Frequency of operation	1.2 MHz	20 MHz
Level shifting Range	0.3 to 1.2 V	0.25 to 1.2 V
Total Power	1.5 μW	490.71 pW
Delay	23.34 nS	64.18 pS
Power Delay Product (PDP)	35.01×10^{-15} J	31.493×10^{-21} J

suggested voltage Level shifter (LS2) is better than the CMOS-based offered voltage level shifter (LS2). TFET-based proposed voltage level shifter (LS2) circuit provides better power in the range of pW as compared to the CMOS-based proposed voltage level shifter. In the TFET-based proposed voltage level shifter (LS2), power due to V_{DDH} is 484.43 pW & V_{DDL} is 6.273 pW. TFET-based voltage level shifter Circuit is faster than the CMOS-based voltage level shifter Circuit. In the TFET-based voltage Level shifter circuit, we obtained Delay in the range of pS. In the TFET-based proposed voltage level shifter (LS2), we received a delay is 64.180 pS. But in the CMOS-based proposed voltage level shifter (LS2), we obtained a postponement of 23.34 nS.

From Table 15.2, we observe that TFET-based proposed voltage level shifter (LS2) has a better Level shifting range as compared to the CMOS-based proposed level shifter (LS2); we obtain 0.25 to 1.2 V in TFET-based proposed voltage level shifter (LS2).

In the TFET-based Improved Level shifter, we obtained less grow-bar current (current from power supply to GND). The current limiting Transistor and splitting type inverter helps to reduce the current due to VDDH and present due to VDDL in an improved Cross coupled Level shifter circuit.

15.4.1.3 TFET-based Proposed Voltage Level Shifter (LS3)

Cross-coupled voltage Level shifter Pullup & Pulldown networks have unequal strength. The different voltage at the gate means PTFET has more Power because V_{DDH} voltage comes across the entrance of PTFET. To reduce this problem, we use MUX in the pull-up network, and with the help of MUX, we provide less Voltage (V_{DDM}) to the Gate of PTFET. This V_{DDM} voltage comes from the TFET-based Voltage divider network.

15.4.1.3.1 TFET-based Multiplexer for Proposed Voltage Level Shifter (LS3)

We design the TFET-based MUX from the transmission gate-type pass transistor Logic. A 2:1 multiplexer in Figure 15.9. This gate selects either input I0 or I1 based on the control signal 'SELECT' value. When the control signal SELECT is logic

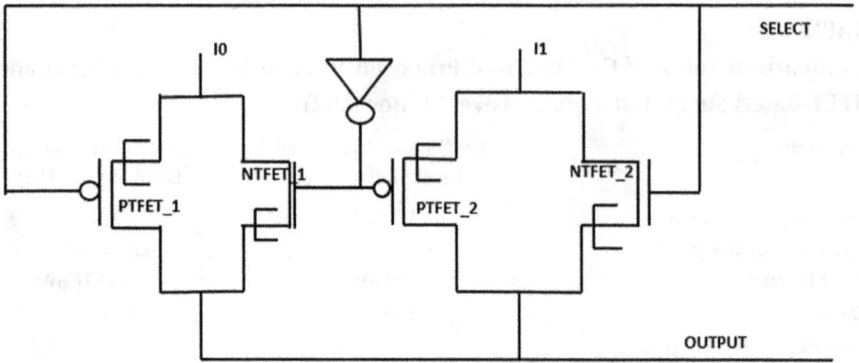

FIGURE 15.9   TFET-based 2:1 MUX 1.

TABLE 15.3
TFET-Based 2:1 MUX

I0	I1	SELECT	OUTPUT
0	0	0	0
0	0	1	0
0	1	0	0
0	1	1	1
1	0	0	1
1	0	1	0
1	1	0	1
1	1	1	1

low, the output equals the input I0, and when it is logic high, the result equals the input I1.

When the control signal SELECT is low, then the upper transmission gate turns OFF, and it will not allow I0 to pass through it; at the same time, the lower transmission gate is 'ON', and will enable I1 to pass through it, so the output = I1.

Table 15.3 shows that whenever SELECT is zero, it will select I0. If SELECT is one, then it will choose I1.

$$Output = \overline{Select} . I_0 + Select . I_1$$

15.4.1.3.2   TFET-based Voltage Divider Network

TFET-based Voltage divider network is the series connection of 5 gates to drain shorted PTFET. We calculate the V_{DDM} voltage from the Drain of the PTFET_3 Transistor. We obtained V_{DDM} voltage 5.38 mV voltage. Here supply voltage V_{DDH} is 1.2V (Figure 15.10).

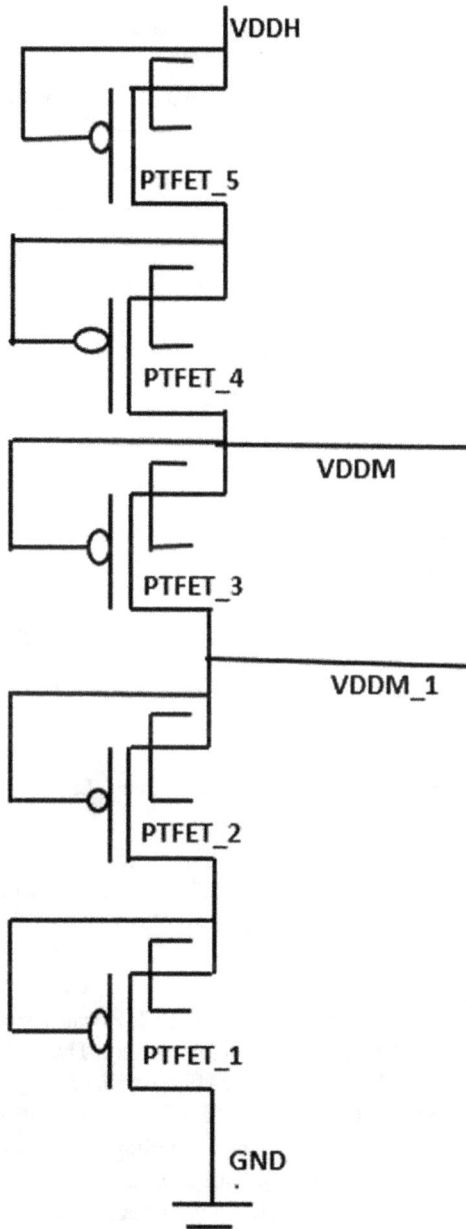

FIGURE 15.10 TFET-based voltage divider network.

The upper cross-coupled PTFET is fast in a cross-coupled circuit because the VDDH potential comes across the PTFET. So Decrease the node voltage of PTFET, we use the MUX, and the input to MUX is Vddm (That generates from the voltage divider network) and D or DBar. Mux-based voltage Level shifter circuit Improves cross-coupled LS in terms of Delay (Figure 15.11).

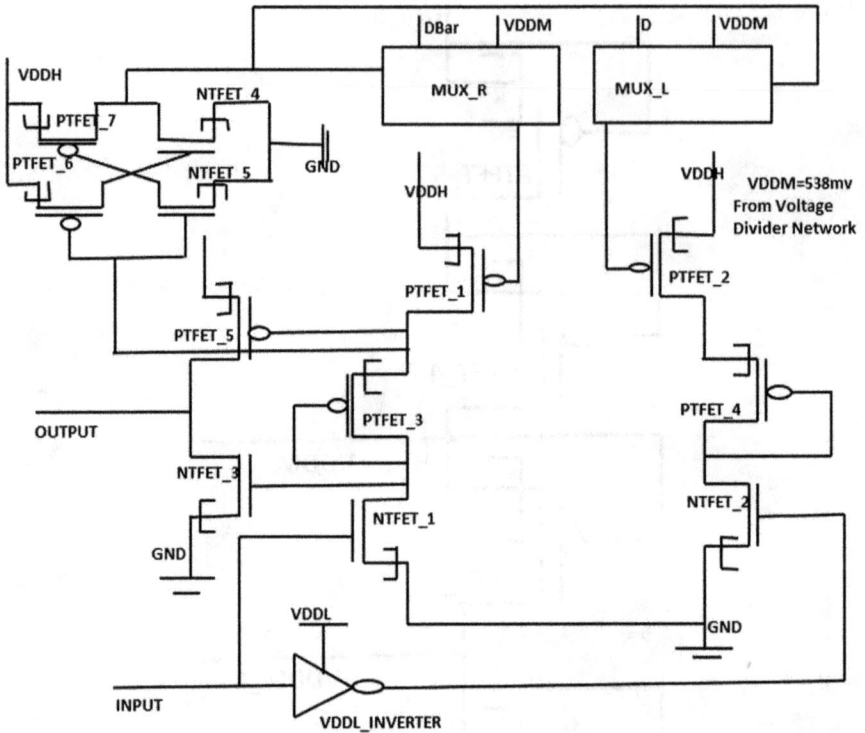

FIGURE 15.11 TFET-based proposed level shifter circuit (LS3).

For proper operation of the SELECT of multiplexers, we provide the cross-coupled network between the source of PTFET_1 & SELECT. Because of the appropriate recovery of the level so that the multiplexer gate properly works, we can say that too fast up the operation.

TFET-based Proposed Level shifter Circuit (LS3) width of the NTFET_1 & NTFET_2 is comparable to PTFET_1 & PTFET_2. Because of the decrease in the gate potential of PTFET, current strength decreased. That's because we require comparable width of the NTFET transistor (Figure 15.12).

Table 15.4 compares the CMOS and TFET implementation of the proposed voltage Level Shifter (LS3). In the proposed voltage level shifter (LS3) circuit, we observe that Delay is less as compared to all topologies. In the CMOS-based proposed Voltage level shifter (LS3) we obtained a delay of 744 pS and in TFET-based proposed voltage level shifter (LS3) we obtained a postponement of 3.874 pS.

From Table 15.4, we observe that the Frequency of operation of the CMOS-based proposed voltage level shifter (LS3) is 4.6 MHz, and TFET-based Propose voltage Level shifter (LS3) is 20 MHz. TFET-based proposed level shifter (LS3) provides a Level shifting range from 0.2V to 1.2V. But CMOS-based proposed voltage level shifter gives a level shifting range from 0.28 to 1.2V. The Circuit

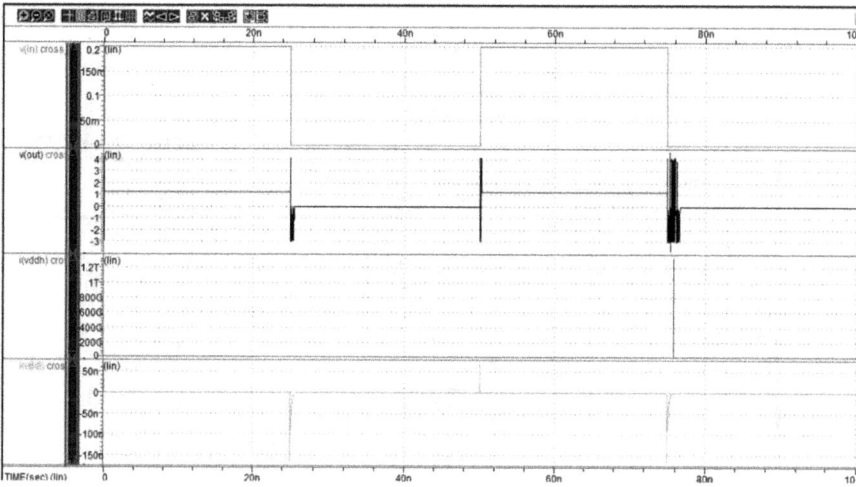

FIGURE 15.12 Wave of input and output voltage and current due to V_{DDH}.

TABLE 15.4

Comparison Table of CMOS-Based Proposed Voltage Level Shifter (LS3) and TFET-Based Proposed Voltage Level Shifter (LS3)

Parameter	CMOS-Based Proposed Level Shifter (LS3)	TFET-Based Proposed Level Shifter (LS3)
Frequency of operation	4.6 MHz	20 MHz
Level shifting Range	0.28–1.2 V	0.2–1.2 V
Total Power	129 μW	2.963 nW
Delay	744 pS	3.874 pS
Power Delay Product (PDP)	95.976×10^{-15} J	11.478×10^{-21} J

offers better level shifting compared to the simple cross-coupled & improved Circuit. TFET-based proposed voltage level shifter (LS3) has more power consumption because of the multiplexers and extra circuitry. In the TFET-based proposed voltage level shifter (LS3), we obtained power due to VDDH being 2.963 nW & VDDL being 4.84 pW. In terms of PDP, this Circuit is better. Because of the use of multiplexers LS3 circuit is faster. We can use this level shifter where speed is my primary purpose.

15.4.1.4 Proposed TFET-based Voltage Level Shifter (LS4)

TFET-based Proposed has a voltage Level shifter (LS4). It has more power than we remove the multiplexer of this level shifter circuit. Insert a cross-coupled stage between the upper cross-coupled stage and the Current Limiting Transistor

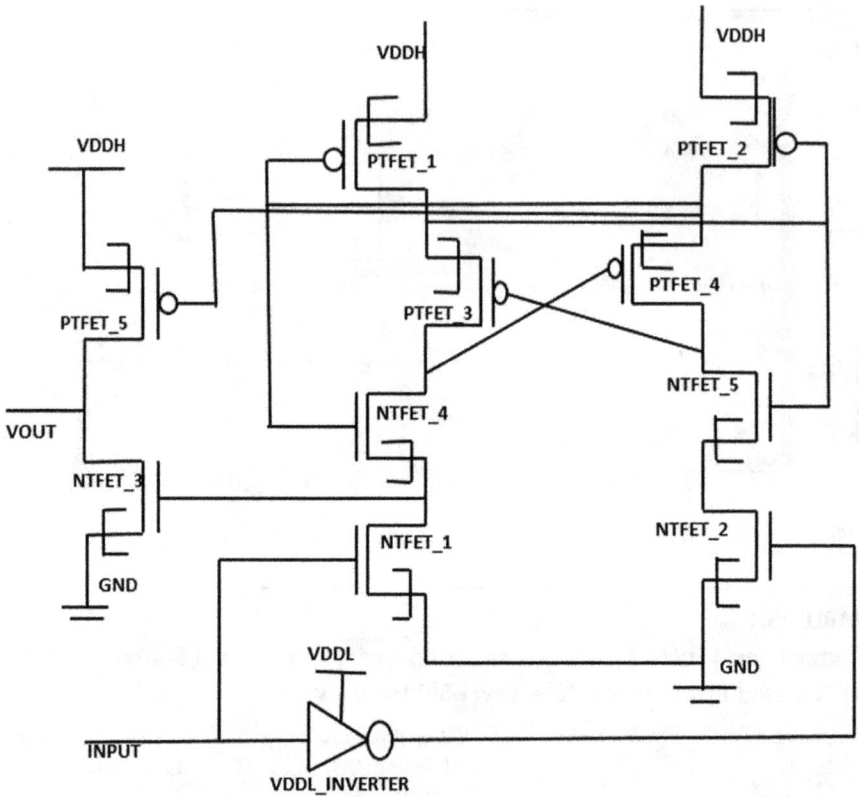

FIGURE 15.13 Proposed TFET-based level shifter (LS4).

(Here, we do not use a diode-connected transistor, We use NTFET Transistor and bias from the Upper Cross coupled PTFET). We use a splitting-type inverter and input to splitting type transistor from the opposite branch (Figure 15.13).

When the input is high, NTFET_1 is on, and GND to drain of NTFET_1:

At bias-condition, GND is passed to drain of NTFET_4. So that's by less amount of leakage current is flow from VDDH & GND that's by power is less (Figure 15.14).

Table 15.5 compares CMOS and TFET implementation of the proposed voltage Level Shifter (LS4). The proposed voltage level shifter (LS4) has two cross-coupled stages that help reduce the Power and Delay of the Circuit. In the CMOS-based presented voltage level shifter (LS4), we obtained power of 666.231 nW; in the TFET-based proposed voltage level shifter (LS4), we received power of 424.483 pW. TFET-based Circuits are faster because the switching in the TFET device is fast. We received a 9.192 pS delay in the TFET-based proposed voltage level shifter (LS4).

We obtained less current due to VDDH & VDDL because of the fast off of the PTFET due to Cross coupled structure. TFET-based proposed level shifter (LS4) has better Delay and, finally better Power delay product (PDP).

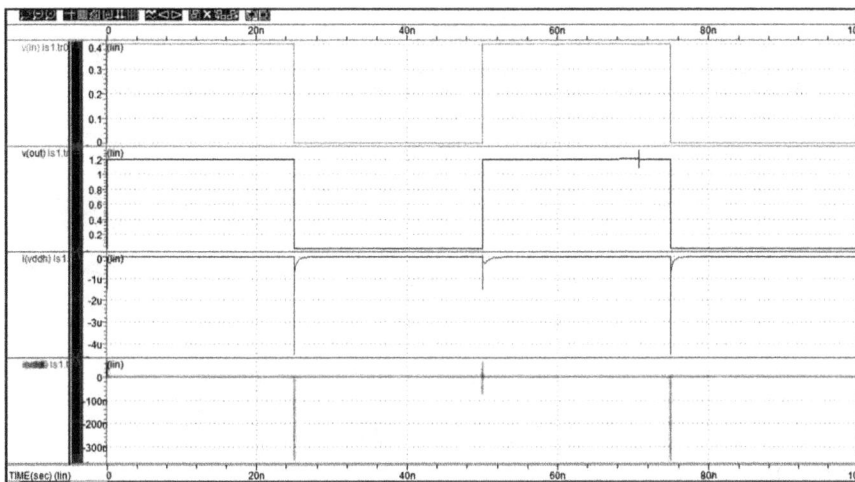

FIGURE 15.14 Waveform of V_{IN}, V_{OUT}, IV_{DDH}, and IV_{DL} of proposed TFET-based level shifter (LS4).

TABLE 15.5

Comparison Table of CMOS-Based Proposed Voltage Level Shifter (LS4) and TFET-Based Suggested Voltage Level Shifter (LS4)

Parameter	CMOS-Based Proposed Level Shifter (LS4)	TFET-Based Proposed Level Shifter (LS4)
Frequency of operation	8 MHz	20 MHz
Level shifting Range	0.35–1.2 V	0.28–1.2 V
Total Power	666.231 nW	424.483 pW
Delay	4.14 nS	9.192 pS
Power Delay Product (PDP)	2.758×10^{-15} J	3.901×10^{-21} J

15.4.2 CURRENT MIRROR LEVEL SHIFTER (CMLS)

The Level shifter in [1] enables a robust and rapid voltage conversion from the lower sub-threshold to the voltage more significant than the threshold region by efficiently level shifting to remove the Problem of Unequal strength of PTFET & NTFET. It is like a contention mitigate type level shifter, but this topology has a vast area and a high static power. The level shifter in [7] uses techniques in the pull-up structure to convert from very low input voltages even lesser than the threshold to the nominal high voltage of 1.8V while consuming very little static power, but the problem with high power dissipation.

15.4.2.1 TFET-based Current Mirror Voltage Level Shifter (LS5)

Cross-coupled voltage Level shifters have strong contention between pull-up and pull-down networks. That's by we go for the current mirror-type level shifter.

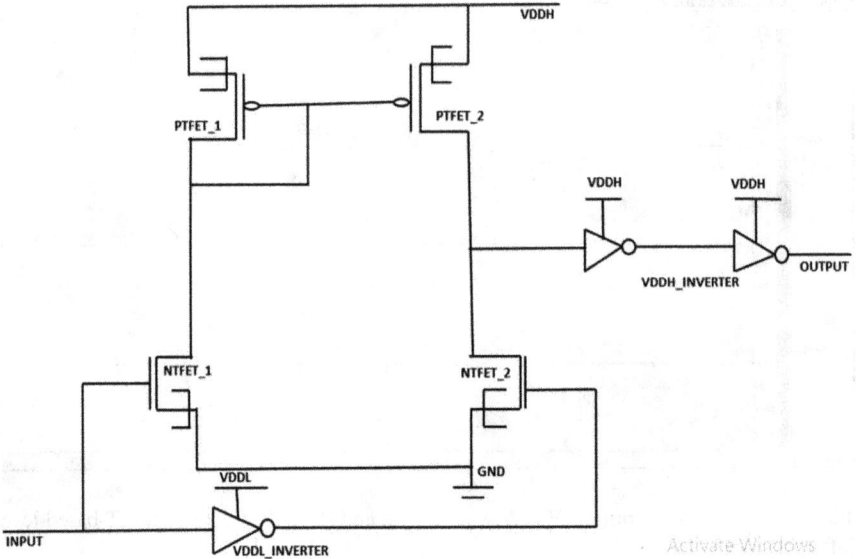

FIGURE 15.15   TFET-based current mirror voltage level shifter (LS5).

By efficiently regulating the conflict, the Current mirror Level shifter delivers robust and quick voltage conversion from the Lower sub-threshold to the above threshold domain. However, it has a lot of static power because due to the current mirror large amount of current is flowing from V_{DDH} to GND (Figure 15.15).

From the below waveform, we observe that the current due to VDDH & VDDL is higher than all level shifter topology because of the current mirror load at the pull-up network. TFET-based current mirror voltage level shifter (LS5) shifts from 0.17 to 1.2V. TFET-based existing mirror voltage level shifter provides a wide conversion range (Figure 15.16).

Table 15.6 compares the CMOS and TFET implementations of the proposed current mirror voltage Level Shifter (LS4). In the Current Mirror Level shifter, because the current mirror pull-up network current is present, that is also a force on the driving Transistor that's by we get the better Frequency of operation in the Current mirror-type voltage level shifter. We obtained an 18MHZ Frequency of operation in the CMOS-based current mirror level shifter, and we received a 33MHZ Frequency of operation in the TFET-based existing mirror voltage level shifter. We can use this Circuit for high-frequency applications.

But the problem Current mirror level shifter is that in the current mirror pull-up network, the current is present every time. When the input is VDDL, NTFET_1 is on, and current flows from VDDH to GND. So due to the current flow, high power dissipation happens. In the TFET-based Current mirror voltage level shifter, we obtained Total Power is 873.52 pW.

Because the Transistor's current mirror structure driving strength is good, we obtained Delay in the CMOS-based existing mirror voltage level shifter is 17.5nS and the TFET-based Current mirror level shifter is 53.914pS.

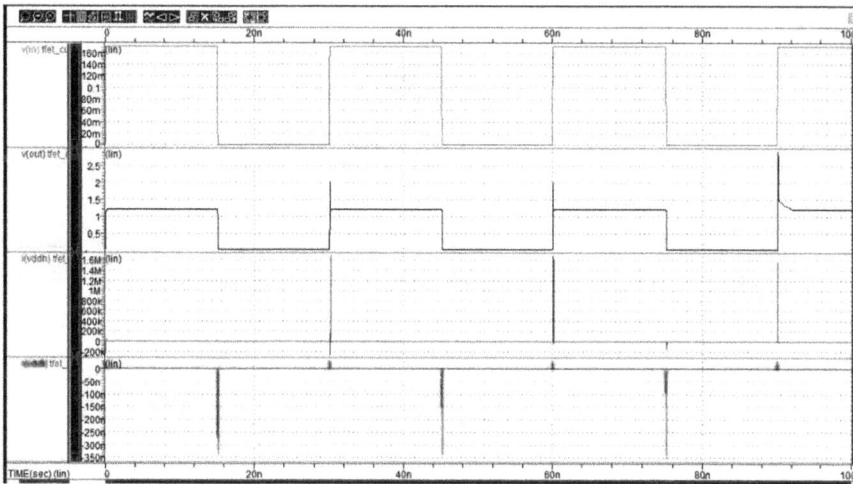

FIGURE 15.16 Waveform of V_{in}, V_{out}, IV_{DDH}, and IV_{DDL} of TFET-based current mirror level shifter (LS5).

TABLE 15.6
Comparison Table of CMOS-Based Current Mirror Voltage Level Shifter (LS5) and TFET-Based Current Mirror Voltage Level Shifter (LS4)

Parameter	CMOS-Based Current Mirror Level Shifter (LS5)	TFET-Based Current Mirror Level Shifter (LS5)
Frequency of operation	18 MHz	33 MHz
Level shifting range	0.25–1.2 V	0.17–1.2 V
Total Power	38.65 μW	873.47 pW
Delay	17.5 nS	53.914 pS
Power Delay Product (PDP)	676.375×10^{-15} J	47.088×10^{-21} J

15.4.2.2 TFET-based Improved Current Mirror Level Shifter (LS6)

When the input is HIGH in a TFET-based current mirror voltage level shifter, the NTFET_1 turns on, and current flows between VDDH and GND. To remove this problem, we use the current Limiting Type transistor between the current mirror and the driving Transistor. In the current limiting Transistor, we use the simple gate to drain shorted Transistor, and after that drain potential of this transistor pass to the gate of NTFET_3, and the source potential pass to the gate of PTFET_3. NTFET_3 & PTFET_3 combination form the splitting type inverter. We assign the proper sizing of the splitting type inverter for proper level shifting operation (Figure 15.17).

In the below waveform, we observe that the TFET-based Improved Current Mirror Level shifter circuit (LS6) has less power than the Current mirror level

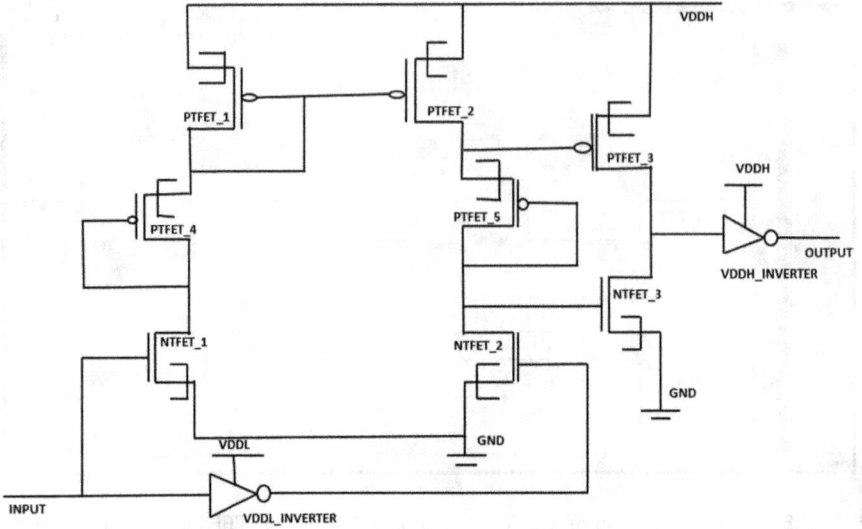

FIGURE 15.17   TFET-based improved current mirror level shifter.

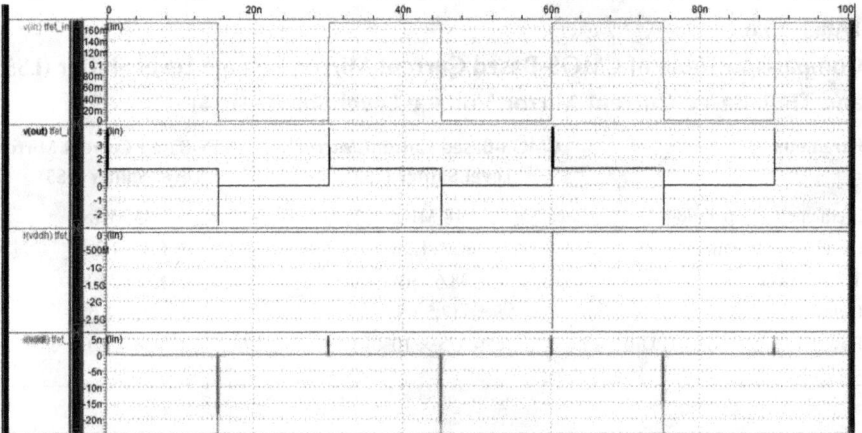

FIGURE 15.18   Waveform of V_{IN}, V_{OUT}, IV_{DDH}, and IV_{DDL} of TFET-based improved current mirror level shifter (LS6).

shifter because of the current limiting Transistor. A current-limiting transistor helps to reduce the crowbar current flow from the V_{DDH} to GND (Figure 15.18).

Table 15.7 compares the CMOS and TFET implementations of the improved Current Mirror voltage Level Shifter (LS6). From the Table 15.7, we observe that the Frequency of operation of CMOS-based improved current mirror-voltage level shifter circuits is 17 MHz and the Frequency of operation of TFET-based improved existing mirror voltage level shifter is 33 MHz.So this Improved current mirror level shifter is helpful for high-speed working modules where the Frequency is in the range of MHz.

TABLE 15.7

Comparison Table of CMOS-Based Improved Current Mirror Voltage Level Shifter (LS6) and TFET-Based Improved Current Mirror Voltage Level Shifter (LS6)

Parameter	CMOS-Based Improved Current Mirror Level Shifter (LS6)	TFET-Based Improved Current Mirror Level Shifter (LS6)
Frequency of operation	17 MHz	33 MHz
Level shifting Range	0.25–1.2 V	0.17–1.2 V
Total Power	4.36 µW	570.876 pW
Delay	4.96 nS	24.749 pS
Power Delay Product (PDP)	21.625×10^{-15} J	14.128×10^{-21} J

In the CMOS-based improved current mirror voltage level shifter, we obtained power of 4.36 µW. and TFET-based improved existing mirror voltage level shifter; we received less energy (570.876 pW). We received a delay of 24.794 pS in the TFET-based improved current mirror voltage level shifter. In a CMOS-based improved existing mirror voltage level shifter, we obtained 4.96 nS.

The improved current mirror voltage level shifter has a wide conversion range because of the current mirror load. We obtained a 0.25 to 1.2V Level shifting range in CMOS-based improved and 0.17 to 1.2V in TFET-based improved current mirror level shifter.

15.4.3 TFET-BASED WILSON CURRENT MIRROR LEVEL SHIFTER (WCMLS)

The feedback transistor PTFET_3 uses the TFET-based Wilson Current Mirror voltage level shifter to prevent static current. The voltage swing at the input node of the first V_{DDH} type inverter minimizes when the static turns off; this results in a sizeable static current from V_{DDH} to GND of the first Inverter. Furthermore, due to the input First V_{DDH} type inverter and charge transfer from the gate of PTFET_1 to Drain of PTFET_3 via PTFET 3, the speed of the fall transition degrades dramatically. Instead of a high static current, the LS in [16] employs a logic error correction circuit to provide low power dissipation. However, it has a problem of low considerable latency and area. To circumvent the voltage swing issue in WCMLS and eliminate the static current, the improved LS in [17] suggests an input-controlled diode chain. However, this results in a considerable area and sluggish transition speed (Figure 15.19).

In [3] a drain-to-source connected current limiter is employed to lower the standby current successfully. However, this comes at a cost: a long delay and a vast area. The LS in [17–19] uses a new topology based on a Current Mirror structure and a level-shifting capacitor to achieve low power. Despite this, there is still a significant amount of Delay and overhead (Figure 15.20).

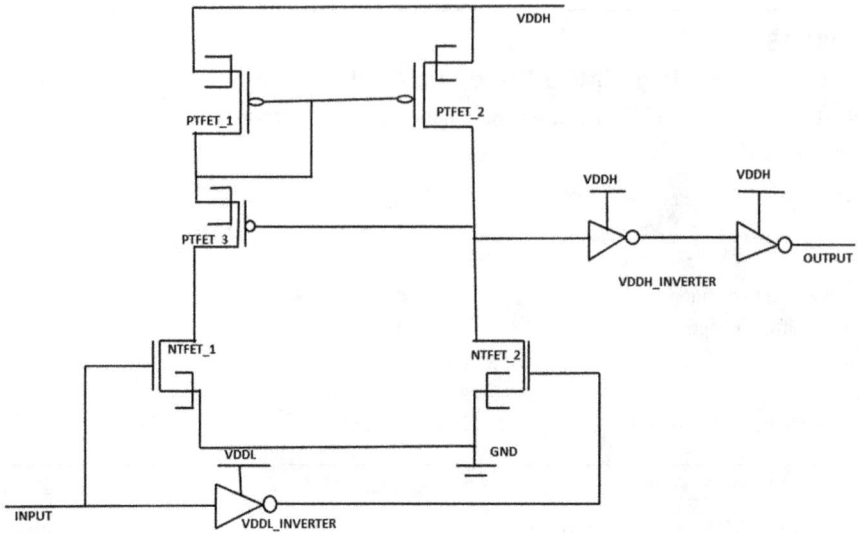

FIGURE 15.19   TFET-based Wilson current mirror voltage level shifter (LS7).

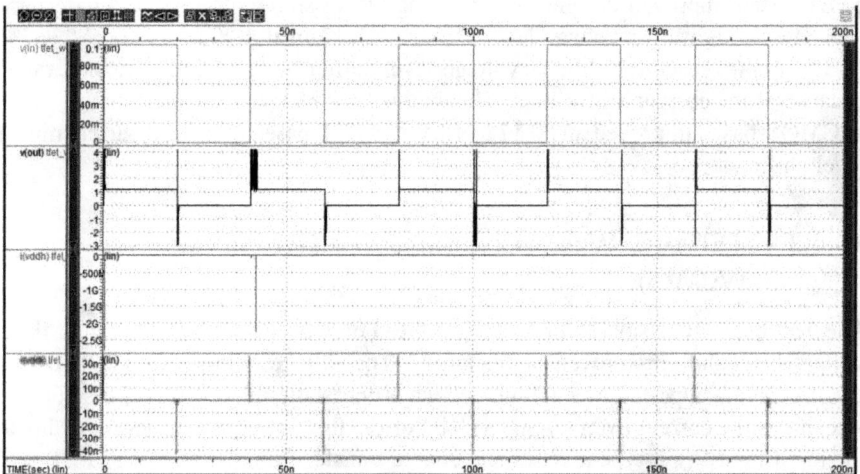

FIGURE 15.20   Waveform of V_{IN}, V_{OUT}, IV_{DDH}, and IV_{DDL} of TFET-based Wilson Current Mirror Level Shifter (LS7).

The TFET-based Wilson current voltage level shifter from the above waveform is the level shift from 0.1 to 1.2V. By seeing the waveform, we observe that at the transition point, Wilson current mirror circuit has leakage because the feedback transistor decreases the potential of input of the V_{DDH} type inverter.

Table 15.8 compares the CMOS and TFET implementations of the Wilson current mirror voltage level shifter (LS7). From Table 15.8, we observe that the CMOS-based Wilson existing mirror Level shifter gives a better result at 10 MHz frequency and the TFET-based Wilson current mirror gives a better response at

TABLE 15.8

Comparison Table of CMOS-Based Wilson Current Mirror Voltage Level Shifter (LS7) and TFET-Based Wilson Current Mirror Voltage Level Shifter (LS7)

Parameter	CMOS-Based Wilson Level Shifter (LS7)	TFET-Based Wilson Level Shifter (LS7)
Frequency of operation	10 MHz	25 MHz
Level shifting Range	0.2 to 1.2 V	0.1 to 1.2 V
Total Power	8.365 nW	636.29 pW
Delay	16.9 nS	42.39 pS
Power Delay Product (PDP)	141.368×10^{-15} J	26.972×10^{-21} J

25 MHz. From the CMOS-based Wilson current mirror level shifter, we got a level shifting range of 0.2 to 1.2V; from the TFET-based Wilson existing mirror level shifter, we obtained 0.1 to 1.2V.

By observation of Table 15.8, we see that the TFET-based Wilson current mirror voltage level shifter has less Power, less Delay, and less PDP.

15.4.4   TFET-BASED IMPLEMENTATION OF EXISTING LEVEL SHIFTER

In the TFET-based implementation of the existing Level shifter, we use the diode-connected Transistor in series with an inverter that reduces the current flow from VDDH to GND. We used one PTFET parallel with this to improve the swing at the input of the VDDH type inverter. Furthermore, the previously proposed voltage level shifter eliminates the VDDL type of Inverter. We use only one Transistor, NTFE_4, to decide the transition from VDDH to GND, resulting in a considerable increase in the growth from VDDH to GND. The feedback transistor NTFET_1, which connects NTFET_3 and the ground, prevents the current mirror from generating a sizeable static current during high output and overcomes the charge transfer problem in the Wilson current mirror level shifter [2]. In the previously proposed LS for Low to high transition operation. When IN is high, NTFET_3 is activated, causing PTFET_2 to generate the mirror current. The Mirror current charges the gate potential of NTFET_2 & PTFET_3, then the input node of the VDDH type inverter is discharged through NTFET_2, and the output goes high (Figure 15.21).

The discharged input node of a VDDH type inverter turns off NTFET_1 to reduce the static current between PTFET_1 and NTFET_3, which reduces the voltage swing at NTFET_2's gate. In the reduced-swing buffer, the Drain to gate shorted Transistor PTFET_5 reduces the voltage swing at PTFET_3's source to the same level as the voltage swing at NTFET_2's gate terminal. As a result, the excessive static current flowing via PTFET_3 reduces, resulting in low standby power for the previously planned LS.

From the waveform shown in Figure 15.22, we observe that TFET-based implementation of the existing voltage-level shifter level shift from 0.1 to 1V. Leakage current is less in this topology.

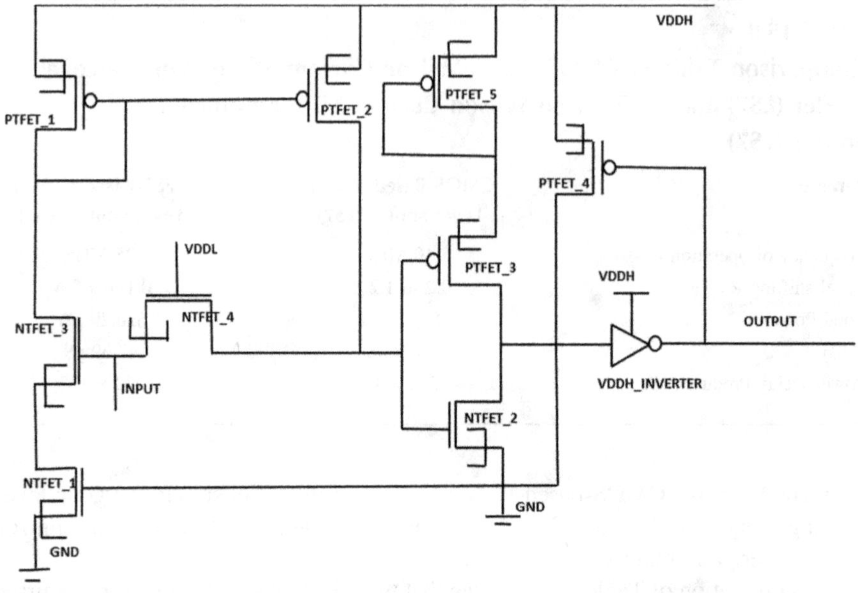

FIGURE 15.21   TFET-based implementation of existing voltage level shifter (LS8).

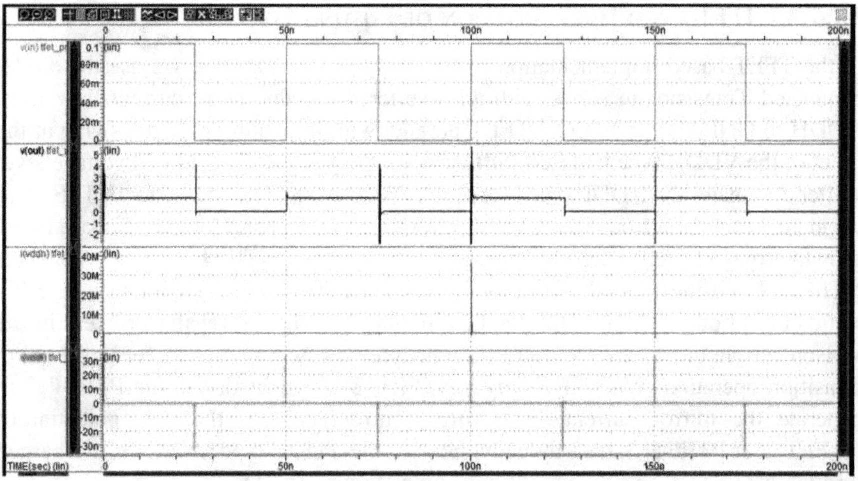

FIGURE 15.22   Waveform of V_{IN}, V_{OUT}, IV_{DDH}, and IV_{DDL} of TFET-based previously proposed level shifter (LS8).

Table 15.9 compares the CMOS and TFET-based implementation of the existing voltage Level Shifter (LS8).

From Table 15.9, we observe that the Conversion range of the existing implementation level shifter is from 0.1 to 1.2 V. The Frequency of operation of CMOS-based implementation of the existing voltage level shifter is 12 MHz, and the

TABLE 15.9

Comparison Table of CMOS-Based Implementation of Existing Voltage Level Shifter (LS8) and TFET-Based Performance of Existing Voltage Level Shifter (LS8)

Parameter	CMOS-Based Proposed Level Shifter (LS4)	TFET-Based Proposed Level Shifter (LS4)
Frequency of operation	12 MHz	20 MHz
Level shifting Range	0.2–1.2 V	0.1–1.2 V
Total Power	12.438 nW	677.398 pW
Delay	7.5 nS	28.93 pS
Power Delay Product (PDP)	93.285×10^{-15} J	19.597×10^{-21} J

Frequency of process of TFET-based implementation of current level shifter gives a better result at 20 MHz.

The main advantage is to use this topology to remove the VDDL type inverter and decrease the leakage current. So the existing implementation of the Voltage level shifter (LS8) has better power consumption and Delay than Wilson's current mirror voltage level shifter (LS7). We obtained less Power and Delay in this topology.

15.5 RESULTS AND DISCUSSION

The proposed TFET-based voltage Level shifter circuits have simulations for the analysis and comparison using HSPICE and 20 nm Double gate InAs Model. Different performance parameters are measured and compared with the existing cells implemented in 20 nm technology—the device parameters and simulation conditions for this analysis are in Table 15.10.

15.5.1 DEVICE PARAMETERS AND SIMULATION CONDITIONS

TABLE 15.10

TFET Parameter Used In Level Shifter

Process Parameter	Value
Gate Length	20 nm
Eto	0.2 nm
Tch	5 nm
Vth	0.145 V
Eg	0.354 eV
V_{DDH}	1.2 V
Temperature	25°C

15.5.2 Frequency of Operation

The Frequency of operation gives information regarding the optimum chosen voltage level shifter operation, which offers a better result. As shown in Table 15.11, it is clear that TFET-based voltage level shifters have a better operation frequency than CMOS-based voltage level shifters. In all topologies of voltage level shifter, the previously proposed current mirror voltage level shifter works on a high frequency. In the TFET-based current mirror circuit, we obtained a better response at 33MHz. The proposed Circuit is working on a 20 MHz frequency. But the proposed Circuit has better Power & Delay.

15.5.3 Level Shifting Range

TFET-based voltage level shifters have a better level shifting range than CMOS-based. Because the lower value of the level shifting range depends upon the V_{DDL} type inverter, we can design less V_{DDL} type inverter with the help of TFET because the TFET device's threshold is less than CMOS. In Table 15.12, we obtained a comprehensive level shifting range of 0.1V to 1.2V in the TFET-based Wilson current mirror-based voltage level shifter (LS7).

15.5.4 Power Dissipation of Voltage Level Shifter

In Table 15.13, we observe that TFET-based voltage Level shifters have less power consumption than CMOS-based voltage level shifters. The proposed level shifter (LS4) has less power than all topologies. Also, in Table 15.13, all

TABLE 15.11
Frequency of Operation of Level Shifter

Level Shifter Topology	Frequency of Operation in TFET-Based Voltage Level Shifter	Frequency of Operation of CMOS-Based Voltage Level Shifter
Cross coupled voltage level shifter (LS1)	20 MHz	1.2 MHz
proposed voltage level shifter (LS2)	20 MHz	1.2 MHz
Proposed voltage level shifter (LS3)	20 MHz	4.6 MHz
Proposed voltage level shifter (LS4)	20 MHz	8 MHz
Current Mirror voltage level shifter (LS5)	33 MHz	18 MHz
Improved current mirror voltage level shifter (LS6)	33 MHz	17 MHz
Wilson current mirror voltage level shifter (LS7)	25 MHz	10 MHz
Existing implementation voltage level shifter (LS8)	20 MHz	12 MHz

TABLE 15.12
Level Shifting Range of all TFET-Based Level Shifter Topology

Level Shifter Topology	Level Shifting Range of TFET-Based Voltage Level Shifter	Level Shifting Range of CMOS-Based Voltage Level Shifter
Cross coupled voltage level shifter (LS1)	0.25–1.2 V	0.3–1.2 V
Proposed voltage level shifter (LS2)	0.25–1.2 V	0.3–1.2 V
Proposed voltage level shifter (LS3)	0.2–1.2 V	0.28–1.2 V
Proposed voltage level shifter (LS4)	0.28–1.2 V	0.35–1.2 V
Current mirror voltage level shifter (LS5)	0.17–1.2 V	0.25–1.2 V
Improved current mirror voltage level shifter (LS6)	0.17–1.2 V	0.25–1.2 V
Wilson current mirror voltage level shifter (LS7)	0.1–1.2 V	0.2–1.2 V
Existing implementation voltage level shifter (LS8)	0.1–1.2 V	0.2–1.2 V

TABLE 15.13
Power Due to Supply of all Topology

Level Shifter Topology	Total Power (pW) of TFET-Based Voltage Level Shifter	Total Power (µW) of CMOS-Based Voltage Level Shifter
Cross coupled voltage Level shifter (LS1)	556.23	16.543
Proposed voltage level shifter (LS2)	490.71	1.5
Proposed voltage level shifter (LS3)	2963	129
Proposed voltage level shifter (LS4)	424.483	0.6662
Current mirror voltage level shifter (LS5)	873.47	38.65
Improved current mirror voltage level shifter (LS6)	570.876	4.36
Wilson current mirror voltage level shifter (LS7)	636.29	8.365
Existing implementation voltage Level shifter (LS8)	677.398	12.438

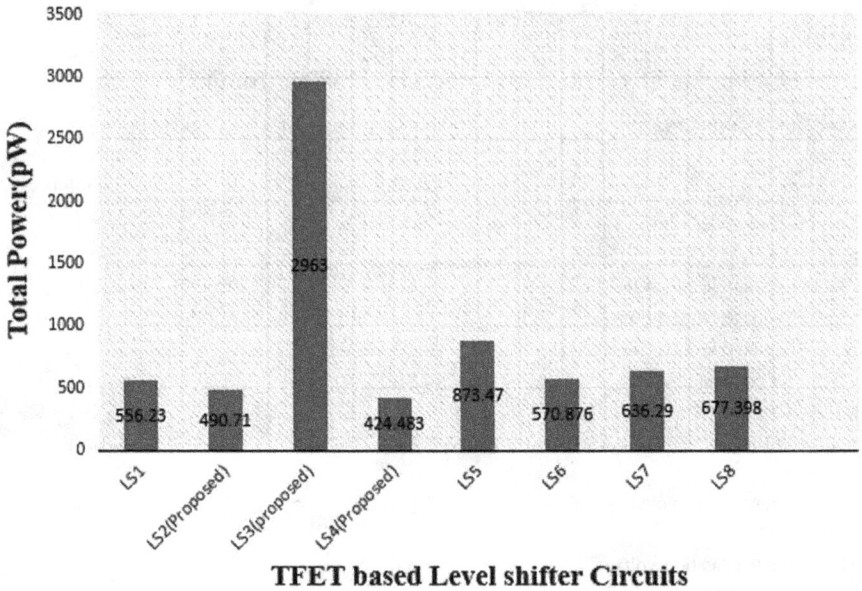

FIGURE 15.23   Power of all topology of TFET-based voltage level shifter.

TFET-based voltage level shifter circuits except the Proposed level shifter (LS3) have power dissipation in the range of pW, and CMOS-based voltage level shifters have power consumption in the field of µW.

Figures 15.23 and 15.24 show the power dissipation of TFET and CMOS implementation of various Level Shifter circuits.

15.5.5   DELAY OF TFET-BASED VOLTAGE LEVEL SHIFTER

The Delay of the TFET-based voltage level shifter circuits is less than CMOS-based voltage level shifter because the Ion/Ioff ratio of the TFET device is higher than the CMOS device. So due to the higher Ion/Ioff ratio, the parasitic capacitor is charged and discharged fast, which, 's by TFET-based voltage level shifter, has less Delay.

In the TFET-based voltage level shifter, we get a delay in the picosecond range. In the proposed TFET-based voltage Level shifter circuit (LS3), we obtained a postponement of 3.874 pS. The proposed Circuit (LS3) is faster because we use a multiplexer (Table 15.14; Figures 15.25 and 15.26).

From the above-shown graph, we see that the Delay of the proposed voltage level shifter (LS3) is a less because of the use of the multiplexer. We use the two cross-coupled inverters in the proposed voltage level shifter (LS3) for the proper voltage level of the select line of the multiplexer so that the multiplexer works appropriately.

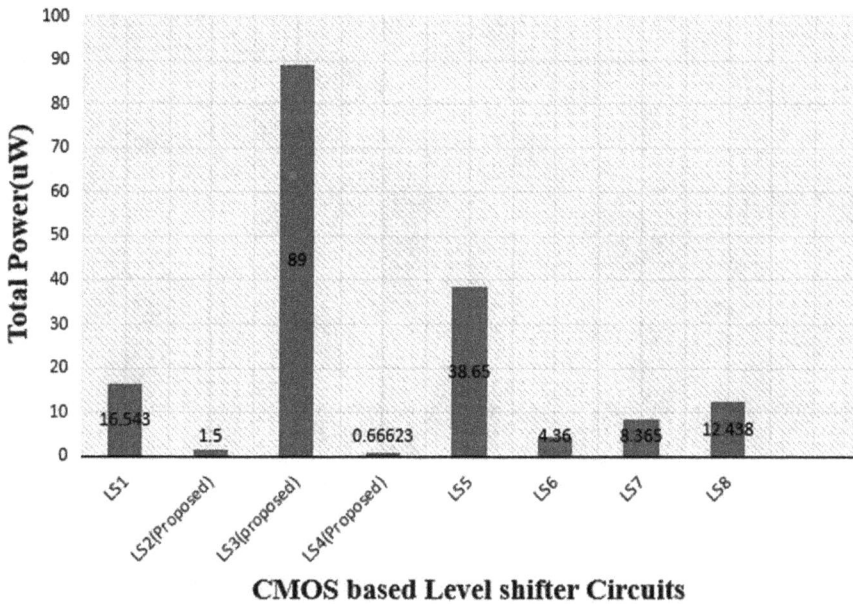

FIGURE 15.24 Power of all topology of CMOS-based voltage level shifter.

TABLE 15.14
Delay of All Level Shifter Topology

Level Shifter Topology	Delay (pS) of TFET-Based Voltage Level Shifter	Delay (nS) of CMOS-Based Voltage Level Shifter
Cross coupled voltage Level shifter (LS1)	71.389	31.7
Proposed voltage Level shifter (LS2)	64.180	23.34
Proposed voltage Level shifter (LS3)	3.874	0.744
Proposed voltage Level shifter (LS4)	9.1921	4.14
Current Mirror voltage Level shifter (LS5)	53.914	17.5
Improved current mirror voltage level shifter (LS6)	24.749	4.96
Wilson current mirror voltage level shifter (LS7)	42.3931	16.9
Existing implementation voltage level shifter (LS8)	28.93	7.5

Delay vs. All circuits

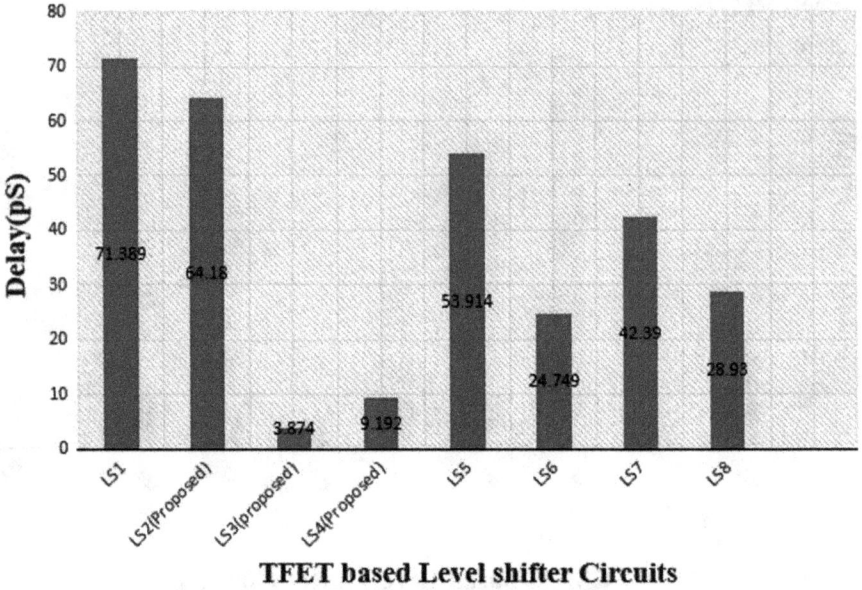

FIGURE 15.25   Delay of all topology TFET-based voltage level shifter.

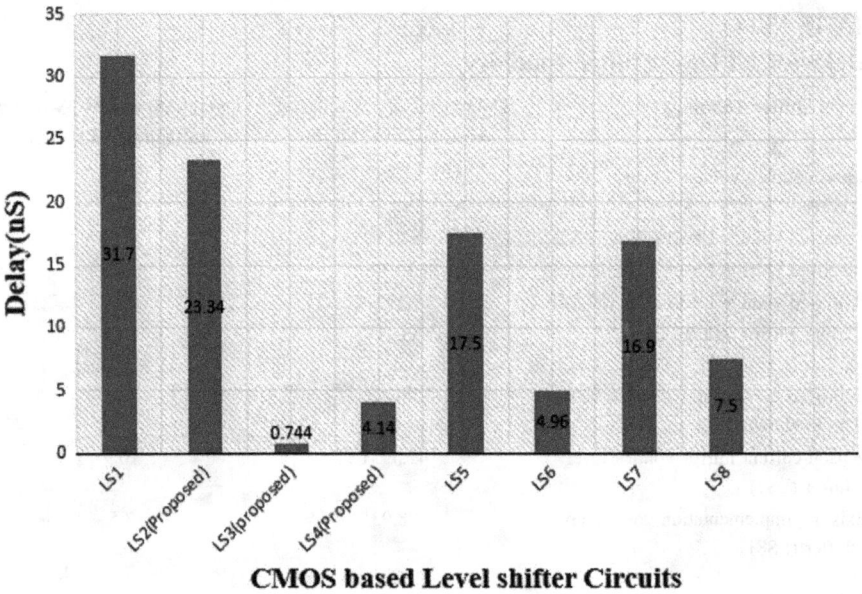

FIGURE 15.26   Delay of all topology CMOS-based voltage level shifter.

15.5.6 PDP OF TFET-BASED VOLTAGE LEVEL SHIFTERS

From Figures 15.27 and 15.28, we observe that the TFET-based voltage level shifter's power delay product is better than CMOS-based voltage level shifters. It is a product of Delay and Power. We calculate PDP in joules. In the TFET-based voltage level shifter, we obtained PDP in the range of $*10^{-21}J$ and CMOS-based voltage level shifters in the field of $*10^{-15}J$. We received better PDP in the TFET-based proposed voltage level shifter (LS4).

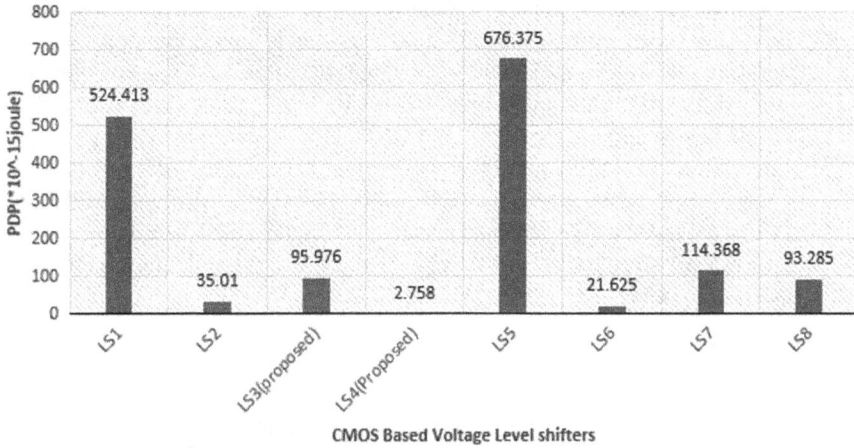

FIGURE 15.27 PDP of all topology CMOS-based voltage level shifter.

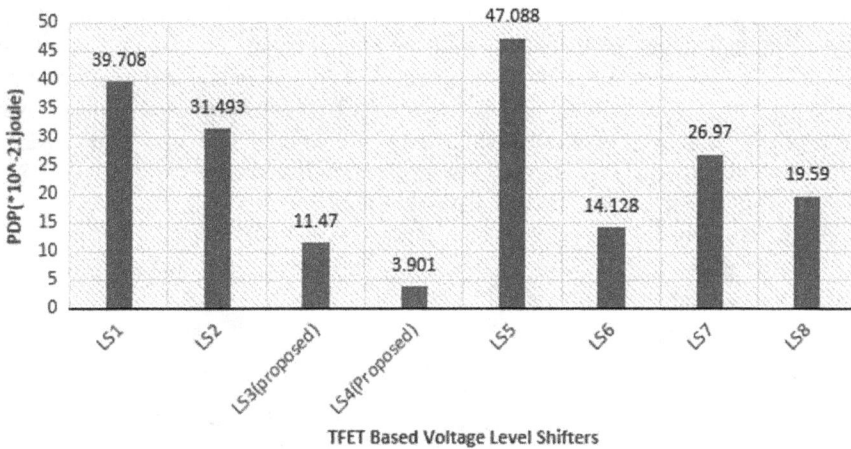

FIGURE 15.28 PDP of all topology TFET-based voltage level shifters.

15.6　IMPORTANCE OF PERFORMANCE METRICS OF LEVEL SHIFTERS

- TFET-based voltage level shifter circuits perform better at higher frequencies (33 MHz). We can use a TFET-based current mirror voltage level shifter (LS5) & TFET-based improved current mirror voltage level shifter (LS6) for Higher frequency applications (33MHz).
- TFET-based circuits have a better level shifting range; in TFET-based Wilson current mirror voltage level shifter (LS7) & existing implementation of TFET-based voltage level shifter (LS8), we obtained a Max range (0.1 to 1.2V). We can use this type of Circuit where the supply voltage is low.
- The TFET-based voltage Level shifter's power is low compared to the CMOS-based voltage level shifter; in TFET-based proposed voltage level shifter (LS4) has minimum power (424.983 pW). We can use this Circuit for low-power applications.
- The Delay of the TFET-based voltage level shifter Circuit is less compared to the CMOS-based voltage level shifter because switching in TFET devices is very fast. We obtained a min delay in the TFET-based proposed voltage level shifter (LS3) (4.84 pS). We can use this Circuit where we require min delay.

15.7　CONCLUSION

TFET-based circuits have a lower threshold that's by a lower conversion range of the TFET-based voltage level shifter is less as compared to CMOS circuits. In the existing TFET-based voltage level shifter implementation, we obtained the level shifting range from 0.1 to 1.2V. TFET-based proposed voltage level shifter (LS2) is the Improvement of the TFET-based cross-coupled voltage level shifter in terms of power. In TFET-based suggested voltage (LS2), we use a current limiting Transistor, and output takes from a splitting type inverter. TFET-based proposed voltage level shifter (LS3) is an improvement in terms of Delay, and this Circuit is the fastest in all voltage level shifter circuits. TFET-based proposed voltage level shifter (LS4) is an improvement both in terms of Power & Delay. Overall, PDP is Better in LS4 Circuit. TFET-based Wilson current mirror voltage level shifter (LS7) is the Improvement of the Current mirror-type voltage-level shifter by inserting the feedback transistor So that current is not flowing from VDD to GND. In this work, we have proposed three topologies of level shifter circuits with the advantages of Delay, Power, and PDP, respectively.

REFERENCES

[1] M. Lanuzza, et al., "An Ultralow-Voltage Energy-Efficient Level Shifter," *IEEE Transactions on Circuits and System II: Express Briefs*, vol. 64, no. 1, pp. 61–65, Jan. 2017.

[2] B. Razavi, *Design of Analog CMOS Integrated Circuits*, 2nd edition. New York: McGraw-Hill, 2001.

[3] H. Lu, T. Ytterdal, and A. Seabaugh, *Universal TFET Model Implementation in Verilog - A Table of Contents. January* 2015.

[4] S. N. Wooters, B. H. Calhoun, and T. N. Blalock, "An Energy-efficient Sub-Threshold Level Converter in 130-nm CMOS," *IEEE Transactions on Circuits and Systems II: Express Briefs*, vol. 57, no. 4, pp. 290–294, April 2010.

[5] S. Kabirpour and M. Jalali, "A Power-Delay and Area Efficient Voltage Level Shifter Based on a Reflected-Output Wilson Current Mirror Level Shifter," in *IEEE Transactions on Circuits and Systems II: Express Briefs*, vol. 67, no. 2, pp. 250–254, February 2020, 10.1109/TCSII.2019.2914036.

[6] A. Shapiro and E. G. Friedman, "Power Efficient Level Shifter for 16 nm FinFET Near Threshold Circuits," in *IEEE Transactions on Very Large Scale Integration (VLSI) Systems*, vol. 24, no. 2, pp. 774–778, February 2016, 10.1109/TVLSI.2015 .2409051.

[7] V. L. Le and T. T. Kim, "An Area and Energy Efficient Ultra-Low Voltage Level Shifter With Pass Transistor and Reduced-Swing Output Buffer in 65-nm CMOS," in *IEEE Transactions on Circuits and Systems II: Express Briefs*, vol. 65, no. 5, pp. 607–611, May 2018, 10.1109/TCSII.2018.2820155.

[8] S. Kabirpour and M. Jalali, "A Low-Power and High-Speed Voltage Level Shifter Based on a Regulated Cross-Coupled Pull-Up Network," in *IEEE Transactions on Circuits and Systems II: Express Briefs*, vol. 66, no. 6, pp. 909–913, June 2019, 10.1109/TCSII.2018.2872814.

[9] R. Lotfi, M. Saberi, S. R. Hosseini, A. R. Ahmadi-Mehr, and R. B. Staszewski, "Energy-Efficient Wide-Range Voltage Level Shifters Reaching 4.2 fJ/Transition," in *IEEE Solid-State Circuits Letters*, vol. 1, no. 2, pp. 34–37, February 2018, 10.1109/LSSC.2018.2810606.

[10] E. Maghsoudloo, M. Rezaei, M. Sawan, and B. Gosselin, "A High-Speed and Ultra-Low-Power Subthreshold Signal Level Shifter," in *IEEE Transactions on Circuits and Systems I: Regular Papers*, vol. 64, no. 5, pp. 1164–1172, May 2017, 10.1109/ TCSI.2016.2633430.

[11] S. R. Hosseini, M. Saberi, and R. Lotfi, "A High-Speed and Power-Efficient Voltage Level Shifter for Dual-Supply Applications," in *IEEE Transactions on Very Large Scale Integration (VLSI) Systems*, vol. 25, no. 3, pp. 1154–1158, March 2017, 10.1109/TVLSI.2016.2604377.

[12] S. Lütkemeier and U. Rückert, "A Subthreshold to Above-Threshold Level Shifter Comprising a Wilson Current Mirror," in *IEEE Transactions on Circuits and Systems II: Express Briefs*, vol. 57, no. 9, pp. 721–724, September 2010, 10.1109/ TCSII.2010.2056110.

[13] Z. Yong, X. Xiang, C. Chen, and J. Meng, "An Energy-Efficient and Wide-Range Voltage Level Shifter With Dual Current Mirror," in *IEEE Transactions on Very Large Scale Integration (VLSI) Systems*, vol. 25, no. 12, pp. 3534–3538, Dec. 2017, 10.1109/TVLSI.2017.2748228.

[14] Y. Cao, W. Ye, X. Zhao, and P. Deng, "An Energy-Efficient Subthreshold Level Shifter With a Wide Input Voltage Range," 2016 IEEE International Symposium on Circuits and Systems (ISCAS), pp. 726–729, 2016, 10.1109/ISCAS.2016. 7527343.

[15] M. Lanuzza, P. Corsonello, and S. Perri, "Fast and wide range voltage conversion in multi supply voltage designs," *IEEE Trans. Very Large Scale Integr. (VLSI) Syst.*, vol. 23, no. 2, pp. 388–391, Feb. 2015.

[16] T. Ytterdal and S. Aunet, "An Energy Efficient Level Shifter Capable of Logic Conversion," *IEEE Transactions on Circuits and Systems II: Express Briefs*, vol. 67, no. 11, November 2020.

[17] Z. Lin, P. Chen, L. Ye, X. Yan, L. Dong, S. Zhang, Z. Yang, C. Peng, X. Wu, and J. Chen, "Challenges and Solutions of the TFET Circuit Design," *IEEE Transactions on Circuits and Systems I: Regular Papers*, vol. 67, no. 12, pp. 4918–4931, 2020, 10.1109/TCSI.2020.3010803.

[18] S. M. Turkane and A. K. Kureshi, "Review of Tunnel Field Effect Transistor (TFET)," *International Journal of Applied Engineering Research*, vol. 11.7, pp. 4922–4929, 2016.

[19] Reddy, N. Nagendra, and D. Kumar Panda, "A Comprehensive Review on a Tunnel Field-Effect Transistor (TFET) Based Biosensors: Recent Advances and Prospects on Device Structure and Sensitivity," *Silicon*, pp. 1–16, 2020.

Index

For Product Safety Concerns and Information please contact our EU
representative GPSR@taylorandfrancis.com
Taylor & Francis Verlag GmbH, Kaufingerstraße 24, 80331 München, Germany

www.ingramcontent.com/pod-product-compliance
Lightning Source LLC
Chambersburg PA
CBHW052119230326
41598CB00080B/3889

9 781032 354699